THERMAL CONDUCTIVITY
20

A Continuation Order Plan is available for this series. A continuation order will bring delivery of each new volume immediately upon publication. Volumes are billed only upon actual shipment. For further information please contact the publisher.

THERMAL CONDUCTIVITY 20

Edited by
D. P. H. Hasselman and J. R. Thomas, Jr.
Virginia Polytechnic Institute and State University
Blacksburg, Virginia

PLENUM PRESS • NEW YORK AND LONDON

ISBN-13:978-1-4612-8069-9   e-ISBN-13:978-1-4613-0761-7
DOI: 10.1007/978-1-4613-0761-7

Proceedings of the Twentieth International Thermal Conductivity
Conference, held October 19–21, 1987, in Blacksburg, Virginia

© 1989 by Purdue Research Foundation
Softcover reprint of the hardcover 1st edition 1989

Plenum Press is a Division of Plenum Publishing Corporation
233 Spring Street, New York, N.Y. 10013

FOREWORD

The International Thermal Conductivity Conference was started in 1961 with the initiative of Mr. Charles F. Lucks and grew out of the needs of researchers in the field. The Conferences were held annually from 1961 to 1973 and have been held biennially since 1975 when our Center for Information and Numerical Data Analysis and Synthesis (CINDAS) of Purdue University became the Permanent Sponsor of the Conferences. These Conferences provide a broadly based forum for researchers actively working on the thermal conductivity and closely related properties to convene on a regular basis to exchange their ideas and experiences and report their findings and results.

The Conferences have been self-perpetuating and are an example of how a technical community with a common purpose can transcend the invisible, artificial barriers between disciplines and gather together in increasing numbers without the need of national publicity and continuing funding support, when they see something worthwhile going on. It is believed that this series of Conferences not only will grow stronger, but will set an example for researchers in other fields on how to jointly attack their own problem areas.

Of the first thirteen Conferences, only four published formal volumes of proceedings. However, effective with the Fourteenth Conference, a policy of publishing formal volumes of proceedings on a continuing and uniform basis has been established. Thus, including the present volume, the following formal volumes of proceedings have been published:

Conference (Year)	Title of Volume	Publisher (Year)
7th (1967)	THERMAL CONDUCTIVITY Proceedings of the Seventh Conference	U.S. Government Printing Office (1968)
8th (1968)	THERMAL CONDUCTIVITY Proceedings of the Eighth Conference	Plenum Press (1969)
9th (1969)	NINTH CONFERENCE ON THERMAL CONDUCTIVITY	U.S. Atomic Energy Commission (1970)
13th (1973)	ADVANCES IN THERMAL CONDUCTIVITY Papers Presented at XIII International Conference on Thermal Conductivity	University of Missouri, Rolla (1974)
14th (1975)	THERMAL CONDUCTIVITY 14	Plenum Press (1976)

15th (1977)	THERMAL CONDUCTIVITY 15	Plenum Press (1978)
16th (1979)	THERMAL CONDUCTIVITY 16	Plenum Press (1983)
17th (1981)	THERMAL CONDUCTIVITY 17	Plenum Press (1983)
18th (1983)	THERMAL CONDUCTIVITY 18	Plenum Press (1985)
19th (1985)	THERMAL CONDUCTIVITY 19	Plenum Press (1988)
20th (1987)	THERMAL CONDUCTIVITY 20	Plenum Press (1988)

Professors D. P. H. Hasselman and J. R. Thomas, Chairmen of the Twentieth Conference, are to be congratulated for their excellent leadership in conducting the Conference and for their painstaking efforts which made the present volume possible.

CINDAS looks forward to working with future host institutions and conference chairmen to ensure that future Conferences continue to produce high-quality volumes of proceedings in this important, specialized field.

C. Y. Ho
Director
Center for Information and Numerical
   Data Analysis and Synthesis
Purdue University

West Lafayette, Indiana
July 1988

PREFACE

The 20th International Thermal Conductivity Conference (ITCC) was listed at the Virginia Polytechnic Institute and State University, Blacksburg, Virginia and sponsored by CINDAS of Purdue University, the EXXON Foundation and the Thermophysical Research Laboratory of the Department of Materials Engineering at VPI. The general chairmen of the conference were Professors D. P. H. Hasselman and J. R. Thomas. A listing of the previous ITTC is given in the subsequent pages.

The 20th ITCC was attended by 66 people, representing 12 different countries. Over forty papers were presented. Unfortunately, some of the contributors from abroad were unable to attend. Nevertheless, some of their papers are included in these proceedings. The content of the papers was presented under main subjects including: Insulation, Liquids, Metals, High-Temperature Materials, Other Materials and Effects, Methods and Composites.

At the banquet, the Thermal Conductivity Award was presented to Dr. A. Cezairliyan. Drs. H. J. Goldsmid, R. S. Graves and D. W. Yarbrough were made Fellows of the ITTC. Mr. L. F. Johnson, graduate student at VPI, was awarded the first Lucks Award.

The chairmen wish to acknowledge all those who have helped make the 20th ITTC a success. The 21st ITTC will be held at the University of Kentucky in Lexington in 1989.

Blacksburg, Virginia
October, 1987

D. P. H. Hasselman
J. R. Thomas
Co-Chairman, 20th ITTC

PREVIOUS THERMAL CONDUCTIVITY CONFERENCES

Conf.	Year	Host Organization and Site	Chairman
1	1961	Battelle Memorial Institute (Columbus, Ohio)	C. F. Lucks
2	1962	National Research Council (Canada) (Ottawa, Canada)	M. J. Laubitz
3	1963	Oak Ridge National Laboratory (Gatlinburg, Tennessee)	D. L. McElroy
4	1964	U. S. Naval Radiological Defence Lab (San Francisco, California)	R. L. Rudkin
5	1965	University of Denver (Denver, Colorado)	J. D. Plunkett
6	1966	Air Force Materials Laboratory (Dayton, Ohio)	M. L. Minges G. L. Denman
7	1967	National Bureau of Standards (Gaithersburg, Maryland)	D. R. Flynn B. A. Peavy
8	1968	Thermophysical Properties Research Center, Purdue University (West Lafayette, Indiana)	C. Y. Ho R. E. Taylor
9	1969	Ames Laboratory and Office of Naval Research (Ames, Iowa)	H. R. Shanks
10	1970	Arthur D. Little, Inc. and Dynatech R/D Co. (Boston, Massachusetts)	A. E. Wechsler R. P. Tye
11	1971	Sandia Laboratories, Los Alamos Scientific Laboratories and University of New Mexico (Albuquerque, New Mexico)	R. U. Acton R. Wagner A. V. Houghton, III
12	1972	Southern Research Institute and University of Alabama (Birmingham, Alabama)	W. T. Engelke S. G. Bapat M. Crawford
13	1973	University of Missouri - Rolla (Lake of the Ozarks, Missouri)	R. L. Reisbig H. J. Sauer, Jr.
14	1975	University of Connecticut (Storrs, Connecticut)	P. G. Klemens

15	1977	Dept. of Energy, Mines And Resources (Ottawa, Canada)	V. V. Mirkovich
16	1979	IIT Research Institute (Chicago, Illinois)	D. C. Larsen
17	1981	National Bureau of Standards (Gaithersburg, Maryland)	J. G. Hust
18	1983	South Dakota School of Mines and Technology (Rapid City, South Dakota)	T. Ashworth D. R. Smith
19	1985	Tennessee Technological University (Cooksville, Tennessee)	D. W. Yarbrough
20	1987	Virginia Polytechnic Institute and State University (Blacksburg, Virginia)	D. P. H. Hasselman J. R. Thomas

CONTENTS

OTHER MATERIALS AND EFFECTS

SESSION 6

METHODS

SESSION 7

COMPOSITES

SESSION 1

INSULATION

LOAD-BEARING EVACUATED FIBROUS SUPERINSULATIONS -

IMPROVEMENTS WITH PEG-SUPPORT AND METAL-COATED FIBERS

J. Fricke,   Physikalisches Institut
D. Büttner,   der Universität Würzburg
R. Caps,   Am Hubland
G. Döll,   D-8700 Würzburg, W.-Germany
E. Hümmer and
A. Kreh

H. Reiss   Brown, Boveri & Cie
                          D-6900 Heidelberg, W.-Germany

ABSTRACT

Optimization of load-bearing evacuated thermal superinsulations requires a detailed investigation of two thermal loss channels: The solid conduction via contacting fibers, and the radiative heat transfer. In this paper we report two important findings which allow to further improve thermal superinsulations: (i) with metal-coated fibers we obtained extinction coefficients of several hundred m^2/kg; this corresponds to a drastic reduction of radiative losses compared to the case of non-coated fibers; (ii) with peg-supported insulation systems the solid conductivity was reduced to about $0.5 \cdot 10^{-3}$ W/(m·K) at 1.15 bar external pressure; this could lead to an improvement by more than a factor of three compared to not-segmented fibrous insulations. The measurements were performed with evacuable and load-controlled guarded hot plate devices and with a FTIR spectrometer.

INTRODUCTION

Load-bearing evacuated fiber systems have been shown to provide good thermal insulation at high temperatures. Best values (figure 1, curves b and c) for the total thermal conductivity are below $5 \cdot 10^{-3}$ W/(m·K) at an external load of 1.3 bar and mean radiative temperatures of about 500 K (corresponding to boundary temperatures of approximately 650 K and 300 K, respectively) [1]. About 50 % of the thermal losses are caused by solid conduction via the contacting fibers, the other 50 % are from infrared (IR) radiative transport. Attenuation of radiative transport is provided by scattering and/or absorption. Typical specific extinction coefficients for plain fiber insulations are of the order of 50 m^2/kg.

Data for such insulations are measured with evacuable guarded hot plate devices under stationary conditions. Such systems have to be evacuated to pressures below 10^{-3} mbar in order to suppress thermal conduction by residual gas. In addition the load onto the insulating specimen has to be adjusted for a controlled variation of the solid conductivity through the fibrous layer. IR transmission and reflection measurements are necessary in order to understand and quantify the radiative heat transfer. The spectral range to be investigated extends from about 2 μm to 50 μm.

The main topic in this report will be the question how and to which extent both solid conductivity, λ_s, and radiative conductivity, λ_r, can be reduced. Two measures pursued in this work are

● Use of a peg-supported system with a covered area fraction, a ≈ 0.18, instead of a full-area load support.

● Application of metal-coated fibers in order to reduce the radiative heat transfer.

EXPERIMENTAL EQUIPMENT

All calorimetric data were retrieved using the two evacuable guarded hot plate machines LOLA I and LOLA II [1].

LOLA I has a circular metering section with ∅ ≈ 480 mm and two guard rings 84 and 15 mm wide. Two reference plates of size 780 x 780 mm² serve as cold boundaries and also as side walls of the vacuum chamber. The attachment of the reference plates to the frame of the vacuum chamber is flexible. This allows to exert an external pressure load p_{ext} between 0 and 1.3 bar onto the specimens. A large metering section is an absolute prerequisite for the investigation of peg-supported insulations.

LOLA II is much smaller with a metering section of ∅ ≈ 120 mm and two guard rings 18 and 20 mm wide. The external pressure load is provided by a hydraulic press, the piston of which penetrates the top part of the vacuum chamber.

Spectral IR reflection and transmission experiments were performed with a Perkin Elmer FTIR spectrometer 1700 within the wavelength range 2 to 45 μm.

MEASUREMENTS WITH PEG-SUPPORTED INSULATIONS

Measured total thermal conductivities λ and thermal loss coefficients k, respectively, generally are presented as a function of a radiative temperature $T_r = [(T_1^2+T_2^2)(T_1+T_2)/4]^{1/3}$. We use the diffusion model expression for the radiative heat flow

$$\dot{q}_r = 4 \cdot n^2 \cdot \sigma (T_1^4 - T_2^4)/(3 \cdot \tau_o) \ . \tag{1}$$

n denotes the effective index of refraction of the porous medium, σ the Stefan Boltzmann constant, τ_o the optical thickness, and T_1 and T_2 are the absolute temperatures of hot and cold wall, respectively. With

$$\dot{q}_r = \lambda_r \cdot (T_1 - T_2)/d \tag{2}$$

a radiative conductivity follows (d is the thickness of the insulation):

$$\lambda_r = (16/3) \cdot n^2 \cdot \sigma \cdot T_r^3 / E_{eff} \ . \tag{3}$$

4

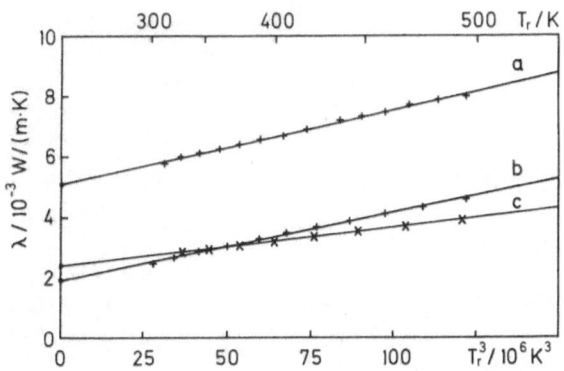

Figure 1: Thermal conductivities for several evacuated porous insulations under a pressure load of 1.3 bar as a function of T_r^3; curve a: fumed silica insulation, ρ = 270 kg/m³, opacified with Fe_3O_4; curve b: borosilicate glass fiber insulation, ρ = 300 kg/m³, the paper was thermally treated at 500°C at an external pressure load of about 1 bar; curve c: same type of glass fiber insulation opacified with Fe_3O_4, ρ = 330 kg/m³.

Here $E_{eff} = \tau_{eff}/d$ denotes an effective value of the extinction coefficient. τ_{eff} is an effective optical thickness. In the following, λ_s describes the solid conduction part of the total thermal conductivity $\lambda = \lambda_s + \lambda_r$. The corresponding loss coefficients are k, k_s and k_r, with k = λ/d.

Figure 1, curve b, shows measured values of the total thermal conductivity λ as a function of T_r^3 of an evacuated, load-bearing glass fiber insulation (borosilicate glass, fiber Ø ≈ 1 to 5 µm, density of the insulation ρ = 300 kg/m³, porosity 0.885) for an external pressure load p_{ext} ≈ 1.3 bar. The observed linear dependence of λ with T_r^3 allows the extraction of the specific extinction e = E_{eff}/ρ ≈ 50 m²/kg as well as an extrapolation of $\lambda(T_r)$ to $\lambda(T_r=0)$ ≈ λ_s ≈ $1.9 \cdot 10^{-3}$ W/(m·K). This value is considerably below the λ_s-value of an evacuated, load-bearing powder insulation (figure 1, curve a). Addition of an opacifier (Fe_3O_4) to the fiber insulation reduces the slope in the (λ, T_r^3)-plot (curve c) and thus the radiative conductivity.

Figure 2 shows the loss coefficient k and the thickness d for the same type of insulation as a function of the external pressure load p_{ext} at constant temperature. Upon compression, the loss coefficient increases. This increase can be attributed to solid conduction, which depends on the number of fiber contacts and their thermal resistance. The radiative loss coefficient is expected to be independent of p_{ext}, as long as the ir scattering processes are not altered upon compression. From $k(p_{ext})$ and $d(p_{ext})$ the total conductivity λ and its components, λ_r and λ_s, can be extracted (see figure 3). The most important result from this plot is the small increase of $\lambda_s < 2 \cdot 10^{-3}$ W/(m·K) at 1 bar to $\lambda_s < 4 \cdot 10^{-3}$ W/(m·K) at 5 bar. In other words: a five-fold increase of the load only doubles the solid conductivity.

This finding initiated a series of measurements with peg-supported insulations. The covered area fraction in these investigations was about ≈ 0.18, thus 35 x 35 mm² large glass fiber pegs were loaded with about 5.6 bar if the load on the total surface of the insulation was 1 bar. The number of pegs within the metering section of LOLA I was 29, thus proper spatial averaging of the thermal losses was guaranteed.

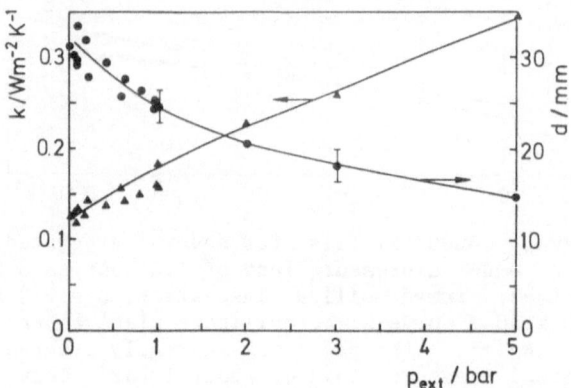

Figure 2: Total heat loss coefficient k (triangles) and thickness of samples d (circles) versus load p_{ext} at radiative temperature T_r = 498 K for glass fiber paper insulation.

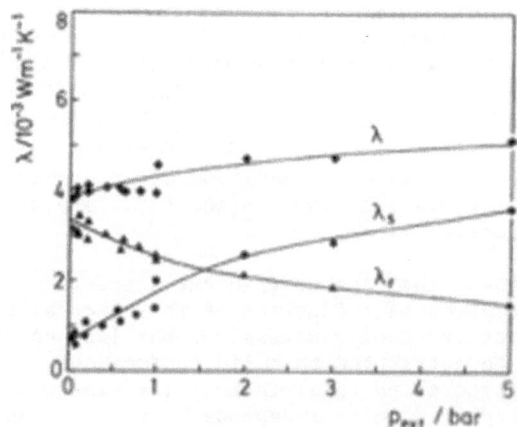

Figure 3: Total thermal conductivity λ (rhombi), and solid conductive (circles) and radiative (triangles) components λ_s and λ_r versus load p_{ext} at radiative temperature T_r = 498 K for glass fiber paper insulation.

First measurements were performed simply with glass fiber pegs installed in the gap between the hot plate and the reference plates, all of which had low emissivity surfaces ($\varepsilon \approx 0.03$). No radiation shields were used to cover the area around the pegs. Due to the strong increase of radiative losses with temperature (see figure 4), this system would be superior to the non-segmented system (figures 2 and 3) only at $T_r < 200$ K. In order to suppress radiative losses, 10 aluminum foils were inserted into the spacing between the pegs (see figure 5). Thermal contact between the foils was prevented by separating them with thin layers of glass silk (mass per unit area and layer $m'' = 2.3 \cdot 10^{-2}$ kg/m^2). As can be seen from figure 4 (curve c), firstly the losses due to solid conduction are extremely low ($\lambda_s \approx 0.5 \cdot 10^{-3}$ W/(m·K)), secondly the radiative transport is largely diminished, even with respect to the non-segmented case.

In a zero-order approximation we assume that the solid conduction is caused only by the glass fiber pegs. The thermal radiation proceeds via the fiber pegs and the remaining space. The total effective radiative conductivity then can be written approximately as

$$\lambda_r = 4 \cdot \sigma \cdot T_r^3 \cdot d \cdot \left\{ \frac{(1-a)}{(2/\varepsilon_{eff}-1)(N+1)} + \frac{4 \cdot n^2 \cdot a}{3 \cdot \tau_o} \right\} . \qquad (4)$$

With $N = 10$, $d = 0.015$ m, $a = 0.18$ and $\tau_o \approx 160$ (from figure 4, curve a) a value $\varepsilon_{eff} \approx 0.07$ is derived if equation 4 is fitted to curve c.

As the surface emissivity for the foils is about 0.03, we have to

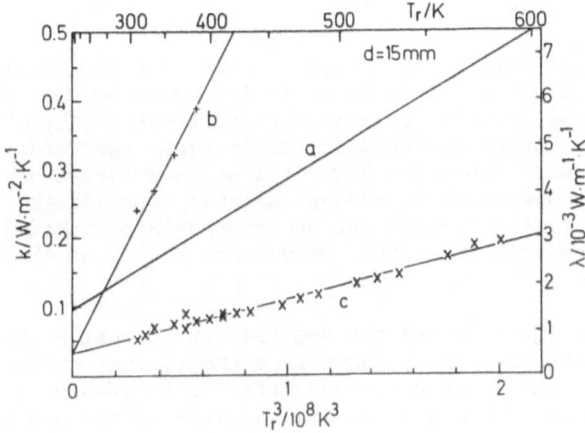

Figure 4: Loss coefficient k for various 15 mm thick insulation systems versus radiative temperature T_r at an external pressure load $p_{ext} \approx 1.15$ bar; curve a: non-segmented glass fiber paper system with $m'' = 3.46$ kg/m^2; curve b: peg-supported glass fiber paper system, covered area fraction $a \approx 0.18$, peg size 35 x 35 mm^2, $m'' \approx 7.0$ kg/m^2 within the pegs, emissivity of boundary $\varepsilon \approx 0.03$; curve c: same peg support plus additional 10 layers of aluminum foil ($\varepsilon \approx 0.03$) as radiation shields.

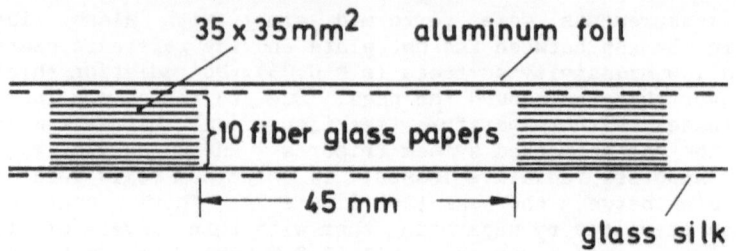

Figure 5: Segmented arrangement with glass fiber paper stacks and aluminum foils as radiation shields. The glass silk prevents direct
thermal contact between adjacent Al-foils.

conclude that the glass silk is thermally coupled to the Al-foils. This
leads to an increased effective emissivity of the layered structure. This
effect has been discussed earlier in optically thin insulations like
aerogels [2]. In order to diminish this effect, instead of the glass silk
an insulating material with less mass per unit area has to be used. A
second effect leading to an increased effective emissivity could be the
absorption of radiation, transferred between the foils, by the peg surfaces.
Which of these possibilities is the dominating one will have to be
clarified by additional measurements.

    In an earlier investigation [3] various peg-supported insulations
have been studied. For the "IDLAS" and the "Multifoil" insulations, thermal
conductivities of about 1.6 and $1.7 \cdot 10^{-3}$ W/(m·K), respectively, at $T_r \approx$
500 K are reported. These data, however, were derived using a rather small
heat flux meter (5 x 5 cm²) covering not even 2 pegs. Furthermore gas
conduction, solid conduction and radiative heat transport were treated as
being independent of each other (which they are not). The radiative
component was considered to be detectable from the slope of the
conductivity versus temperature curve. Such a procedure is not in
accordance with equation 3.

MEASUREMENTS WITH COATED FIBERS

    Plain fibrous insulating materials have specific extinction coeffi-
cients of about 50 m²/kg. The extinction is caused by scattering (which is
predominant in the forward direction) and by absorption. The wavelength
dependence of the spectral specific extinction coefficient $e(\Lambda)$ for a
microglass board with fiber $\emptyset \approx 0.5$ to 1 μm is depicted in figure 6. The
extinction can be improved by adding suitable opacifiers, like Fe_3O_4.
Though the radiative transport can be reduced (see figure 1), the added
weight is non-negligible and the solid conductivity of the system usually
increases.

    Therefore we investigated the possibility to coat a microglass board
with thin metal layers. It is known from theory that extremely thin metal
fibers act like small antennae and effectively absorb thermal radiation
[4]. Ni-fibers with $\emptyset \approx 0.3$ μm in diam provide huge specific extinction
coefficients $e \approx 1500$ m²/kg (see figure 7). The transmission measurements
were performed with Ni-fibers embedded in KBr. For comparison another
calculation using Mie theory for the extinction cross section was added. As
such metal fibers are not readily available, we tried to produce metal-
coated thin fibrous layers by using standard sputter technique instead.
The coating process was applied to one side of microglass paper, and the
metal penetrated into about ¼ of the paper. The specific spectral
extinction derived from IR transmission measurements is shown in figure 6
(curve b). The reflectivity of the metal-coated side was about 15 %, and

nearly independent of wavelength. These results underline that the optical properties of glass fibers no longer could be "seen" through the coating. The corresponding Rosseland mean $e_R \approx 300$ m²/kg is quite encouraging. The

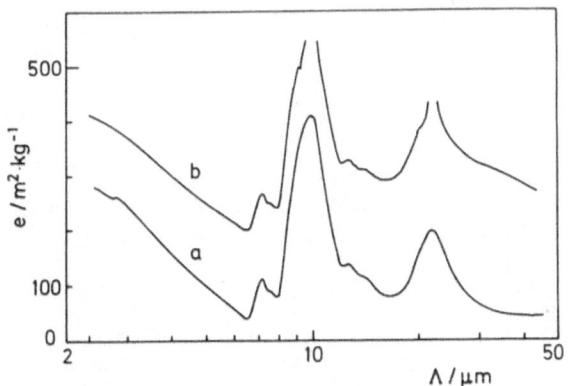

Figure 6: Specific spectral extinction coefficient e(Λ) for a microglass board (a) before and (b) after sputtering with Cr.

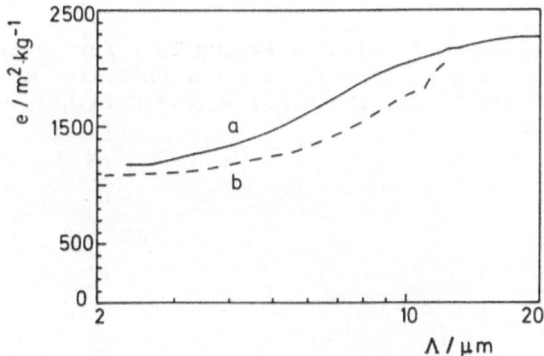

Figure 7: Spectral specific extinction coefficient e(Λ) for Ni-fibers (⌀ ≈ 0.3 µm) embedded in KBr, (a) measured, (b) Mie theory calculation.

following calorimetric measurements (figure 8) using a stack of 20 paper layers with 15.6 g/m² each, however, revealed two problems which have not been solved up to now: (i) The proximity of the coated fibers caused an increase of the solid conductivity within the metal-coated zones. (ii) The paper layers apparently act like a compressed stack of high-emissivity ($\varepsilon_{eff} \approx 0.85$) opaque high-conductivity foils and insulating fiber layers in-between, both being in close solid thermal contact. Thus a considerably smaller extinction coefficient of $e_m \approx 120$ m²/kg results from figure 8 if $n^2 = 1$ is assumed.

Both problems possibly could be eliminated if the coated fibers were finely dispersed in a non-coated environment.

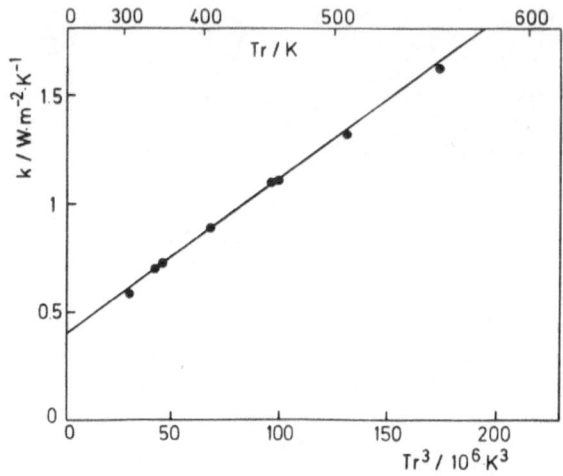

Figure 8: Thermal loss coefficient k versus T_r^3 for 20 layers of metal-coated microglass paper; mass per unit area m" = 0.312 kg/m², boundary emissivity $\varepsilon \approx 0.03$; $p_{ext} \approx 0.1$ bar (solid conduction suppressed).

OUTLOOK

We have shown that both the solid conduction and the radiative losses in evacuated superinsulations can be reduced considerably. By using segmented arrangements values of $\lambda_s \approx 0.5 \cdot 10^{-3}$ W/(m·K) are feasible. Likewise, with properly coated and dispersed fibers the radiative conductivity could be lowered to $\lambda_r \leq 1 \cdot 10^{-3}$ W/(m·K) at 500 K. Thus we expect the realization of superinsulations with total thermal conductivities in the order of $\lambda \approx 1 \cdot 10^{-3}$ W/(m·K) in the near future. The experimental investigation of such insulations will have to be performed with extreme care, however, as the decreasing thermal flux along the temperature gradient becomes difficult to detect and to control (especially with a relative increase of lateral flux component).

ACKNOWLEDGEMENTS

This work was supported by the German Bundesministerium für Forschung und Technologie (BMFT).

REFERENCES ·

[1] D. Büttner, J. Fricke and H. Reiss, Analysis of radiative and solid conduction components of the total thermal conductivity of an evacuated glass fiber insulation, measurements with a 700 x 700 mm² variable load guarded hot plate device, AIAA 20th Thermophysics Conf., June 1985, Williamsburg, Va., paper No. 85 – 1019

[2] P. Scheuerpflug, R. Caps, D. Büttner and J. Fricke, Internat. J. Heat Mass Transfer 28, 2299 (1985).

[3] Report Union Carbide Corporation – Linde Division Tonawanda, N.Y., Contract No. EM-78-C-01-5160 (1979).

[4] K. Y. Wang, C. L. Tien, J. Quant. Spectrosc. Radiat. Transfer 30, 213 (1983).

MODELLING OF TRANSIENT TESTS TO DETERMINE THERMAL PROPERTIES OF FIBERGLASS INSULATIONS

J. R. Thomas

Mechanical Engineering Department
Virginia Polytechnic Institute and State University
Blacksburg, Virginia 24061

ABSTRACT

Transient test methods for fibrous insulations require a sophisti-cated analysis technique because of the more rapid response of radiation heat transfer as compared to conduction, and because of the interaction between the two modes. A high-order combined-mode heat transfer model has been developed for this purpose, and benchmarked against experimental results. The model is capable of reproducing the experimental data to within 1.2°C throughout the duration of a transient test when the sample specific heat is corrected for moisture content. The model has yielded several interesting insights into transient testing techniques.

INTRODUCTION

Transient test methods have recently been introduced [1-4] to deter-mine the thermal resistance and conductivity of thermal insulations. Such methods offer the potential for considerable savings as compared to steady-state methods, since the long wait for steady conditions may be avoidable. These methods also introduce some challenging questions for thermal analysts, however, particularly for low-density fibrous insula-tions in which radiation is known to make a considerable contribution to the total energy transfer [5]. Since radiation propagates at the speed of light, the early stages of a transient test are likely to be dominated by radiative transfer, possibly leading to misleading results for the conductivity. Attempts to analyze the results of such experiments with independent calculations of radiative and conductive components or with low-order models have not been totally successful, so a more sophisti-cated approach was considered necessary. The problem is complicated by the interaction of radiation and conduction through the absorption and re-emission of radiant energy, and by the highly anisotropic scattering properties of glass fibers [6].

The model to be described below was developed as an attempt to ac-count for these effects. Coupled radiative-conductive heat transfer is accounted for through a simultaneous solution of the equation of radia-tive transfer and the energy equation [7]. The anisotropic scattering kernel is represented by a 20-term Legendre polynomial expansion and the equation of transfer is solved by a high-order P_N approximation [9]. The energy equation is solved by implicit finite differences.

ANALYSIS

The specific objective of the project was to develop a model for the flat-screen test apparatus at Oak Ridge National Laboratory (ORNL) [10]. This system is shown schematically in Figure 1. It consists of a screen wire heater (labelled "hot boundary") of thickness ℓ, the insulation specimen of thickness L, and the cold boundary, which is formed of a thick slab of copper to enforce isothermal conditions. A typical experiment proceeds as follows [3]: The system begins at a uniform equilibrium temperature; at time t = 0 a constant voltage is applied to the screen, producing a current and corresponding heat generation at the rate $\dot{q}$. Thermocouples attached to the screen monitor its temperature, and an average value is logged every 30 or 60 seconds. This continues until the screen reaches a new equilibrium temperature.

Problem Formulation

It is assumed that the heating element may be represented by a lumped-capacitance model in which energy is generated by the electrical current at the rate $\dot{q}$ (W/m^3). The screen loses heat by conduction and radiation into the specimen, and gains heat by radiation from the specimen. Thus an energy balance on the screen yields [4]

$$\rho_h \, C_h \ell \, \frac{dT_h}{d\xi} = \dot{q}\ell + k \left. \frac{\partial T}{\partial x} \right|_{x=0} - \varepsilon_h \, \sigma \, T_h^4 + \alpha_h \, \bar{q}_r \,, \tag{1}$$

in which T_h represents the screen (hot boundary) temperature, ρ_h and C_h its density and specific heat, and ε_h and α_h the total emissivity and absorptivity of the screen surface, respectively. The symbols k and T represent the thermal conductivity and temperature of the insulation, so that the term $k \left. \frac{\partial T}{\partial x} \right|_{x=0}$ represents conduction away from the screen. ξ represents time and $\bar{q}_r$ the radiative flux incident on the screen from the medium.

Within the insulation specimen, we have heat transfer by combined radiation and conduction. Thus $T(x,\xi)$ satisfies the equation [7]

$$k \, \frac{\partial^2 T}{\partial x^2} - \frac{\partial q_r}{\partial x} = \rho_p \, C_p \, \frac{\partial T}{\partial \xi} \,, \tag{2}$$

where ρ_p and C_p are the density and specific heat of the insulation, and q_r is the radiative flux. The radiative flux is computed from the radiation intensity $I(x,\mu)$, where μ is the direction cosine, according to [7]*

$$q_r(x,\xi) = 2\pi \int_{-1}^{1} I(x,\mu)\mu d\mu \,. \tag{3}$$

---

*The time dependence of $I(x,\mu)$ is suppressed in the notation since it is only parametric. See Ozisik [7], p. 252.

The radiation intensity satisifes the integrodifferential equation of transfer [3,7]

$$\mu \frac{\partial}{\partial x} I(x,\mu) + \sigma_e I(x,\mu) = \frac{\sigma_s}{2} \int_{-1}^{1} p(\mu,\mu') I(x,\mu') d\mu' \qquad (4)$$
$$+ \kappa \frac{\sigma}{\pi} T^4(x,t) \, ,$$

where κ and σ_s represent the absorption and scattering coefficients, respectively, $\sigma_e = \kappa + \sigma_s$ is the extinction coefficient, σ is the Stefan-Boltzmann constant, and $p(\mu,\mu')$ is the phase function [11]. The phase function, which represents the fraction of the radiation originally traveling in direction μ which is scattered into directions within $d\mu'$ about μ', is usually represented by an expansion in Legendre polynomials [10]:

$$p(\mu,\mu') = \sum_{m=0}^{L_s} \beta_m P_m(\mu) P_m(\mu') \, . \qquad (5)$$

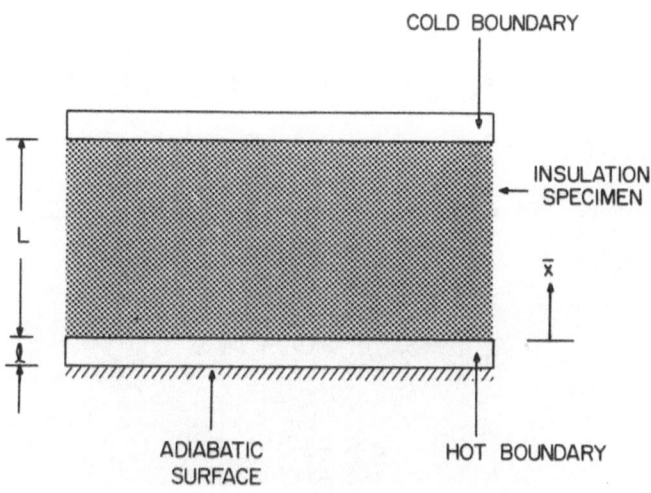

Figure 1. System Geometry

Following Tong [4], the equations are nondimensionalized as follows. The optical length variables τ and τ_o are defined as

$$\tau = \sigma_e x \, , \qquad 6a)$$

and

$$\tau_o = \sigma_e L \, ; \qquad (6b)$$

the nondimensional time

$$t = \frac{k \xi}{\rho_p C_p L^2} \, , \qquad (6c)$$

and temperature

$$\theta = \frac{T}{T_i} \, . \qquad (6d)$$

The heat generation rate in the screen is nondimensionalized as

$$\dot{Q} = \frac{\dot{q}\ell}{4\sigma T_i^4} . \tag{6e}$$

It is also convenient to define a nondimensional radiation intensity as

$$\psi(\tau,\mu) = \frac{I(\tau,\mu)}{\sigma T_i^4} , \tag{6f}$$

and heat flux as

$$Q_r = \frac{q_r}{4\sigma T_i^4} . \tag{6g}$$

Here T_i is the initial temperature of the system (K). Using Eqs. (5) and (6), the problem is reformulated as

$$\frac{N H_R}{\tau_o} \frac{d\theta_h}{dt} = \dot{Q} + N \frac{\partial\theta}{\partial\tau}\bigg|_{\tau=0} - \frac{\varepsilon_h}{4}\theta_h^4 + \varepsilon_h Q_r^- , \tag{7}$$

$$\frac{\partial^2\theta}{\partial\tau^2} - \frac{1}{N}\frac{\partial Q_r}{\partial\tau} = \frac{1}{\tau_o^2}\frac{\partial\theta}{\partial t} , \tag{8}$$

$$\mu\frac{\partial\psi}{\partial\tau} + \psi(\tau,\mu) = \frac{\omega}{2}\sum_{m=0}^{L}\beta_\ell P_\ell(\mu)\int_{-1}^{1} P_\ell(\mu')\psi(\tau,\mu')\,d\mu' + (1-\omega)\theta^4 , \tag{9}$$

$$Q_r(\tau) = \frac{1}{2}\int_{-1}^{1}\psi(\tau,\mu)\,\mu\,d\mu , \tag{10}$$

and

$$Q_r^- = -\frac{1}{2}\int_{o}^{1}\psi(0,-\mu)\,\mu\,d\mu . \tag{11}$$

In Eq. (7),

$$N = \frac{k\sigma_e}{4\sigma T_i^3} . \tag{12}$$

is the conduction-to-radiation parameter, and

$$H_R = \frac{\rho_H C_H \ell}{\rho_p C_p L} , \tag{13}$$

and following Kirchoff's law [7], α_h was replaced with ε_h.

The initial and boundary conditions for these equations are

$$\theta_h(0) = \frac{T_h(0)}{T_i} = 1 , \tag{14}$$

$$\theta(x,0) = 1 , \tag{15}$$

$$\theta(\tau_o,t) = 1 , \tag{16}$$

$$\theta(0,t) = \theta_h(t) , \tag{17}$$

for the conduction equations. For the radiative transfer equation both surfaces are assumed opaque but reflecting and emitting. For generality, the reflection is allowed to be partly specular and partly diffuse. This leads to the equations [7]

$$\psi(0,\mu) = \varepsilon_h \frac{\theta_h^4}{\pi} + \rho_1^s \psi(0,-\mu) + 2\rho_1^d \int_0^1 \psi(0,-\mu')\,\mu'\,d\mu' , \quad \mu > 0 ; \tag{18}$$

$$\psi(\tau_0,-\mu) = \frac{\varepsilon_c}{\pi} + \rho_2^s \psi(\tau_0,\mu) + 2\rho_2^d \int_0^1 \psi(\tau_0,\mu')\,\mu'\,d\mu' , \quad \mu > 0 . \tag{19}$$

Eqs. (7-19) constitute the complete formulation of the coupled radiation and conduction heat transfer problem.

Solution method

Following Benassi et al. [8], the P_N approximation for the solution of the radiative transfer equation is written in the form

$$\psi(\tau,\mu) = \sum_{\ell=0}^{N} \frac{2\ell + 1}{2} P_\ell(\mu) \sum_{j=1}^{J} \left[A_j e^{-\tau/\xi_j} + (-1)^\ell B_j e^{-(\tau_0-\tau)/\xi_j} \right] g_\ell(\xi_j)$$
$$+ \psi_p(\tau,\mu) \tag{20}$$

The polynomials $g_\ell(\xi)$ were introduced by Chandrasekhar [10], and obey the recurrence

$$(\ell + 1)\, g_{\ell+1}(\xi) = h_\ell\, \xi\, g_\ell(\xi) - \ell\, g_{\ell-1}(\xi) , \tag{21}$$

with $g_0(\xi) = 1$ and $h_\ell = 2\ell + 1 - \omega\beta_\ell$. We assume N is odd so that the eigenvalues ξ_j, $j = 1, 2, \ldots, J$, where $J = (N + 1)/2$, are the J positive zeros of $g_{N+1}(\xi)$. $\psi_p(\tau,\mu)$ represents a particular solution of Eq. (9) corresponding to the source term $(1 - \omega)\theta^4$. The arbitrary constants A_j and B_j must be chosen so as to force $\psi(\tau,\mu)$ to satisfy the boundary conditions (18) and (19). Thus Eq. (20) is substituted into Eqs. (18) and (19) and the two resulting equations are multiplied successively by $P_{2\alpha+1}(\mu)$, $\alpha = 0, 1, \ldots, (N-1)/2$, and integrated over $0 < \mu < 1$ to yield N + 1 linear algebraic equations for the A_j and B_j. The particular solution $\psi_p(\tau,\mu)$, evaluated at the boundaries, appears on the right-hand side of these equations, so it must be determined, as described below, before the linear equations can be solved for the A_j and B_j. With these contants determined, the heat flux $Q_r(\tau)$ is determined from Eqs. (10) and (20):

$$Q_r(\tau) = (1 - \omega) \sum_{j=1}^{J} \left[A_j e^{-\tau/\xi_j} - B_j e^{-(\tau_0-\tau)/\xi_j} \right] + \int_{-1}^{1} \psi_p(\tau,\mu)\,\mu\,d\mu . \tag{22}$$

The heat conduction equation (8) is solved by an implicit finite-difference technique: the interval $0 < \tau < \tau_0$ is divided into M subintervals of length $\Delta\tau = \tau_o/M$ and equation (8) is written

$$\frac{\theta_{m+1}^{(p+1)} - 2\theta_m^{(p+1)} + \theta_{m-1}^{(p+1)}}{(\Delta\tau)^2} - \frac{1}{N}\frac{\partial Q_r^{(p)}}{\partial\tau}\bigg|_{\tau=\tau_m} = \frac{1}{\tau_o^2}\frac{\theta_m^{(p+1)} - \theta_m^{(p)}}{\Delta t} , \qquad (23)$$

where

$$\theta_m^{(p)} \triangleq \theta(\tau_m, t_p) , \qquad (24)$$

and

$$\Delta t = t_{p+1} - t_p .$$

When Eq. (23) is evaluated at the $M + 1$ points $\tau_1, \tau_2, \ldots, \tau_{M+1}$, we have a system of $M + 1$ linear algebraic equations for the $\theta_m^{(p+1)}$, $m = 1, 2, \ldots, M + 1$. The value of θ at $\tau = 0$ must be determined from Eq. (7); this is accomplished through the forward-difference approximation

$$\theta_h^{(p+1)} = \theta_h^{(p)} + \Delta t\left(\frac{\tau_o}{N\,H_R}\right)\left\{\dot{Q} + N\left(\frac{\theta_2^{(p)} - \theta_1^{(p)}}{\Delta\tau}\right) - \frac{\epsilon_h}{4}[\theta_h^{(p)}]^4 + \epsilon_h\, Q_r^-\right\} . \qquad (25)$$

The particular solution $\psi_p(\tau,\mu)$ corresponding to the source term $(1 - \omega)\,\theta^4(\tau,t)$ is determined as follows: at a given time step t_p, $\theta^4(\tau,t_p)$ is represented by a spline approximation within each sub-interval,

$$\theta^4(\tau,t_p) = \sum_{k=0}^{K} a_{km}(\tau - \tau_m)^k, \quad \tau_m < \tau < \tau_{m+1} . \qquad (26)$$

The particular solution is then expressed as an expansion in Legendre Polynomials:

$$\psi_p(\tau,\mu) = \sum_{\ell=0}^{N} \left(\frac{2\ell + 1}{2}\right) c_{\ell m}(\tau)\, P_\ell(\mu) , \quad -1 < \mu < 1 ; \tau_m < \tau < \tau_{m+1} . \qquad (27)$$

When Eqs. (26) and (27) are substituted into Eq. (9) and the resulting equation projected onto the Legendre polynomial basis, a set of linear, first-order, ordinary differential equations is obtained in each interval $\tau_m < \tau < \tau_{m+1}$. These equations may be solved explicitly for the $c_{\ell m}(\tau)$, [11].

Calculation procedure

Starting from the initial temperatures, θ_h is calculated at the end of the first time step, $t = \Delta t$, from Eq. (25). This is used as the boundary condition for Eqs. (23); these equations are then solved to yield a new temperature distribution throughout the medium. This temperature distribution is then used in Eq. (26) to evaluate the coefficients a_{km} in the spline approximation. The coefficients $c_{\ell m}(\tau)$ in the particular solution are next determined [11]. These, in turn, are used in the equations for the P_N constants, A_i and B_i. Once these are determined, a new heat flux distribution is calculated from Eq. (22), Q_r^- is computed from Eq. (11), and Eq. (25) is used to start the next time step. This procedure is repeated until a new equilibrium temperature distribution is attained.

RESULTS

Numerical checks

The two principal parameters which control the accuracy of the numerical results are the order, N, of the P_N approximation, and the number of spatial intervals, M. The minimum value of N for meaningful results is L_s, the order of the Legendre expansion of the scattering phase function (Eq. 5). For low-density fiberglass insulation, a 20-term expansion derived from experimental data is available [6] and was chosen for this study. Thus, since N must be odd, we must use $N \geqslant 21$ when the full anisotropic scattering kernel is used. Using this value, a study of the effect of number of spatial intervals on the accuracy of the solution was performed. The maximum difference between the calculated and measured results was chosen as an appropriate measure of solution accuracy. Figure (2) shows this quantity plotted as a function of M for one data set. It is clear that a minimum occurs at approximately M = 100. Thus, for all subsequent calculations a value of $M \geqslant 100$ was used.

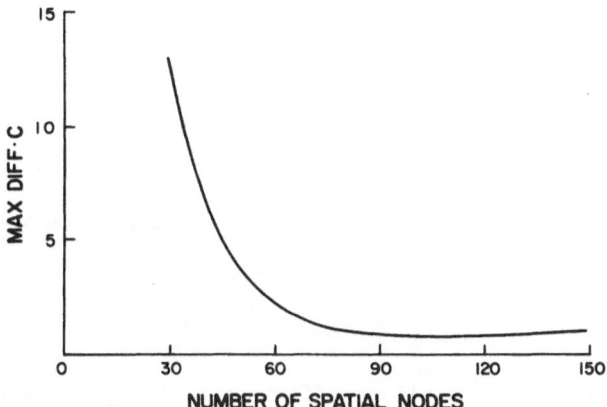

Figure 2. Effect of number of mesh intervals on the accuracy of the combined-mode model.

The P_N-order was next increased to values as high as N = 149. It was found that the accuracy of the result was little affected by this parameter, so a value of N = 29 was used for all subsequent calculations.

A similar study of the effect of time-step size revealed a broad minimum in the error in the vicinity of $\Delta t = 30$ secs, so this value was used for all calculations.

Table 1. Input data for Equations (12) and (13)

k	0.027 W/mK
ρ_H	2115 kg/m^3
c_H	435 J/kgK
ℓ	0.32 mm
L	76.26 mm

Extinction coefficient

The value of the extinction coefficient for the fiberglass samples used in the experiment is not well known. As a means of determining this parameter, several simulations were run using different values of the extinction coefficient for comparison with the experimental data. Typical results of such an exercise are shown in Figure 3. It is seen that accurate knowledge of the extinction coefficient is needed to obtain good agreement between the model and the data. By choosing the extinction coefficient to match the final equilibrium temperature, a good fit to the data is obtained throughout the experiment. This is illustrated in Figure 4; in this simulation, agreement was obtained to within ±1 C throughout the 50 minute experiment. Since the extinction coefficient varies with temperature, a different value must be used for each experimental run, however. Following this scheme, agreement was obtained to within ±1.2 C for all six data sets provided by ORNL.

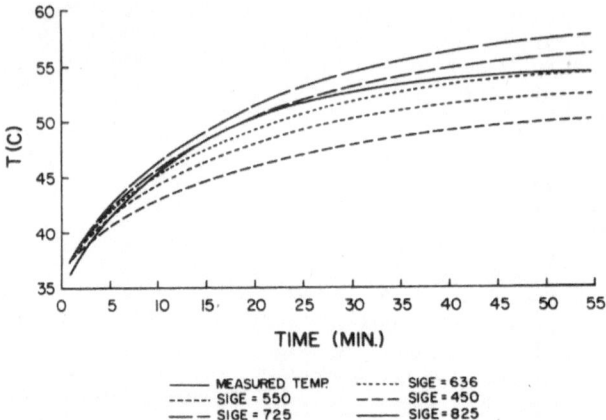

Figure 3.   Effect of extinction coefficient on the calculated temperature history; data set F4.

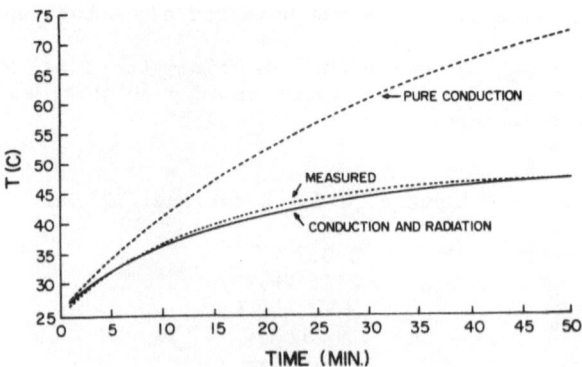

Figure 4.   Comparison of measured and computed screen temperatures with a pure conduction model.

Temperature-dependent specific heat was also incorporated into the model through correlations of Hust [12] and McElroy [11]. The correlation of McElroy apparently includes the effect of moisture in the sample [11].

Apparent Thermal Conductivity

One of the main objectives of the experiments is the determination of the thermal conductivity of the insulation samples. The _apparent_ conductivity is defined through the equation

$$q = k_{app} \frac{\Delta T}{L} , \tag{28}$$

where q is the total heat flux (conduction and radiation). Results for k_{app} as computed by the present model are compared to those predicted by the widely-used equation of Rennex [13] in Figure 5. The P-29 results agree with Ref. [13] to within 2.5% at low temperature, but differ by 12% at 50 C. Considering the simplicity of the Rennex equation, this is very good agreement, and suggests that this equation should be quite useful for predicting the apparent conductivity of radiation samples.

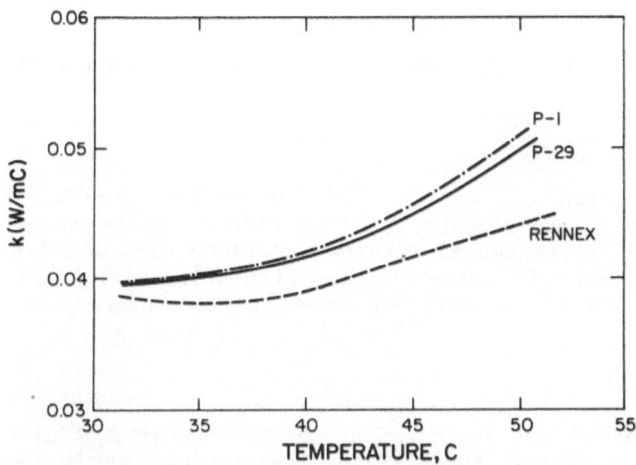

Figure 5. Apparent thermal conductivity of low-density fiberglass as computed by three different models.

Surface emissivity/reflectivity

The boundary conditions shown in Eqs. (18) and (19) are quite general and afford the possibility of exploring the effects of surface reflectivity. Assuming Kirchoff's law holds, and setting $\rho_1^s = \rho_2^s$, $\varepsilon_1 = 1 - \rho_1^d$, $\varepsilon_2 = 1 - \rho_2^d$, $\varepsilon_1 = \varepsilon_2 = \varepsilon$, the effect of varying ε is shown in Figure 6. These results were computed for the end-of-transient equilibrium conditions of the first data set supplied by ORNL. The increase in conduction is a little misleading; it results from the increase in temperature of the hot surface as its reflectivity decreases, for fixed $\dot{q}$.

Clearly, reflective surfaces can be of considerable importance in reducing heat transfer through low-density insulations.

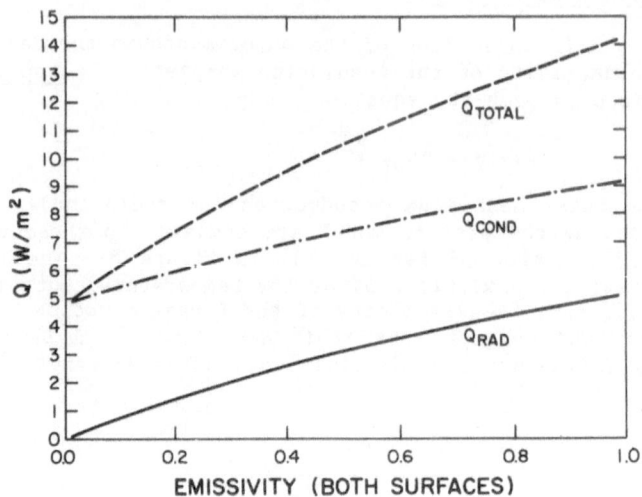

Figure 6. Effect of boundary emissivity on predicted heat flux.

CONCLUSIONS

A new coupled-mode heat transfer model has been developed for analyzing transient experiments in insulations. It is accurate to within ±1.2 C for all experimental results provided. The model can be used to study the effects of scattering and absorption properties, surface reflectivities, and temperature dependent specific heat.

ACKNOWLEDGEMENT

This research was sponsored by the U.S. Department of Energy as part of the National Program for Building Thermal Envelope Systems and Materials.

REFERENCES

1. McElroy, D. L., Graves, R. S., Yarbrough, D. W., and Tong, T. W., "Non-Steady State Behavior of Thermal Insulations," 9th European Conference on Thermophysical Properties, R. Taylor (Editor), 1986.

2. Tong, T. W., McElroy, D. L. and Yarbrough, D. W., "Transient Radiation and Conduction Heat Transfer in Porous Thermal Insulations," ASME Winter Annual Meeting, Paper No. 84-WA/HT-102, 1984, also published in Journal of Thermal Insulation, Vol. 9, 1985, pp. 13-29.

3. Yarbrough, D. W., McElroy, D. L., Tong, T. W., and Wood III, J. K., "Analysis of Transient Thermal Measurements on Fibrous Insulations Obtained with an Unguarded Flat Tester," 4th International Chemical Engineering Conference, CHEMPOR '85, Coimbra, Portugal, 1985.

4. Tong, T. W., "Analysis of Transient Behavior and Radiation Measurements of Commercial Thermal Insulation," Oak Ridge National Laboratory Report, ORNL/SUB/83-43366/1, 1985.

5. H. A. Fine, S. H. Jury, D. W. Yarbrough, and D. L. McElroy, "Analysis of Heat Transfer in Building Thermal Insulation," ORNL/TM-7481 (December, 1980).

6. M. S. Schuetz, "Heat Transfer in Foam Insulation," M.S. Thesis, Massachusetts Institute of Technology (1982).

7. M. N. Ozisik, _Radiative Transfer_, John Wiley and Sons, New York (1973).

8. M. Benassi, R. M. Cotta, and C. E. Siewert, "The P_N Method for Radiative Transfer Problems with Reflective Boundary Conditions," J. Quant. Spectrosc. Radiat. Transfer _30_, 547 (1983).

9. R. S. Graves, D. W. Yarbrough, and D. L. McElroy, "Apparent Thermal Conductivity Measurements by an Unguarded Technique," Eighteenth International Thermal Conductivity Conference, Rapid City, S.D. (1983).

10. S. Chandrasekhar, _Radiative Transfer_, Oxford University Press, London (1950).

11. J. R. Thomas, Jr., "Calculational Model Development for Fibrous Thermal Insulation Transient Test Procedures," ORNL/SUB-27494/1 (in press).

12. J. G. Hust, J. E. Callanan, and S. A. Sullivan, "Specific Heat of Insulations," in Thermal Conductivity 19 (in press).

13. B. G. Rennex, "Thermal Parameters as a Function of Thickness for Combined Radiation and Conduction Heat Transfer in Low-Density Insulation," _Journal of Thermal Insulation._

APPARENT THERMAL CONDUCTIVITY OF HIGH

DENSITY AND LOW DENSITY FIBERGLASS INSULATIONS

S. Yajnik* and J. A. Roux**
University of Mississippi
University, MS 38677
U.S.A.

ABSTRACT

A numerical model for coupled conduction and radiation heat transfer in fibrous insulations was used to predict the steady-state heat transfer in two standard reference materials (SRM); SRM 1451 (low density fiberglass - 9.25 kg/m^3) and SRM 1450(b)(high density fiberglass - 129 kg/m^3). Gray isotropic radiative properties were input into the transient numerical model which marches to steady state. Theoretical apparent thermal conductivity values were computed as a function of mean temperature for both density fiberglasses and as a function of thickness and density for the low density fiberglass insulation. These results are compared to the experimental apparent thermal conductivity values obtained from tests performed at the Oak Ridge National Laboratory (ORNL). In addition, the relative importance of radiation and conduction for different density fiberglass insulations is also discussed.

INTRODUCTION

Heat transfer in fibrous insulations with coupling of combined modes of heat transfer has been investigated by several researchers[1-5]. The two predominant heat transport mechanisms in fibrous insulations are conduction and volumetric radiation. Radiation accounts for up to 50 percent of the overall heat transfer for most fiberglass insulations at moderate temperatures (250-350 K); the radiative component being greater for low density fiberglass.

The present research had a two-fold purpose: the determination of the spectral radiative properties (extinction coefficient and albedo) of the two fiberglass insulations[6] and the investigation of their heat transfer characteristics using the computed gray isotropic radiative properties. The objective of this paper is to present steady-state heat transfer results for the two standard reference (fiberglass) materials. Another purpose is to compare the theoretical heat transfer predictions with the experimental heat transfer results from the ORNL data and to quantify the relative importance of the radiative component of the

---

* Graduate Assistant
** Professor, Department of Mechanical Engineering

overall heat transfer for different density fiberglass insulations. The numerical model was formulated using a control volume approach[7] for solving the energy equation in conjunction with the one-dimensional radiative transport equation which is solved using the method of Chandrasekhar[8].

PROBLEM STATEMENT

The physical geometry of a single layer of fiberglass insulation is depicted in Fig. 1. The origin of the coordinate system was taken at the substrate of the insulation. The angle between the positive y direction and the propagated direction of a radiation ray defines the polar angle, θ. The top interface temperature, T_0, and the substrate temperature, T_S, completely define the energy equation boundary conditions.

The single-layer, one-dimensional insulation problem is modeled to have coupled conductive and radiative heat transfer. The fiberglass insulation is treated as a layer of air containing a matrix of glass fibers and is taken to be an absorbing, emitting and scattering medium. Scattering and absorption account for radiation attenuation, while emission and scattering result in an augmentation process. The orientation of the fibers and the density account for scattering in fiberglass. The basic assumptions made in modeling the heat transfer characteristics of the fiberglass insulations are listed as follows: 1) one-dimensional, coupled conductive and radiative heat transfer, 2) axisymmetric scattering, 3) isotropic scattering, 4) the top insulation interface is transparent to radiation, and 5) constant bulk density.

The radiative transport model described above was also employed to determine the spectral radiative properties[6] which are used for the present heat transfer calculations. The gray isotropic extinction coefficients (β) and albedo values (ω) for the SRM 1451 and SRM 1450(b) which are used in the present case are summarized in Table 1. The temperatures for the top and bottom surfaces were the same as the experiments performed at the ORNL for the two fiberglass materials[9,10].

ANALYSIS

The one-dimensional axisymmetric integro-differential equation of radiative transfer and the one-dimensional energy equation constitute the governing equations for the present heat transfer analysis. The iterative solution process for this problem involves two major steps: 1) solving the radiative transport equation using a guessed temperature field, and 2) updating the temperature field by solving the energy equation in an iterative fashion employing a control-volume based finite difference approach; these steps are repeated until convergence.

The one-dimensional radiative transport equation for a plane parallel medium which absorbs, emits, and isotropically scatters is given by,[6]

$$\mu \frac{dI(\tau,\mu)}{d\tau} = -I(\tau,\mu) + \frac{\omega}{2} \int_{-1}^{1} I(\tau,\mu')d\mu'$$

$$+ n^2(1-\omega) I_b(T(\tau)) \qquad (1)$$

with boundary conditions

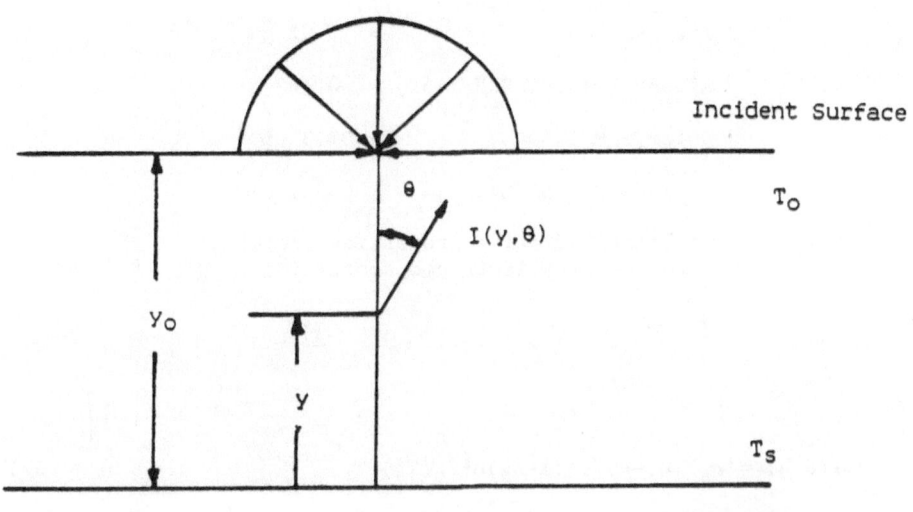

Fig. 1 Geometry of single layer of insulation with coordinate
system.

Table 1. Thermal and radiative properties
for fiberglass

Density of 1451, low density
fiberglass (γ_{f1})   = 9.25 kg/m^3
Density of 1450(b), high density
fiberglass (γ_{fh})   = 129.0 kg/m^3

True thermal conductivity
(k_{true} for fiberglass[12])

$$= a + bT + 8.55374 \times 10^{-5} \, \gamma_f$$

$$a + bT = k_{air}$$
$$= 4.976 \times 10^{-3} + 7.000 \times 10^{-5} \, T$$
(k_{true} in W/mK ; T in degrees K)

Extinction coefficient* (β)
for SRM 1451   =   3.92 cm^{-1}

Albedo (ω) for SRM 1451   = 0.162

β for SRM 1450(b)   = 37.3 cm^{-1}

ω for SRM 1450(b)   = 0.201 cm^{-1}

*All the radiative properties mentioned
are for gray isotropic scattering

$$I(0,\mu) = \rho_s I(0,-\mu) + (1-\rho_s)n^2 I_b(T_s) \qquad (2a)$$

$$I(\tau_0,-\mu) = \rho_n I(\tau_0,-\mu) + (1-\rho_n)n^2 I_b(T_0) \qquad (2b)$$

Here, τ is the optical depth, μ is the cosine of the polar angle θ,
$I(\tau,\mu)$ is the radiative intensity at depth τ in direction μ, n is the
refractive index of the medium (here n=1), I_b is Planck's blackbody
intensity function, and ρ_s and ρ_n are the substrate and top surface
reflectivities respectively. The net radiative flux at depth τ is
obtained by integrating the intensity from:

$$q_r(\tau) = 2\Pi \int_0^\infty \int_{-1}^1 I(\tau,\mu)\mu d\mu d\lambda \qquad (3)$$

where λ refers to the wavelength integration. The one-dimensional
energy equation for transient coupled conduction and radiation in a
planar medium is given by

$$\frac{\partial}{\partial y}(k \frac{\partial T}{\partial y}) - \frac{\partial q_r}{\partial y} = \frac{\partial}{\partial t}(\gamma \, C_p T) \qquad (4)$$

where, γ is the density of the fiberglass, C_p is the specific heat of
the medium, k (or, k_{true}) is the true thermal conductivity of the

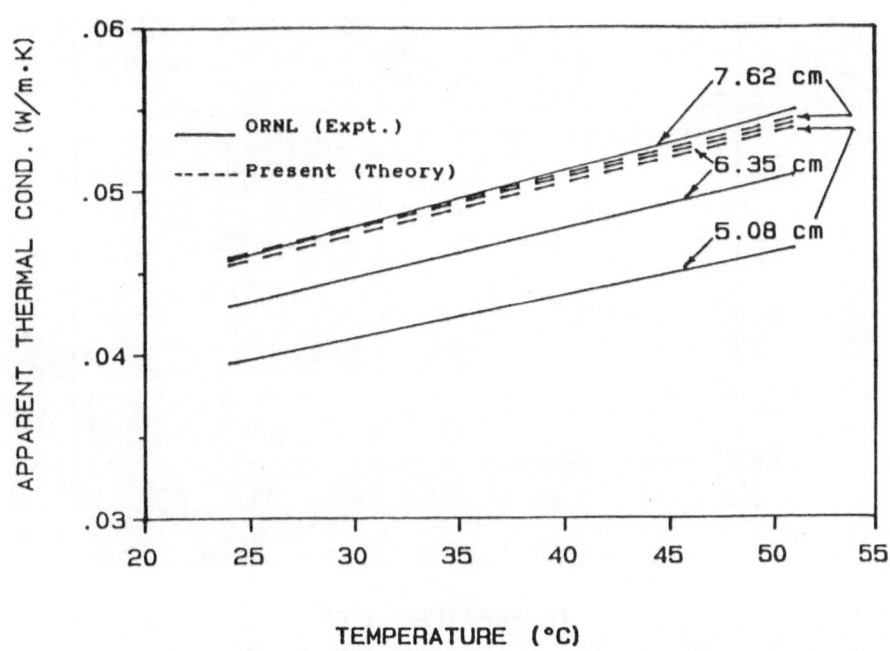

Fig. 2 Comparison of temperature dependence of theoretical and experimental apparent thermal conductivity for low density fiberglass with constant density (= 10 kg/m^3).

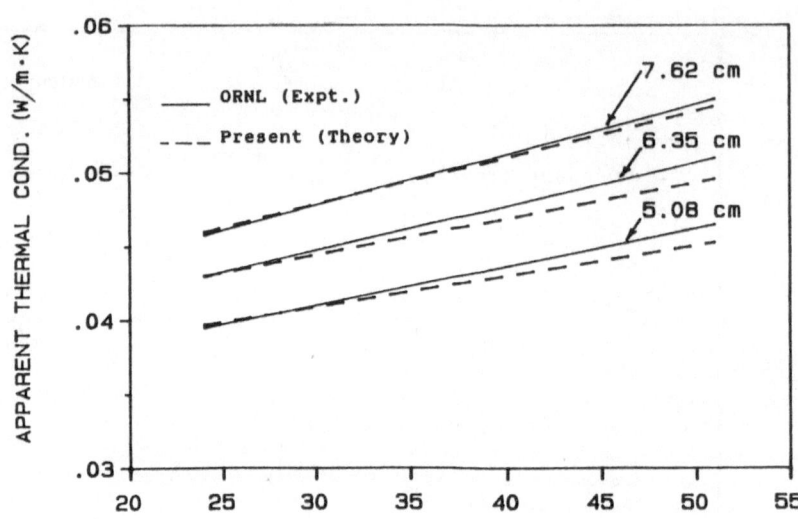

TEMPERATURE (°C)

Fig. 3   Comparison of temperature dependence of theoretical and
experimental apparent thermal conductivity for low density
fiberglass with variable density.

fiberglass. The fiberglass top interface temperature (T_0) and substrate temperature (T_s) comprise the boundary conditions for Eq. (4) and were known constants for a steady-state analysis from the actual experimental measurements at ORNL.

Discretization of the energy equation was accomplished by using a control-volume based finite difference scheme. Temperature nodes were located within each control volume; the grid size being $0.0033y_0$ (fine) near the boundaries and $0.2y_0$ (coarse) away from the boundaries. The transient computer program marched to steady-state to give the total heat flux (q_{total}) which is the sum of the conductive component (q_c) and the radiative component (q_r) of the overall heat transfer.

RESULTS AND CONCLUSIONS

The gray radiative properties listed in Table 1 were input into the heat transfer program for solving the energy equation in a one-dimensional plane insulation layer with coupled radiation and conduction. The k_{true} values for fiberglass were obtained from Houston[12]. From the theoretical steady-state heat transfer values, the theoretical "apparent thermal conductivity" was computed for both insulation materials. "Apparent thermal conductivity" is defined as follows:

$$k_{app} = \frac{q \cdot d}{\Delta T} \qquad (5)$$

where, q is the steady-state heat flux in a one-dimensional plane insulation layer; d is the thickness of the layer; and ΔT (T_0-T_s) is the temperature difference between the top and bottom surfaces of the insulation. It should be noted that since k_{app} is proportional to the overall heat transfer, percentage changes in k_{app} and q are the same.

The theoretical "apparent thermal conductivity" predictions were compared to experimental "apparent thermal conductivity" values obtained from measurements performed at the Oak Ridge National Laboratory (ORNL) and the National Bureau of Standards (NBS). The experiments to determine the experimental "apparent (or effective) thermal conductivity" values, performed at the ORNL used a large unguarded hot plate apparatus. This apparatus uses an electrically heated nichrome wire screen as the heat source and the hot boundary for vertical, one-dimensional steady-state heat flow[13].

A comparison of the experimental and theoretical k_{app} values for the low density fiberglass (1451) is shown in Fig. 2. From Fig. 2 it is clear that, although the theoretical temperature dependence of the apparent thermal conductivity (k_{app}) agrees to within 3% of the experimental values for the 7.62 cm fiberblanket, the "thickness effect" (the shift in the k_{app} with change in thickness) is not observed in the theoretical results. This is physically realistic, since a small change in the thickness should not result in a large change in k_{app}. The "thickness effect" was found to be really a "density effect" since the same 7.62 cm thick sample was compressed to a thinner fiberblanket in the experimental measurements. Changing the density by the same factor as the thickness resulted in good agreement between the theoretical and experimental k_{app} (Fig. 3). The three experimental k_{app} curves in Fig. 3 agree very well with the theoretical k_{app} curves, the largest error of $\approx$4% was observed for the 6.35 cm thick fiberblanket at 50°C (122°F). It is also observed from Fig. 3 that the apparent thermal conductivity increased by about 15 % from low mean temperatures (75°F or $\approx$24°C) to

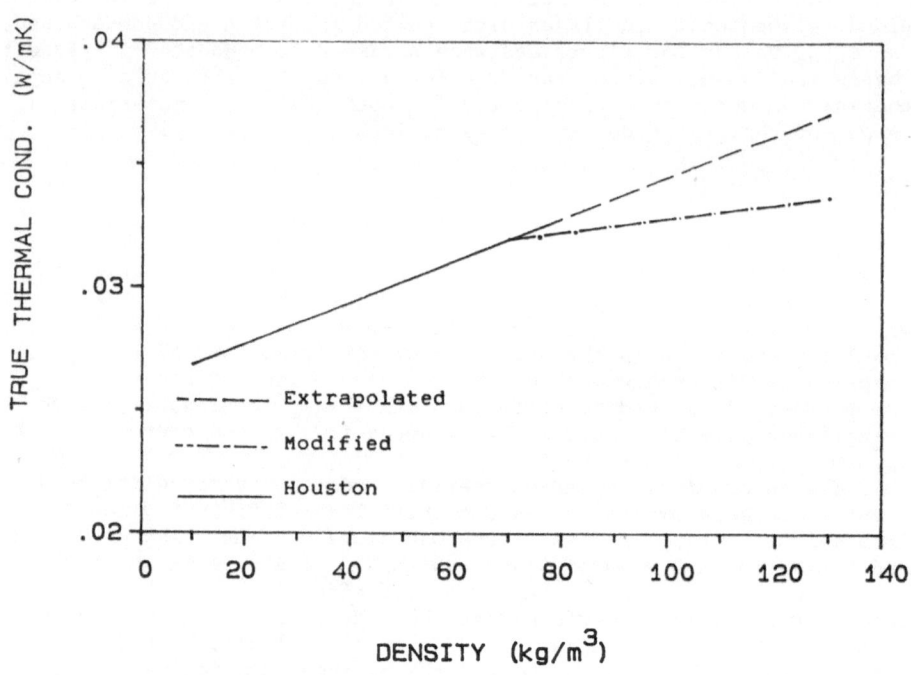

Fig. 4 Density dependence of true thermal conductivity up to 130 kg/m^3 for fiberglass at 24 ^{o}C.

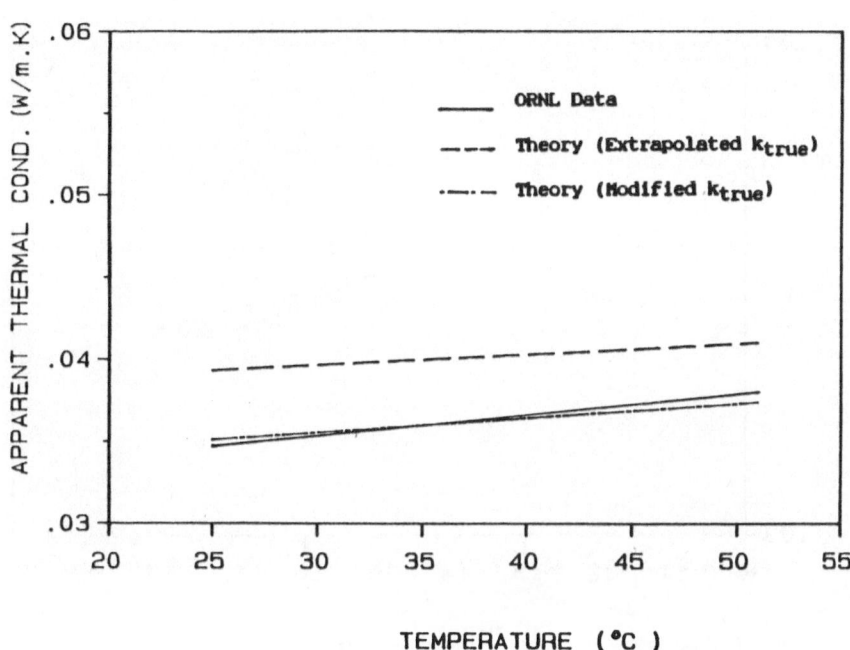

Fig. 5 Comparison of temperature dependence of theoretical and experimental apparent thermal conductivity for high density fiberglass (density = 129 kg/m^3).

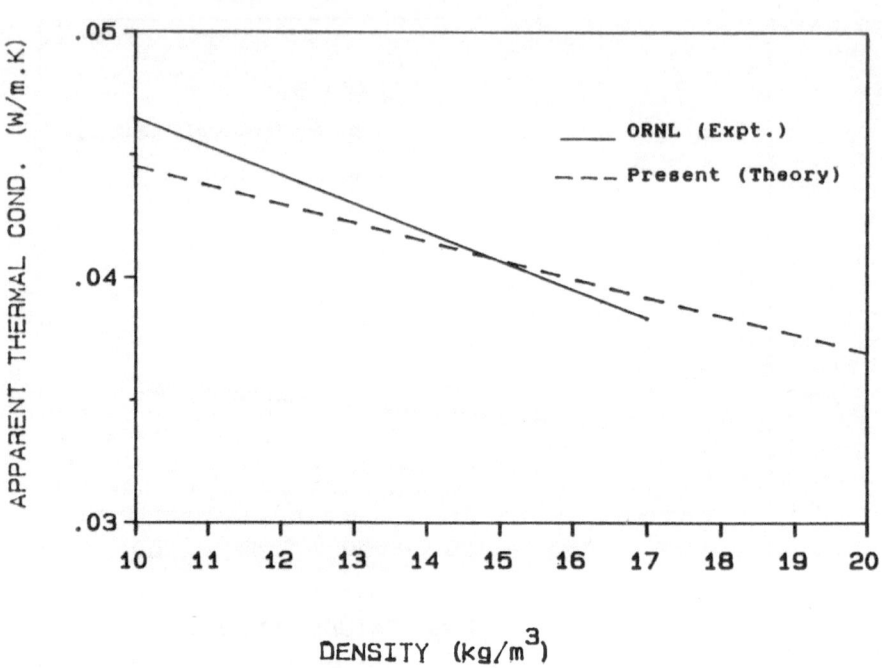

Fig. 6 Comparison of density dependence of theoretical and
experimental apparent thermal conductivity upto 20 kg/m^3
for low density fiberglass at 24 °C.

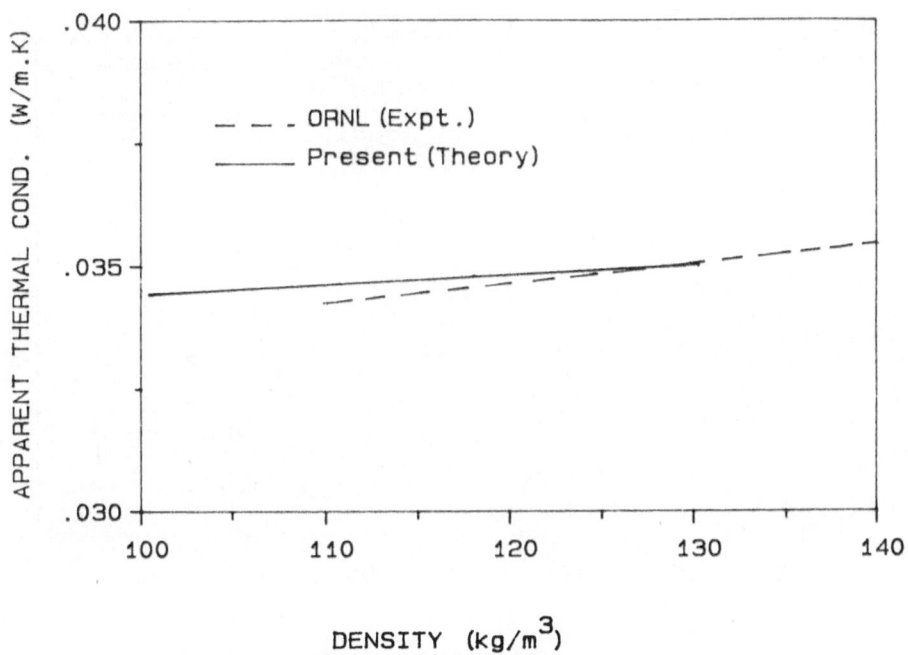

Fig. 7 Comparison of density dependence of theoretical and experimental apparent thermal conductivity for high density fiberglass (100–140 kg/m^3) at 24 °C.

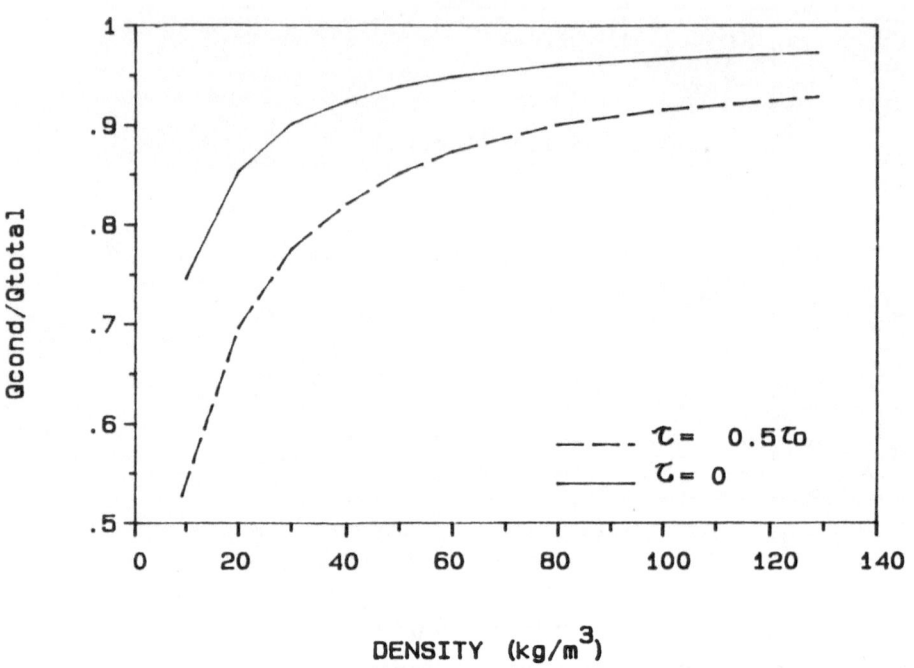

Fig. 8   Relative importance of conduction and radiation as a
function of fiberglass density at 24 °C.

higher mean temperatures (122°F or 50°C). This is a significant result since it implies that the low density fiberglass (1451) is more effective in winter than in summer. In fact this was found to be true for both fiberglass insulations. This temperature dependence is essentially the same as the temperature dependence of the air thermal conductivity.

As in the low density fiberblanket, determination of k_{app} for the high density fiberboard (129 kg/m^3) required gray isotropic radiative properties and k_{true} values for the SRM 1450(b). The radiative properties used are listed in Table 1; the k_{true} for fiberglass obtained from Houston[12] was, however, valid only up to fiberglass densities of ≈70 kg/m^3. Since extrapolation beyond 70 kg/m^3 resulted in unrealistically high k_{true} values, a modified k_{true} curve was used for the high density fiberglass. The modified k_{true} curve as a function of density is shown in Fig. 4. The dashed and dotted line (modified k_{true}) in Fig. 4 shows that the extrapolation (dashed line in Fig. 4) of the original[12] k_{true} (solid line in Fig. 4) is indeed very high. A comparison of the theoretical and experimental k_{app} values for the SRM 1450(b) (high density fiberglass) is shown in Fig. 5. The difference between the theoretical (dashed-line in Fig. 5) and the experimental curve was found to be considerable due to the high k_{true} value (for the extrapolated k_{true}). It was found that using the modified linear interpolation for k_{true} from 70-129 kg/m^3 reduced the difference between the modified theoretical and experimental k_{app} values to about 2% (dashed and dotted line in Fig. 5).

Experimental data from the ORNL showed a strong density dependence of the k_{app} values for the low density fiberglass insulations. This has also been indicated by the "density effect" of Fig. 3. The experimental k_{app} values for the high density fiberglass were a weak increasing function of density. Using the gray β (extinction coefficient) and ω (albedo) values for the low and high density fiberglass, theoretical k_{app} values were determined for fiberglass densities from 10 to 20 kg/m^3 and from 110 to 130 kg/m^3. A comparison of the theoretical and experimental k_{app} values for densities between 10 and 20 kg/m^3 is given in Fig. 6 and for densities between 110 and 130 kg/m^3 in Figure 7. Agreement to within 3% is observed between the experimental and theoretical curves for the low density fiberglass; the high density fiberglass k_{app} curves agreed to within 2% (Fig. 7). It is important to note the steep negative slope of the k_{app} values with respect to the fiberglass density as compared to the slight rise (less than 5%) in the k_{app} values for the high density fiberglass. This is a significant result since, a small density change (10 to 20 kg/m^3) improves the effectiveness of the low density fiberglass by more than 20% for a mean temperature of 24°C.

The extinction coefficient, β, of the high density fiberglass (129 kg/m^3) was about 10 times that for low density fiberglass (9.25 kg/m^3). This implies that the radiative component of the overall heat transfer should be a strong function of the fiberglass density. An increase in β implies a decrease in the radiative component of the overall heat transfer. This is evident from the Rosseland equation which is applicable to the optically thick limit (a large central region of the insulation); this equation is given as,

$$Q_{total} = (k_{true} + 16\sigma T^3/3\beta)dT/dy \qquad (6)$$

The first term in the brackets corresponds to the conduction component and the second term in the brackets corresponds to the radiative component. Here σ is Planck's constant. Equation (6) implies a decrease in the radiative component with increasing β values. This is also observed in Fig. 8 which gives the relative heat transfer importance of the radiative component in fiberglass insulations. Figure 8 shows the ratio of the predicted conduction component of the heat transfer to the overall heat transfer as a function of fiberglass density. Figure 8 was obtained using interpolated β values and the modified k_{true} (Fig. 4) for fiberglass. It is evident from Fig. 8 that for densities above 80 kg/m^3 the radiative component decreases to less than 10% of the overall heat transfer. Also, for fiberglass insulations, an increase in albedo (ω) implies an increase in the radiative component of the overall heat transfer. However, the overall heat transfer was not found to be sensitive to changes in the albedo values. Moreover, the gray albedo was found to be almost independent of the fiberglass density (≈ 0.20).

It is recommended that heat transfer at intermediate density (between 9.25 kg/m^3 and 129 kg/m^3) fiberglass insulations be investigated for two reasons: 1) The k_{app} values for high density fiberglass (Fig. 7) were found to be greater than the projected k_{app} value (at 20 kg/m^3) in Fig. 6. This implies that there should be a minimum in the k_{app} versus density curve for fiberglass. Optimization of the fiberglass insulations should be pursued. 2) The predictions pertaining to the relative importance of radiation (Fig. 8) for fiberglass densities can be confirmed by investigating the heat transfer characteristics of various density fiberglass insulations.

ACKNOWLEDGEMENT

This work was funded by the Department of Energy and the Oak Ridge National Laboratory under Contract No. 19X-55930C.

REFERENCES

1. Verschoor, J.D., and Greebler, P., "Heat Transfer by Gas Conduction and Radiation in Fibrous Insulations," Trans. of ASME, Vol. 74, 1952, pp. 961-968.
2. Pellane, C.M., "Heat Flow Principles in Thermal Insulations," J. Therm. Ins., Vol. 1, July 1977, pp. 48-80.
3. Houston, R.L., and Korpela, S.A., "Heat Transfer Through Fiberglass Insulations," Proceedings of the 7th Int. Heat Transfer Conference, Vol. 2, 1982, pp. 499-504.
4. Tong, T.W., and Tien, C.L., "Insulations — Part I: Analytical Study," J. Heat Transfer, Vol. 105, No. 1, 1983, pp. 70-75.
5. Rish, III, J.W., and Roux, J.A., "Heat Transfer Analysis of Fibrous Insulations With and Without Radiant Barriers for Summer Conditions," Journal of Thermophysics and Heat Transfer, Vol. 1, No. 1, January 1987, pp. 43-49.
6. Yajnik, S., "Determination of Radiative Properties of Fiberglass and Foam Insulations," Masters Thesis, University of Mississippi, University, MS., May 1987.
7. Patankar, S.V., Numerical Heat Transfer and Fluid Flow, Hemisphere Publishing Corp., Washington, D. C., 1980.
8. Chandrasekhar, S., Radiative Transfer, Dover, New York, 1960.
9. Hust, J.G., "Glass Fiberboard SRM for Thermal Resistance", NBS Special Publication 260-98, August 1985, pp. 1-20.
10. Hust, J.G., "Glass Fiberblanket SRM for Thermal Resistance", NBS Special Publication 260-103, Sept. 1985, pp. 1-15.

11. Yeh, H.Y., "Radiative Properties and Heat Transfer Analysis of Fibrous Insulations," _Ph.D. Dissertation_, 1985, University of Mississippi, University, MS., May 1986.
12. Houston, R.L., "Combined Radiation and Conduction in a Non-gray Participating Medium that Absorbs, Emits, and Anisotropically Scatters," _Ph. D. Dissertation_, Ohio State University, 1980.
13. McElroy, D.L., Graves, R.S., Yarbrough, D.W., and Moore, J.P.,"A Flat Insulation Tester that Uses an Unguarded Nichrome Screen Wire Heater," _American Society for Testing and Materials_, 1985, pp. 121-139.

THERMAL CONDUCTIVITY OF PVC FOAMS
FOR SPACECRAFT APPLICATIONS

D. D. Burleigh and J. Gagliani

Materials and Processes Engineering
General Dynamics Space Systems Division
P.O. Box 85990
San Diego, CA 92138

ABSTRACT

The thermal conductivity of 13 closed-cell PVC foams has been measured at nominal mean specimen temperatures of 100K, 215K, and 310K (ambient). The densities of these foams vary from 35 to 330 kg/m³. The relationship between density and thermal conductivity is shown.

INTRODUCTION

Polyvinyl chloride (PVC) foam is widely used in many applications. Its most desirable property is its low thermal conductivity, which it has in common with most other polymeric foams. PVC foam is semi-rigid so it may be used structurally. It is commonly used as a core material in sandwich panels in buildings, railroad cars, cargo trailers, and aircraft, spacecraft, and marine applications.

Other attributes include low water absorption and flame resistance. Special grades of PVC can be used for short periods of time up to a temperature of 500F, but when heated above 650K (700F), they decompose and exude HCL.

MATERIAL

The materials tested in this work were CO_2 blown, closed-cell PVC foams made by Diab Barracuda under the trademark of Divinycell. Two different grades of foam were tested, the H series and the HT series. The H series material has a maximum use temperature of 345K (160F); the HT series material will tolerate temperature up to 395K (250F). Another PVC foam, Cellulaire H920A, was also tested for comparison.

TEST METHOD

The test method used in the evaluation of the thermal conductivity of the foams is based in principle on ASTM C177. The same method was used previously for evaluation of the thermal conductivity of polyimide foams.[1]

Two thin circular concentric electrical heaters were sandwiched between two specimen panels whose dimensions are approximately 15 by 15 by 1 centimeter thick. The temperatures of the specimen surfaces and of the center and guard heaters were measured by twelve 36-gauge chromel-constantan (Type E) thermocouples.

At equilibrium, the electrical power to the center heater and the average hot and cold face temperatures were recorded. Thermal conductivity, λ, was calculated from the relationship:

$$\lambda = C\,(EI)t/A\Delta T$$

where: C = geometry and unit factor

(EI) = center heater power

t = average center-section thickness

A = effective center-section area

ΔT = temperature difference between hot and cold faces

The heat sinks used in this testing were a pair of aluminum plates with interior channels through which coolant can be pumped. For the lowest temperatures (100K), liquid nitrogen was the coolant. Chilling the outer surfaces of the plates with blocks of dry ice produced a specimen temperature of 210K. Circulating water through the plates resulted in a specimen temperature of near ambient, roughly 310K.

A plastic bag surrounded the specimen stack and a dry nitrogen gas purge flowed around the specimen at a rate of about 0.05 liters/second. Figure 1 shows the test stack with the purge bag and cooling plates.

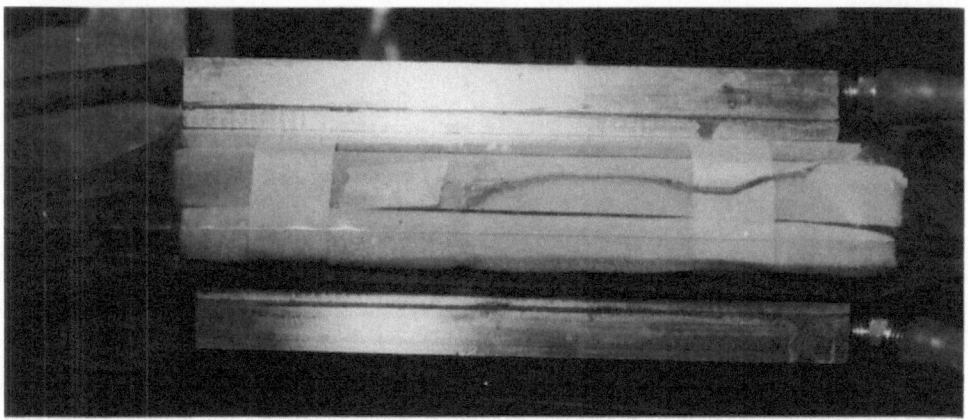

Figure 1. Detail of apparatus showing cooling plates, purge bag, and specimens.

The test apparatus is shown in Figure 2. The apparatus has been calibrated by measuring the thermal conductance of NBS standard reference material (SRM) 1450B. Based on these measurements, the accuracy is estimated to be 2% at ambient temperatures. The densities of the foams were measured by standard methods.

TEST RESULTS

It was found that foams selected from random manufacturers' lots do not always follow a coherent relationship between thermal conductivity (λ) and density (ρ). When foams are purchased, conductivity is rarely specified, and the foam production processes are not designed for controlling this property. We originally tested a set of H foams that were taken from manufacturers' stock; they had been produced at several factories and possibly with different processes. When we tested these foams, the relationship between λ and ρ was not coherent. We requested that the manufacturer provide us with another uniform set of H foams. These foams were tested, and the data is reported in this paper. In contrast, the HT series material was only manufactured by one factory, and the test results were found to be satisfactory.

We had originally hoped to measure the cell wall thickness and the average cell size and correlate these properties to density and thermal conductivity. Measurement of the first two quantities proved impossible, however, so this effort was abandoned.

Figure 2. Thermal conductivity test apparatus.

It is known that the graph of the relationship between thermal conductivity and density for many foams and fibrous insulations exhibits a curve that is U-shaped (concave up).[2] The high thermal conductivity at low density is the result of heat connection through the discontinuities of the cellular structure and that at high density is determined by the thermal conductivity properties of the solid material. We have found a similar curve for a family of PVC closed-cell foam.

Test data is shown in Table 1 and Figures 3 through 5. In Figure 3 this concave relationship is shown for the H series foams. However, for the HT series we did not find this relationship because foams of lower density were not available.

Table 1. Thermal Conductivity of PVC Foams

MATERIAL	DENSITY (kg/m³)	THERMAL CONDUCTIVITY (W/mK)		
		100K	210K	310K
H30	36.4	—	—	0.031
H45	47.0	—	—	0.031
H60	66.0	—	—	0.030
H100	105.6	—	—	0.030
H130	127.2	—	—	0.033
H200	190.4	—	—	0.035
H250	330.3	—	—	0.063
HT50	49.6	0.015	0.027	0.034
HT70	68.7	0.016	0.029	0.037
HT90	98.6	0.019	0.032	0.041
HT110	94.9	0.018	0.032	0.040
H920A	45.3	0.014	0.028	0.039

NOTE: SPECIMEN TEMPERATURES ARE NOMINAL ($\pm$4K)

CONCLUSION

The relationship between thermal conductivity and density of foam materials can be important in the selection of insulation systems for aerospace structures.

From Figure 3 it is clear that due to a negative slope in a portion of the λ versus ρ curve for the H series foams, the minimum thermal conductivity of this material system occurs at a material density significantly higher than the minimum material density tested.

As a result, a thin wall of dense foam could be an equivalent thermal insulator to a thicker wall of less dense foam. Furthermore, since less of the dense foam is required for equivalent thermal

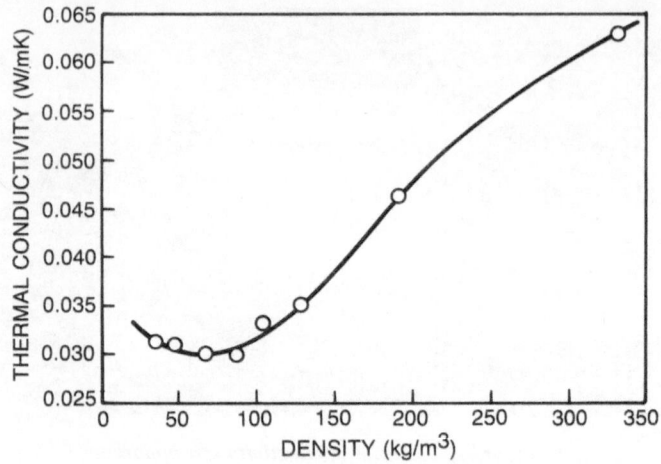

Figure 3. Thermal conductivity of H series; Divinycell PVC foams versus density.

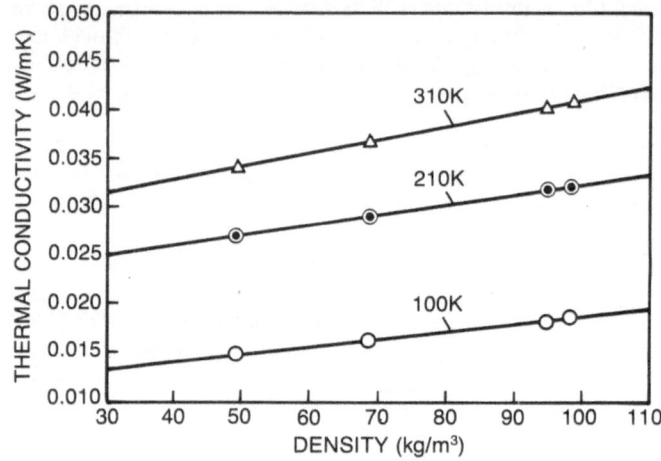

Figure 4. Thermal conductivity of HT series; Divinycell PVC foams versus density.

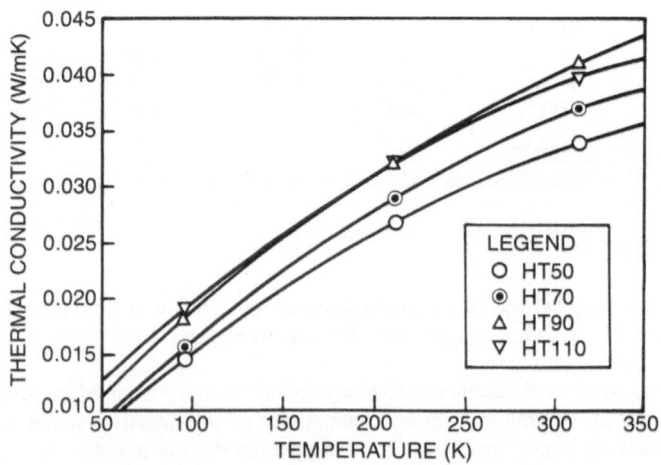

Figure 5. Thermal conductivity of HT series; Divinycell PVC foams versus temperature.

conductance, the weight of the two systems may be equivalent. The selection of the dense foam may carry no weight penalty and would therefore be justified since it would provide benefits such as improved mechanical properties, increased impact resistance, and reduced aerodynamic cross section if used on external surfaces.

REFERENCES

1. D. Burleigh, "Thermal Conductivity of a Polyimide Foam," *Thermal Conductivity 18,* Proceedings of the 18th International Conference on Thermal Conductivity, October 3-5, 1983, pp. 437-433 (1985).

2. J. H. Saunders and K. C. Frisch, *Polyurethanes: Chemistry and Technology, Part II,* Wiley & Sons, p. 252 (1964). Data attributed to F. O. Guenther, SPE Regional Technical Conference, October, 1961.

THERMOPHYSICAL PROPERTY MEASUREMENT IN BUILDING INSULATION MATERIALS

USING A SPHERICALLY-SHAPED, SELF-HEATED PROBE

Brian P. Dougherty and
William C. Thomas

Mechanical Engineering Department
Virginia Polytechnic Institute and State University
Blacksburg, VA

ABSTRACT

A technique for making in situ thermal conductivity and diffusivity measurements of building insulation materials was investigated. The two thermophysical properties are simultaneously determined given the result from a single transient test. The measurement apparatus consists of an effectively spherical, small-volume probe that is interfaced with a computer-based data acquisition and control system. Although different small-volume probes were evaluated, encapsulated bead thermistors received the most interest and so are emphasized in this paper. The results from using the technique to measure the thermal conductivity of five insulation materials are included.

NOMENCLATURE

A = area, m^2

a = probe effective radius, m

e = feedback error, $^\circ C$

i = current, A

K = feedback gain, $mA/^\circ C$

k = thermal conductivity, $W/m \cdot ^\circ C$

q = heat transfer rate, W

R = thermistor resistance, Ω

R^2 = statistical parameter

r = spherical coordinate, m

T = temperature, $^\circ C$

t = time, s

Greek Symbols

α = thermal diffusivity, m^2/s

δ = slope of the $q(t)$ vs $t^{-1/2}$ plot, $W \cdot s^{1/2}$

Subscripts

a = actual
b = thermistor bead, theoretical model
c = correction
d = desired
m = medium
o = output
r = reference
s = surface
ss = steady-state
t = thermistor, physical probe

INTRODUCTION

A portable instrument for making in situ thermophysical property measurements of insulation materials would be a valuable tool for building energy analyses and determining the actual performance of insulation systems. While a practical in situ measurement device is not yet available, contributions to the evolution of such an instrument are the subject of this paper. The in situ measurement of thermal conductivity and thermal diffusivity would have obvious advantages over relying on the typical data tabulations in handbooks or manufacturer's literature. In the case of handbooks, only a small sampling of the commercially available products is documented. Moreover, documented property values are for new, dry materials. In practice, the thermal performance of insulation can degrade from material changes, settling, and moisture accumulation.

A transient technique for measuring the thermal conductivity and diffusivity of insulation materials has been investigated. The technique incorporates a probe equipped with a small-volume sensor element that is interfaced with a computer-based data acquisition and control system. Once inserted into the insulation, the active probe element simultaneously acts as a small heater and as a temperature sensor. The probe is heated suddenly such that its temperature response approximates a step change. The computer controls the temperature response and also monitors the power dissipation vs. time. The computer with the appropriate statistical software is subsequently used to reduce the large amount of data collected. From the statistical analysis, the slope and ordinate intercept of the characteristic linear plot (power vs. the inverse square root of time) are determined. Simple algebraic expressions in terms of these parameters are then solved to give the two thermophysical properties.

The identification of suitable small-volume sensor elements is a major applications consideration. The ideal probe, besides possessing the dual characteristics of a heater and a temperature sensor, would have spherical geometry and negligible mass. The measured temperature would be that of the isothermal surface of the adjacent test material. The operation of a physical probe can only approach that of the ideal instrument. When the investigation began, bead thermistors were apparently the only device that had been identified for this measurement application [1]. Although other probe types were evaluated during the study, the use of an encapsulated bead thermistor as the active probe received the most attention and is emphasized in this paper. (Evaluation of a second probe design, designated the "two-sensor" probe, was initiated in the investigation and reference 2 details the progress with this alternative probe type.)

BACKGROUND

An encapsulated bead thermistor is suited for the small-volume heating and temperature measurement application. A thermistor is a resistance-based sensor conventionally employed as a temperature transducer. Thermistors, which are ceramic semiconductors, can be self-heated to an arbitrary level above the ambient medium temperature by increasing the impressed signal voltage. Of the different thermistor types, bead thermistors more nearly approximate a small sphere. Encapsulation of the active thermistor element prevents electrical shorting when the probe is submerged in an electrically conducting liquid or embedded in a wetted insulating material.

Thermistors have high resolution and acceptable repeatability when used as temperature transducers. An equation of the form

$$T = B + D \ln(R) \tag{1}$$

fits the data well. Figure 1 shows a typical result.

The use of thermistors for thermophysical property measurements was apparently pioneered by researchers in the biomedical field [3,4]. Within the biomedical field, the thermistor-based measurement technique was used to determine thermal conductivity, thermal diffusivity, and perfusion rates of biomaterials [5,6]. The only known application of the technique to materials with conductivities in the range of building insulations was by Woodbury and Thomas [7]. They used the technique as part of an investigation to determine indirectly the moisture content of a glass fiber matrix by measuring the material's thermal conductivity. References 1 and 7 detail the correlations which were used for the determination of moisture distribution.

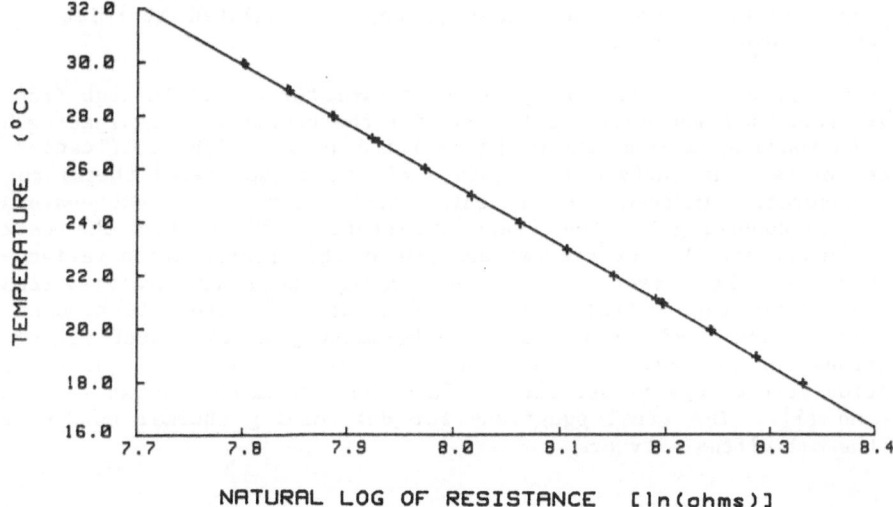

Fig. 1. Curve Fit to the Thermistor Temperature vs. Resistance Data

As compared to the previous researchers, a main innovation of the present work is to implement digital control of the temperature response of the thermistor. Analog controllers were used exclusively in prior investigations. The change to an all-digital system permitted a more reliable and efficient data acquisition and data reduction procedure. Following a typical 30-second test, the thermal conductivity of a sample is determined, using laboratory type equipment, in less than 15 minutes. An all-digital system also advances the prospects for a portable instrument using a dedicated microprocessor.

This paper emphasizes the application of the technique for measuring thermal conductivity. The use of a thermistor for measuring thermal diffusivity of insulation materials has yet to be validated.

THEORY

A theoretical analysis yields the starting point for designing an in situ measurement instrument. The analysis involves solving the one dimensional transient heat conduction equation for an infinite medium with a spherical cavity. The spherical boundary is modelled as experiencing a step temperature rise. The resulting temperature distribution within the infinite medium varies as a function of radial position and time [2,3]. The time-dependent power input required to maintain the boundary temperature is

$$q(t) = -k_m A_s \frac{\partial T_m(a,t)}{\partial r} = k_m \Delta T_s \, 4\pi a\left(1 + \frac{a}{\sqrt{\pi \alpha_m t}}\right) \qquad (2)$$

This equation shows that the ordinate intercept and slope of a plot of heat transfer rate (or, equivalently, the power dissipated) vs the inverse square root of time, $q(t)$ vs $t^{-1/2}$, depend directly on the thermal conductivity, k_m, and thermal diffusivity, α_m, of the medium. The previously described characteristics of the ideal probe are based on this analytical model. The probe, which corresponds to the spherical cavity, effects the boundary condition at the medium-probe interface. Although deviating from the ideal probe, encapsulated thermistors are suitable approximations.

The operation of the encapsulated thermistor differs enough from the ideal probe that the analytical model for the thermistor dissipating heat to a surrounding medium had to be modified [4,5]. The modification results because the surface temperature of the encapsulated thermistor is not measured. Instead, the measured temperature is a volume-averaged value corresponding to the least resistance path within the ceramic bead. Moreover, the surface temperature of the encapsulation varies as a function of time, even though the average bead temperature remains constant (once the initial transients die out). In order to account for the temperature gradient within the thermistor, coupled heat conduction equations for the bead and surrounding medium were solved. The complete solution methodology is detailed by Balasubramaniam and Bowman [4] and by Valvano [5]. The final equations for determining thermal conductivity and thermal diffusivity are

$$\frac{1}{k_m} = \frac{4\pi a \Delta T}{q_{ss}} - \frac{1}{5k_b} \qquad (3)$$

$$\alpha_m = \left[\frac{a}{\frac{\delta}{q_{ss}} \sqrt{\pi} \left\{ 1 + \frac{1}{5} \frac{k_m}{k_b} \right\}} \right]^2 \tag{4}$$

The parameters q_{ss} and δ relate to how the internal heat generation of the bead thermistor varies with time in yielding a volume-averaged temperature response approximating a theoretical step rise.

Both analytical and experimental approaches have been used to show that the power supplied to a thermistor must vary according to

$$q(t) = q_{ss} + \delta \, t^{-1/2} \qquad t > 0 \tag{5}$$

for the thermistor to undergo a temperature response approaching a step change [1,2,5]. Thus, although the analytical model for the physical thermistor differs from that for the ideal probe, the characteristic power vs the inverse square root of time relationship is common to both models. A typical thermistor temperature response curve is given in Fig. 2a. The initial abrupt transient is completed in approximately 3 seconds. Figure 2b shows that once the target temperature rise is obtained, the power varies linearly with $t^{-1/2}$. From this characteristic plot, the ordinate intercept yields q_{ss} while the slope gives δ.

The two thermistor properties, a and k_b, are determined by calibration using at least two materials with known thermal conductivities (k_m). The calibration test procedure is actually the same as for the measurements test. For calibration, equation 3 is rearranged as

$$\frac{1}{k_b} = 5Ca - \frac{5}{k_m} \tag{6}$$

where $C = 4\pi\Delta T/q_{ss}$. The two probe parameters are deduced from a plot of the resulting linear equations for each calibration medium. Although only two calibration materials are required, a minimum of three is recommended. Ideally, the calibration lines (equation 6) should intersect at a point. In practice, however, the intersection of three lines forms a triangle. The centroid is taken as the best approximation for determining the effective properties of the probe. Figure 3 shows the triangle formed from the three calibration curves for the thermistor referenced in this paper.

EXPERIMENTAL APPARATUS

The experimental apparatus is configured to permit two modes of probe operation. The actual measurement process begins by embedding the probe into a test medium sample and allowing enough time for the system to reach thermal equilibrium. Shortly before the actual transient measurements begin, the probe is used in a conventional mode (negligible self-heating) to measure the temperature of the sample. From this ambient temperature measurement and the specified temperature rise (either 2 or 3 °C), the target thermistor temperature is calculated. The temperature of the probe is quickly elevated to this target value by resistively heating the ceramic thermistor bead while digital feedback control is used to regulate the temperature response. The controlled variable is the current supplied to the probe. The thermistor voltage drop, the output current, and the resulting thermistor resistance (temperature) are recorded as functions of time. The test run terminates after a preset time, usually 30 seconds.

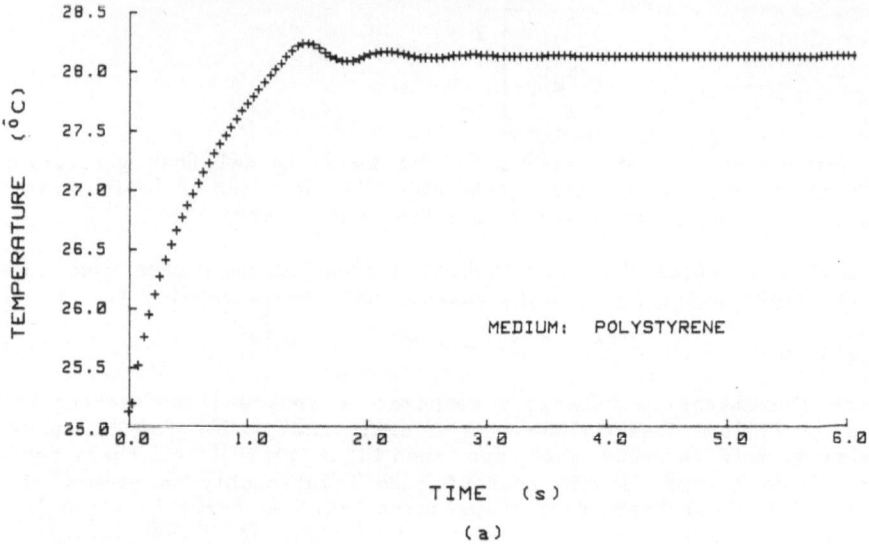

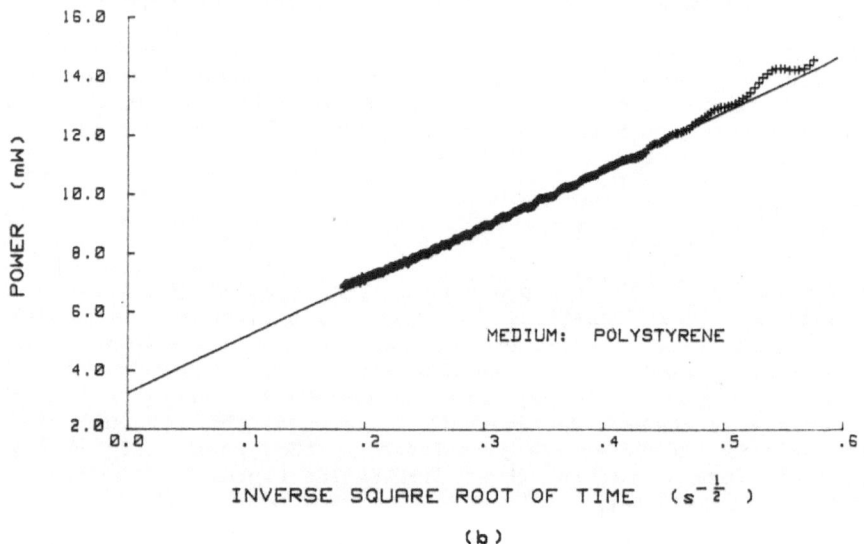

Fig. 2. Thermistor Characteristics:
(a) Approximating a Theoretical Step Temperature Rise
(b) Corresponding Probe Power Dissipation as a Function of $t^{-1/2}$

As shown in Fig. 4, a Hewlett-Packard 9836A Series 200 computer is an integral part of the laboratory apparatus for the two-mode operation. Besides input and output control, the computer:

(1) prompts the user for test parameters (e.g., desired temperature rise, total run time, etc.),

(2) executes the feedback control logic,

(3) records the pertinent run variables as a function of time, and

(4) prints and stores the collected data to a magnetic disc.

The power supply programmer (PSP, HP 59501B) and the current-programmable

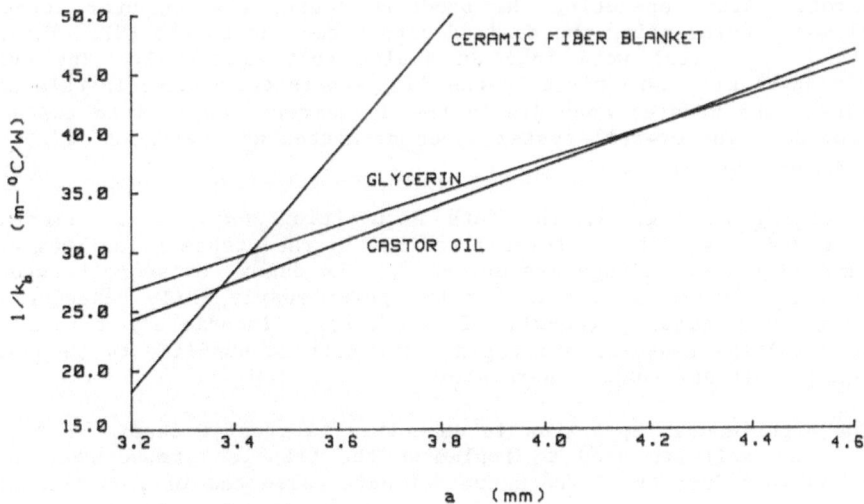

Fig. 3. Determination of Thermistor Effective Properties

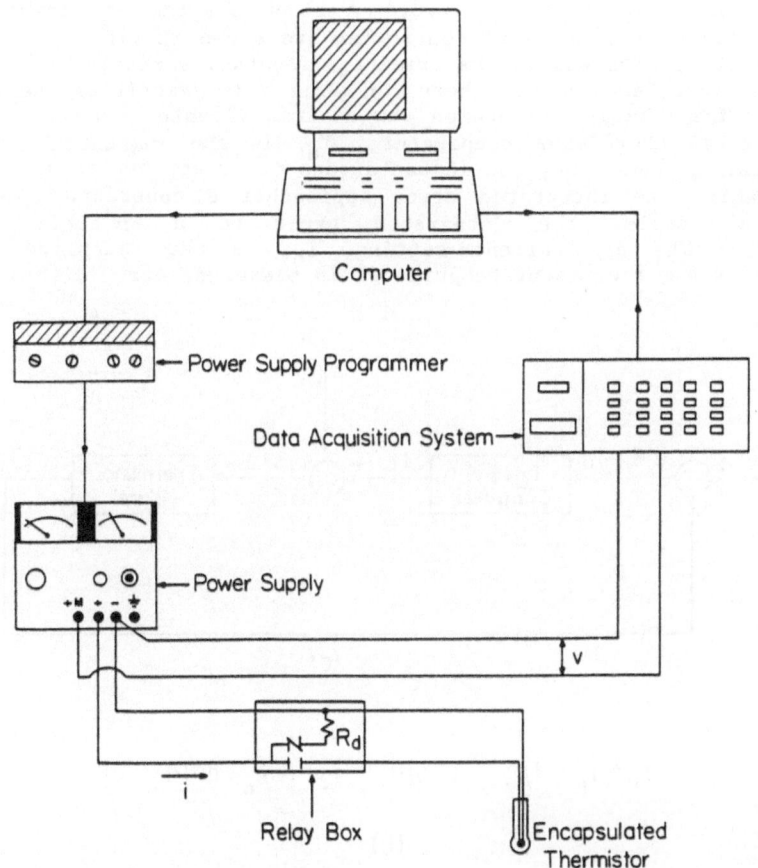

Fig. 4. Configuration of Experimental Apparatus

power supply (CPPS, HP 6186C) supplied the electrical current to the thermistor. After executing the feedback logic, the computer sends a digital word representing the desired output current to the PSP. The PSP converts this digital word into an analog voltage signal. The CPPS, which is physically hard-wired to the PSP, senses the change in this analog voltage and updates accordingly the dc current supplied to the load (thermistor). The overall system speed permitted an average of 21.6 updates per second.

Referring to Fig. 4, the data acquisition and control unit (HP 3497A) is used for both the initial thermistor resistance (i.e., temperature) and transient voltage measurements. The dummy resistor, housed in the relay box, provides a load for the power supply while reaching its initial current setting (usually full-scale). Immediately before the transient voltage measurements begin , the current supplied to the dummy resistor is switched to the thermistor.

The experimental apparatus is described in greater detail in reference 8. The software used to implement the transient measurement procedure and to reduce the large amount of data collected is also discussed in this reference.

An important part of the control algorithm was the implementation of digital-based logic for effecting the desired thermistor response. An integral control strategy was used. A block diagram and mathematical representation of the integral controller are shown in Fig. 5. Although thermistor resistance was the experimental feedback variable, temperature rather than resistance is used here for clarity in describing the control strategy. The difference between the desired elevated temperature, T_d, and the actual thermistor temperature, T_a, is the instantaneous error e. Integrating over time, the instantaneous error approaches zero (i.e., $T_a \rightarrow T_d$) while the integrated error approaches a constant. The gain setting, K, scales the integrated error to a suitable current correction. The new current setting, i_o, is then supplied to the thermistor, a new thermistor temperature is measured, etc.

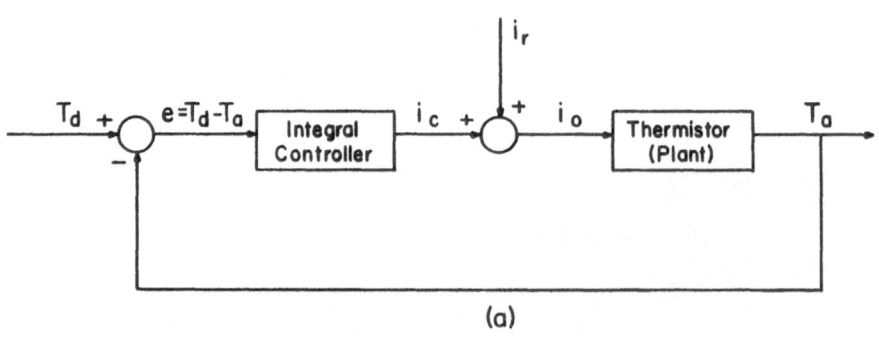

(a)

$$i_o = i_r + i_c \quad \text{where} \quad i_c = K \int_0^t e(t)\, dt$$

(b)

Fig. 5. Integral Feedback Control Stategy:
(a) Block Diagram
(b) Mathematical Representation

RESULTS AND DISCUSSION

For the calibration and measurement application, a teflon-encapsulated precision thermistor (YSI 44105) was used. The thermistor was evaluated as a measuring device with five insulation materials. These materials included a fibrous insulation batt with a density of 100 kg/m^3, a fibrous insulation blanket (50 kg/m^3), an alumina-silicate ceramic fiber blanket, a polystyrene sample, and a polymer foam slab. Multiple runs were conducted on each material. An initial temperature of 25°C was used for all tests.

The thermal conductivities of the insulation materials were also measured using a heat flux thermal conductivity test instrument (Dynatech R-MATIC). The operation of this instrument is based on ASTM C-518 [9]. The uncertainty associated with using the instrument is estimated to be ±5 per cent.

Probe Calibration

The thermistor was calibrated using glycerin, castor oil and the ceramic fiber blanket. Representative plots of power vs $t^{-1/2}$ for the seven materials are given in Fig. 6. The data plotted for the insulation materials is limited to that of the ceramic fiber blanket because the different data sets become indistinguishable. For each of the three calibration media, an expression of the form of equation 6 was obtained. The results from the multiple tests were averaged to yield a lone calibration equation. The thermistor average calibration equations are plotted in Fig. 3. From the centroid of the resulting triangle, the probe effective radius and reciprocal bead thermal conductivity were determined to be 3.66 mm and 32.4 m ^{o}C/W, respectively.

Only one insulation material was used for calibrating because the curves for the five materials are close together. The closeness occurs because the insulation materials all have conductivities in the neighborhood of 0.035 W/m·^{o}C. As the calibration curves converge, the location of the intersection of two such curves is much more strongly dependent on the uncertainties in both the calibration measurement and the calibration

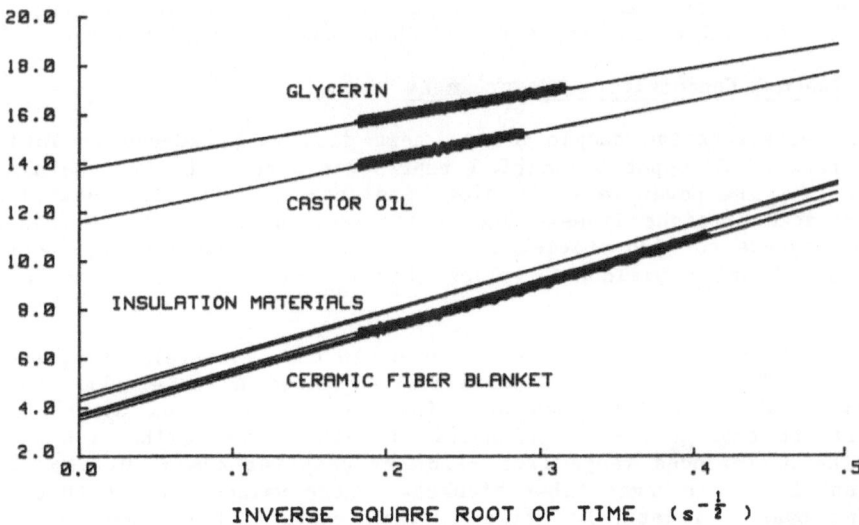

Fig. 6. Thermistor Characteristic Curves

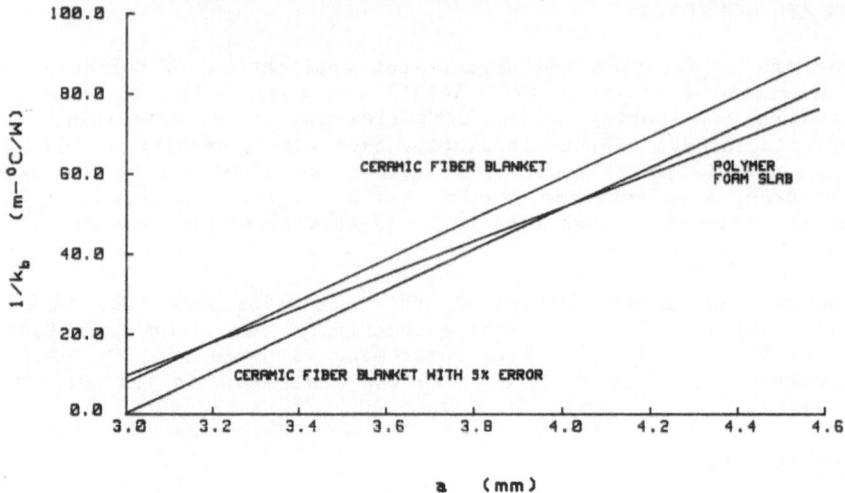

Fig. 7. Effects of Poor Reference Materials on Thermistor Effective
Properties

materials' thermal conductivities. Figure 7 shows the effect of a 5 per
cent error in the reference thermal conductivity of an insulation
material on the intersection point of two potential calibration curves.
The intersection point changes from 3.19 mm and 17.4 m $^{\circ}$C/W to 4.01 mm
and 52.3 m $^{\circ}$C/W. This change corresponds to a 25 per cent increase in
the effective radius and a 200 per cent increase in the reciprocal bead
thermal conductivity. The need for reliable standard reference materials
is obvious.

The choice of the ceramic fiber blanket as the calibration material
bracketing the lower end of the thermal conductivity interval was based
on a different consideration. The ceramic blanket was chosen because it
gave the best closure around the embedded thermistor probe. The reference thermal conductivities of all five insulation materials were measured using the heat flux meter. Consequently, the measured thermal conductivity for all these materials had about the same uncertainty.

Probe Thermal Conductivity Measurements

A representative sample of the test data is presented in Table 1.
Given that an R^2 upper bound of 1 represents a perfect fit, the high R^2
values from the power vs. $t^{-1/2}$ plots indicate that the data set closely
approximated straight lines. The consistency of the C values (equation
6) corresponds to high precision measurements. In particular, identical
test runs 5 and 6 yielded C values that differed by less than 0.1 percent.

The thermal conductivities of the seven test materials are listed in
Table 2. The calibration materials, designated by asterisks, are included to show the influence of using three calibration materials, as
compared to two, to form a triangle. For the three calibration materials, the differences range from −2.8 per cent for castor oil up to 6.1
per cent for the ceramic fiber blanket. These values suggest that calibrating over a relatively large thermal conductivity range and using
materials with significantly different compositions increases the probe

measurement uncertainty. The calibrated probe gave good performance for the majority of the insulating materials. The magnitudes of the differences ranged from 3.3 to 16.8 per cent.

Table 1. Thermistor Test Data

I.D. No.	Medium	T (^oC)	ΔT (^oC)	q_{ss} (mW)	R^2	C ($^oC/W$)
1	$100kg/m^3$	25.02	3	4.2745	0.9990	44097
2	insulation batt	25.07	3	4.2827	0.9992	44012
3		25.06	3	4.3051	0.9982	43783
4		25.04	3	4.2892	0.9993	43947
5		24.99	2	2.8665	0.9990	43838
6		24.99	2	2.8691	0.9990	43800
7		25.01	2	2.8198	0.9991	44565
8	castor oil	25.01	2	7.6979	0.9967	16324
9		25.04	3	11.588	0.9950	16266
10		25.05	3	11.679	0.9858	16139
11	polystyrene	25.09	2	2.3313	0.9976	53904
12		25.11	2	2.3479	0.9979	53522
13		25.11	3	3.4728	0.9981	54278
14		25.12	3	3.4791	0.9982	54180

Table 2. Thermistor Thermal Conductivity Results

Medium	Thermal Conductivity (W/m $\cdot$ ^oC)		Reference No.	Per Cent[@] Difference
	Thermistor k_t	Reference k_r		
castor oil*	0.185	0.180	10	-2.8
glycerin*	0.281	0.292	11,12	3.8
100 kg/m^3 insulation batt	0.0389	0.0333	R-MATIC	-16.8
polymer foam slab	0.0411	0.0429	R-MATIC	4.2
50 kg/m^3 insulation blanket	0.0325	0.0336	R-MATIC	3.3
ceramic fiber blanket*	0.0322	0.0343	R-MATIC	6.1
polystyrene	0.0303	0.0361	R-MATIC	16.1

*designates calibration material

[@]Per Cent Diff. $= \dfrac{(k_r - k_t)}{k_r} \times 100$

CONCLUSIONS

An in situ thermophysical measurement apparatus was investigated and developed further. Although the search for probes approaching the performance of the ideal probe continues, encapsulated thermistors are adequately suited for the measurement of thermal conductivity. The thermistor-based measurement technique was validated for measuring the thermal conductivity of insulation materials. The data acquisition and control portion of the apparatus was successfully developed. The integral control strategy closely controlled the temperature response of the self-heated thermistor. Following a typical 30-second thermal conductivity test, the thermal conductivity of the test sample was available in approximately 15 minutes.

The lack of suitable standard reference materials limits the performance of the calibration-dependent thermistor probe. Reference materials are needed which span a wide enough interval to avoid similar calibration curves. As with the tightly grouped insulation materials, the effective probe properties are strongly influenced by uncertainties in the thermal conductivities of reference materials.

ACKNOWLEDGEMENTS

This project was funded by the National Bureau of Standards (USDOC-NBS Grant No. 60NANB6D0642). The authors are grateful to the NBS Solar Equipment Group for the financial support and to its members for their assistance.

REFERENCES

1. Woodbury, K. A., "An Experimental and Analytical Investigation of Liquid Moisture Distribution in Roof Insulating Systems," Ph.D. Thesis in Mechanical Engineering, Virginia Polytechnic Institute and State University, August 1984.

2. Thomas, W. C., and Dougherty, B. P., "Moisture Distribution Measurement in Solar Insulation Materials," Final Report on USDOC-NBS Grant No. 60NANB6D0642 (to be published Jan. 1988).

3. Chato, J. C., "A Method for the Measurement of Thermal Properties of Biological Materials," Thermal Problems in Biotechnology, American Society of Mechanical Engineers, LCN068-58741, New York, 1968, pp. 16-25.

4. Balasubramaniam, T. A., and Bowman, H. F., "Temperature Field Due to a Time-Dependent Heat Source of Spherical Geometry in an Infinite Medium," Journal of Heat Transfer, Trans. ASME, Vol. 96, Series 3, No. 3, Aug. 1974, pp. 296-299.

5. Valvano, J. W., "The Use of Thermal Diffusivity to Quantify Tissue Perfusion," Ph.D. Thesis in Medical Engineering, Harvard University - MIT Division of Health Sciences and Technology, Aug. 1981.

6. Balasubramaniam, T. A., and Bowman, H. F., "Thermal Conductivity and Thermal Diffusivity of Biomaterials: A Simultaneous Measurement Technique," Journal of Biomechanical Engineering, Trans. ASME, Vol. 99, Aug. 1977, pp. 148-154.

7. Woodbury, K. A., and Thomas, W. C., "Effective Thermal Conductivity of Moisture-Laden Fiber Insulation Matrix," _Proceedings of the National Heat Transfer Conference_, Denver, Colorado, Aug. 4-7, 1985.

8. Dougherty, B. P., "An Automated Probe for Thermal Conductivity Measurements," Masters Thesis in Mechanical Engineering, Virginia Polytechnic Institute and State University, November 1987.

9. ASTM Standard C158-76, "Standard Test Method for Steady-State Thermal Transmission Properties by Means of the Heat Flow Meter," _Annual Book of ASTM Standards_, Vol. 14:01, 1984, pp. 47-78.

10. Touloukian, Y. S., "Thermophysical Properties of Matter," _The TPRC Data Series_, Vol. 3, IFI Plenum Press, New York, 1970.

11. Asher, G. B., "Development of a Computerized Thermal Conductivity Measurement System Utilizing the Transient Needle Probe Technique: An Application to Hydrates in Porous Media," Ph.D. Thesis in Chemical Engineering and Petroleum Refining, Colorado School of Mines, May 1987.

12. Takizawa, S., Murata, H., and Nagashima, A., "Measurement of the Thermal Conductivity of Liquids by Transient Hot-Wire Method," _Bulletin of the JSME_, Vol. 21, No. 152, Feb. 1978, pp. 273-278.

SESSION 2

METALLIC MATERIALS

THEORY OF LORENZ RATIO OF METALS AND ALLOYS

P. G. Klemens

Department of Physics and
Institute of Materials Science
University of Connecticut, Storrs, Connecticut 06268

The theoretical basis of the Wiedemann-Franz-Lorenz law is reviewed. The Lorenz ratio takes the Sommerfeld value L_o provided (a) the thermal conductivity is electronic (b) the electron gas is highly degenerate and (c) the effective relaxation time is the same for electrical and for thermal conduction. The conditions are discussed for high degeneracy, and for deviations from L_o at high temperatures. It is shown that the relaxation time is unique except for pure metals at low temperatures and cases when electron-electron scattering is important. Corrections for the lattice thermal conductivity are not discussed in detail.

INTRODUCTION

Wiedemann and Franz[1] showed that the thermal conductivity λ of metals is approximately proportional to the electrical conductivity σ, and Lorenz[2] showed that this ratio depends on temperature. Although Lorenz's measurements were confined to $0°$ C and $100°$ C, he surmised the ratio to depend on absolute temperature T. Thus one defines a Lorenz ratio

$$L = \lambda / \sigma T \qquad (1)$$

Theoretical models of a gas of mobile electrons inside a metal make L have the form

$$L = n \left(k/e\right)^2 \qquad (2)$$

where k is the Boltzmann constant, e the electronic charge, and n a numerical constant. Drude's simple model[3], which disregards the thermal spread of electron velocities, lead to n=3, in rough agreement with then available data. Lorentz[4] considered a classical distribution of electron velocities and assumed a constant mean free path; he found n=2, in worse agreement with observations. Sommerfeld[5], using Fermi-Dirac statistics, found $n=\pi^2/3$.

Of course, this model is concerned only with the electronic component of the thermal conductivity. The Sommerfeld value of the Lorenz ratio

$$L_o = \lambda_e / \sigma T = (\pi^2/3)(k/e)^2 \qquad (3)$$

requires three conditions to hold:
    (a) the electron gas is highly degenerate,
    (b) the electron relaxation time, which enters in the expression for both λ and σ, should be the same for both properties.
    (c) the lattice component of the thermal conductivity should either be negligible, or the total thermal conductivity can be corrected for the lattice component to give λ_e.

THE BOLTZMANN EQUATION(6)

Let $f(\underline{p})$ be the average number of electrons in each state of momentum $\underline{p}$ in the electron gas. If f departs from its thermal equilibrium value f^o, then the rate of return to equilibrium, in the absence of disturbances such as an electric field F or a temperature gradient grad T, is given by

$$df/dt]_{int} = -(f-f^o)/\tau = -g/\tau \qquad (4)$$

This equation defines the relaxation time τ.

An electric field and a temperature gradient cause a systematic rate of change in the distribution function given by

$$df/dt = e(\underline{F} \cdot \underline{v}) df^o/dE - (\underline{v} \cdot \text{grad} T) df^o/dT \qquad (5)$$

where $\underline{v} = \partial E/\partial p$ is the electron velocity. In the Boltzmann equation for a steady state, these two rates of change add up to zero. Now

$$f^o(E) = (e^\varepsilon + 1)^{-1} \qquad (6)$$

where $\varepsilon = (E-\zeta)/kT$ and ζ is the Fermi energy. In metals ζ is only a weak function of T.

Equating the sum of (4) and (5) to zero, and expressing df^o/dT in terms of $df^o/d\varepsilon$ one obtains

$$g(\underline{p}) = -\tau(\underline{p})(kT)^{-1}(df^o/d\varepsilon) \underline{v} \cdot [e\underline{F} - (k\varepsilon + d\zeta/dT)\text{grad} T] \qquad (7)$$

Both the electric and heat current densities vanish in equilibrium, and can be expressed in the form

$$\underline{j} = e \int \underline{v}(\underline{p}) g(\underline{p}) D \, d\underline{p} \qquad (8)$$

$$\underline{Q} = \int \underline{v}(\underline{p}) [E - \zeta] g(\underline{p}) D \, d\underline{p} \qquad (9)$$

where D is the number of electron states per unit volume of solid in an element d$\underline{p}$ of momentum space.

Since the factor $df^0/d\varepsilon$ in $g(p)$ is a function of E only, one can arrange the integration over $\underline{p}$ as an integration over E and over successive energy contours (surfaces of constant E) in p-space. Thus one can write the isothermal conductivity σ, defined by $\underline{j} = \sigma\underline{F}$, in the form

$$\sigma = \int \sigma(E)\left(-df^0/d\varepsilon\right) d\varepsilon \tag{10}$$

where $\sigma(E)$ is an average, over the energy contour E, of $\gamma v_z^2 D$, and where v_z component of v in the field direction.

Since $df^0/d\varepsilon = - e^\varepsilon(e^\varepsilon + 1)^{-2}$ is an even function of ε, it is advantageous to split the term in (7) within the square brackets into a term in ε and a term independent of ε. The latter, $e\underline{F}' = e\underline{F} - (df^0/dT)\,\mathrm{grad}T$, is the electric field which appears in an external measurement. One also notes that the integrand of $\underline{Q}$ differs from that of $\underline{j}$ by a factor $(kT/e)\,\varepsilon$.

Defining moments of the function $(-df^0/d\varepsilon)\,\sigma(\varepsilon)$ by

$$M_n = \int \sigma(\varepsilon)\,\varepsilon^n \left(-df^0/d\varepsilon\right) d\varepsilon \tag{11}$$

in particular $\sigma = M_0$, then the electric field F' which makes $j = 0$ is

$$\underline{F}' = S\,\mathrm{grad}T = \mathrm{grad}T\,(k/e)\,M_1/M_0 \tag{12}$$

where S is the Seebeck coefficient. The thermal conductivity is $-Q/\mathrm{grad}\,T$ when $j = 0$, and is thus

$$\lambda_e = \left(k^2 T/e\right)\left[M_2 - M_1^2/M_0\right] \tag{13}$$

The Lorenz ratio becomes

$$L = \lambda_e/\sigma T = (k/e)^2\left[M_2/M_0 - (M_1/M_0)^2\right]$$
$$= (k/e)^2 M_2/M_0 - S^2 \tag{14}$$

HIGH DEGENERACY - CASE OF METALS

Equations (12), (13) and (14) hold for electronic conduction in all solids and even in liquids. For metals, a simplification is possible, since the function $\sigma(E)$ varies only slowly near the Fermi energy ζ. One may thus expand $\sigma(E)$ as a Taylor series about ζ, i.e.

$$\sigma(E) = \sigma(\zeta) + (E - \zeta)\sigma'(\zeta) + \tfrac{1}{2}(E - \zeta)^2 \sigma''(\zeta) + \ldots \tag{15}$$

where $\sigma'(\zeta)$, $\sigma''(\zeta)$ denote the first and second derivations at $E = \zeta$.

High degeneracy implies that such an expansion is valid for $E - \zeta$ up to several kT. Since $\sigma(E)$ contains not only the density of states $D(E)$ but also a factor $v^2 \gamma(E)$, this criterion of high degeneracy is more stringent than the one used for equilibrium properties, which only requires $D(E)$ to be slowly varying over an interval of several kT.

Using (15) and retaining terms up to the second derivative, one finds

$$M_0 = \sigma = \sigma(\zeta) + (\pi^2/3)(kT)^2 \sigma''(\zeta) \tag{16}$$

$$M_1/M_0 = S = (\pi^2/3)(k/e) \, kT \, \sigma'(\zeta)/\sigma(\zeta) \tag{17}$$

and

$$L = M_2/M_0 - S^2$$
$$= (k/e)^2 \left[\pi^2/3 + (8\pi^4/45) \, k^2 T^2 \sigma''(\zeta)/\sigma(\zeta) \right] - S^2 \tag{18}$$

Here $\sigma'(\zeta)$, $\sigma''(\zeta)$ are the first and second derivatives with respect to E at $E = \zeta$, keeping ζ constant. They should not be equated with derivatives of the electrical conductivity with respect to the Fermi energy.

The leading term in (18) yields the Sommerfeld value L_0, the other terms give corrections varying as T^2 and important at higher temperature. Equation (18) can be put in the form

$$L = L_0 \left[1 + 5 \, (kT)^2 \sigma''/\sigma \right] - (S/156)^2 \tag{19}$$

where S is expressed in microvolts per degree. The term in S^2 can usually be disregarded; even if S is as large as $40 \, \mu V/K$, this correction is only -6.5%.

Unfortunately when the term in σ'' becomes important, the Taylor expansion (15) may no longer be adequate. However, (18) or (19) can at least provide a criterion for the adequacy of L_0.

Consider, as example, a transition metal at 1000 K. Structure in the density of states $D(E)$ due to the d-band has a characteristic width E_B, typically around 0.5 ev or 6 kT. Thus $(kT)^2 \sigma''/\sigma$ is about $(kT/E_B)^2 = 0.03$, if the Fermi energy falls near a minimum of the density of states. Thus one would expect $\Delta L/L_0$ to be of the order of 15%.

Considerations of lattice stability lead one to expect ζ to fall more frequently near a minimum of the $D(E)$ curve than a maximum. If $\sigma(E)$ parallels $D(E)$, one would expect deviations $\Delta L/L_0$ to be more commonly positive; this seems to be the case. However, negative deviations do occur. If conduction takes place in one band, while an overlapping band has low mobility but provides receptor states for scattering from the first band, the variation of $\sigma(E)$ could be in the opposite sense to that of $D(E)$, so that σ'' and $\Delta L/L_0$ would be negative even though ζ were still at a minimum of $D(E)$. This appears to be the case for some actenides which have overlapping f- bands.

UNIQUENESS OF THE RELAXATION TIME

The treatment of the previous section presupposes a unique

66

relaxation time, and hence a unique function $\sigma(E)$. However, a full treatment of the interaction operator of equation (4) would make the relaxation rate $1/\tau$ depend not only on the magnitude of g, but also on its functional form, and the functional form of g differs in the cases of electrical and thermal conduction.

The functional form of g is given by (7). In the case of isothermal electrical conduction, g reduces to

$$g = e \, F \, \tau \, v \, (-df^o/dE) \, \phi(\cos\theta) \qquad (20)$$

where θ is the angle between the electron momentum and the field direction. The function ϕ is odd in $\cos\theta$; in the case of isotropic energy contours $\phi(\cos\theta)$ is simply $\cos\theta$. Apart from the factor $(-df^o/d\varepsilon)$, which is even in ε, g is independent of ε, but reverses sign as the direction of p is reversed. Hence the electric field causes an excess of electrons moving in the field direction and a deficit in the opposite direction.

In the case of a temperature gradient, together with the small thermoelectric field F' to prevent a net electric current, g has the form

$$g = (-\text{grad } T) \, \tau \, v (-df^o/dE) \, k \, \varepsilon \, \phi(\cos\theta) \qquad (21)$$

where we have neglected the terms due to F'. Note that g differs in functional form from (20) by an additional factor ε. Hence the electrons flowing down the temperature gradient are slightly hotter than the average temperature; those in the opposite direction are slightly colder.

Resistance is produced by processes which take electrons from regions in p-space of large g to regions with small g for either sign of g. We thus find that electrical resistance is produced by interactions which reverse the direction of p, or reverse $\phi(\cos\theta)$, irrespective of whether ε changes or not. Thermal resistance, on the other hand, can be produced in two ways: by processes which change the direction of p but do not change ε, and by processes which change ε but do not change the direction of p.

Processes which change direction without energy have been termed[7] "horizontal"; processes which change energy but not direction are "vertical". We divide processes into these two groups, each of which contribute to the relaxation rate $1/\tau$, and we find that

$$1/\tau_\sigma = 1/\tau_H \qquad \text{for electric conduction}$$

$$1/\tau_\lambda = 1/\tau_H + 1/\tau_V \qquad \text{for thermal conduction} \qquad (22)$$

when $1/\tau_H$ and $1/\tau_V$ are the relaxation rates for horizontal and vertical processes, respectively. If $1/\tau_V$ is significant, $\tau_\sigma > \tau_\lambda$ and the Lorenz ratio L is less than L_o, even when the electron gas is highly degenerate.

The relaxation rate is due to three kinds of interaction processes, each contributing to the electrical resistance: (a) scattering by defects and solute atoms (b) scattering by phonons and (c) electron-electron interactions.

Scattering by defects is elastic, i.e. $1/\tau = 1/\tau_H$, and the corresponding resistances are related by the Sommerfeld Lorenz ratio

$$\rho = L_e W T \tag{23}$$

where ρ is the electrical, W the thermal resistivity.

Scattering of electrons by phonons is, at ordinary and high temperatures, almost elastic and isotropic in direction, so that (23) also holds for this component of resistivity. At low temperatures, however, scattering is inelastic and favors small angles, so that $1/\tau_V$ becomes increasingly important as temperature is lowered, and dominates the thermal resistance at very low temperatures. For this resistivity component

$$L = L_o f(T/\theta) \tag{24}$$

where at low temperatures $f \propto (T/\theta)^2$, but $f \to 1$ as T approaches and exceeds θ.

Electron-electron collisions, which are important in transition metals at low and very high temperatures, are always inelastic, and can be divided into two groups. One group conserves electron momentum, and for that group $1/\tau_H = 0$. The other group does not, and for that group $1/\tau_H$ is comparable to $1/\tau_V$. This partition is more or less independent of temperature, so that $L = \alpha L_o$, where the parameter α is a constant comparable to 1/2. The exact value requires a detailed knowledge of the electronic band structure.

Alloys have, therefore, a unique relaxation time and $L = L_o$ if the electron gas is highly degenerate. For pure metals L is always lower than L_o at low temperatures, but L approaches L_o with increasing temperature, provided high degeneracy is maintained. In pure transition metals at very high temperatures, L may again drop below L_o if electron-electron collisions are important. However, in transition metals at high temperatures there are also deviations from high degeneracy which also cause L to deviate from L_o, making the interpretation of observed Lorenz ratios particularly uncertain.

THE LATTICE COMPONENT

The lattice component of thermal conductivity λ_p, that is the heat transport by phonons, required additional discussion beyond the scope of this review, and has been reviewed in reference(6) and more recently in reference(8). In most cases, this component is comparable at ordinary and high temperatures to the thermal conductivity of dielectric solids of comparable elastic properties and similar concentration of point defects. Its temperature dependence should therefore range from $\lambda_p \propto T^{-1}$ for pure metals to $\lambda_p \propto T^{-1/2}$ for concentrated alloys. These temperature variations should be corrected for thermal expansion, since the theories usually assume constant volume, but the theories are somewhat uncertain, so that these corrections are usually omitted.

Whatever the temperature dependence of λ_p may be, some account must be taken of it at intermediate temperatures in all but the most conducting alloys. Only if the phonons are very strongly scattered by electrons can one assume λ_p to be temperature independent(8).

68

REFERENCES

1. G. Wiedemann and R. Franz, Ann. Physik (2)89:497 (1853).

2. L. Lorenz, Ann Physik (3)13:422 (1881).

3. P. Drude, Ann. Physik (4)1:566 (1900).

4. H. A. Lorentz, Proc. Amsterdam Acad. 7:438,585 (1905).

5. A. Sommerfeld, Z. Physik 47: 1 (1928).

6. P. G. Klemens, Theory of the thermal conductivity of solids, in "Thermal Conductivity" Vol. 1, R. P. Tye, ed., Academic Press, London (1969).

7. P. G. Klemens, Thermal conductivity of solids at low temperatures, in "Handbuch der Physik" Vol. 14, S. Fluegge, ed., Springer, Berlin (1956).

8. P. G. Klemens and R. K. Williams, Thermal conductivity of metals and alloys, Int. Metals Review 31:197 (1986).

CHANGES IN THERMAL DIFFUSIVITY DURING
HEATING OF PM POWDER COMPACTS

Wolfgang Neumann

Austrian Research Centre
Seibersdorf, Austria

Angelica Hallen

Swedish Institut for Metals Research
Stockholm, Sweden

ABSTRACT

The temperature distribution in powder compacts plays an important role in powder metallurgy technology. The thermal diffusivity of a number of materials was measured up to high temperatures. Since during the sintering process different degrees of densification occur, the influence of the density on the diffusivity was measured and the results were compared with data calculated by existing semi-empirical models.

In the low temperature range (up to 400 $^{\circ}$C) the experimental results of the green bodies were influenced by the lubricant. After the burn off of these additions an irreversible increase of diffusivity was obtained. This increase occured far below the sintering temperature and was carefully investigated and discussed.

INTRODUCTION

In powder metallurgy thermophysical properties play an important role, mainly because of two reasons:

- PM materials are often used for high temperature application, where the suitability and life-time of the PM parts strongly depend on the thermophysical properties of the material.

- PM materials are produced by compaction of loose powders followed by heating the compacted parts to a specific temperature for sintering. The temperature distribution in the material during this heat treatment influences the sintering process, therefore it is of great interest to know the thermal diffusivity over the related temperature range.

In both cases the density is an important parameter. In the first case one has to consider that almost all of the PM produced materials have a certain amount of residual porosity. In the second case it is obvious that the density changes during sintering. Therefore, the main task of this study was to determine the thermal diffusivity as a function of density and temperature. Representative samples of different types of materials, such as cemented carbides, light metals and refractory metals were investigated.

METHODS AND RESULTS

The Laser-Flash-Method was used to measure the thermal diffusivity. The method and the apparatus is described in a previous paper (1). The temperature increase of the rear surface of the specimen was detected by an infrared sensor, which enabled measurements from room temperature up to 1400 °C. The specimen dimensions were 10 mm diameter and 1 to 3 mm high. The metallic specimens with a bright surface were grid blasted to improve the absorption of the laser energy.

On some materials, measurement of the thermal conductivity by the use of a comparative method with a steady state heatflow (cylindrical specimens: 25 mm Ø, 25 mm l) were conducted for purposes of comparison with the diffusivity data.

The primary task was to investigate parameters, which influence the conductivity. Density is one of the most important parameters, therefore samples of sintered high speed steel of different densities were measured by the comparative method. The samples were already sintered but still contained a certain amount of pores. Since the differences between the results of the various samples did not change with temperature, a conductivity versus pores ratio plot represents the influence of density over the whole temperature range. To give a more general view the conductivity data are related to the value of full dense material (fig. 1). In this form they can be easily compared with results from an empirical formula reported by Tye (2), which describes the influence of porosity on the thermal conductivity of metals

$$\frac{\lambda}{m} = \frac{1 - f}{1 + 11f^2}$$

λ ... conductivity of the porous material
λ_m ... conductivity of dense material
f ... pores ratio

The results obtained with the high speed steel samples fit this model quite well.

In contrast to the results above, the diffusivity of a molybdenum alloy did not agree with the above function (fig. 2).

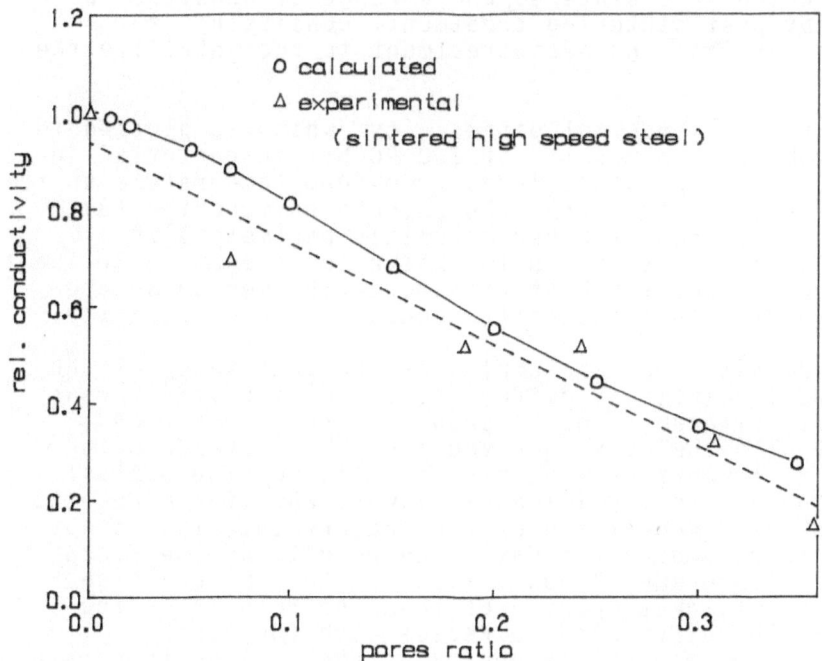

Fig. 1. INFLUENCE OF PORES ON THERMAL CONDUCTIVITY

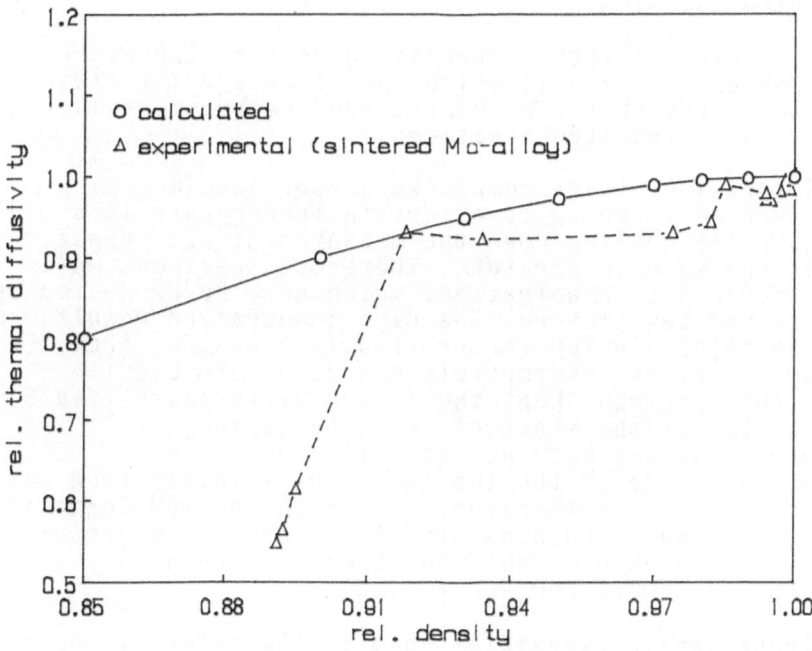

Fig. 2. INFLUENCE OF DENSITY ON THERMAL DIFFUSIVITY

In these experiments the thermal diffusivity was measured on samples, which were sintered, and a range of densities were achieved by post sintering treatment, consisting of sucessive forgings and heat treatment to recrystallize the material.

The relative thermal diffusivity, which is plotted in the diagram are the results at 300 °C but characterize in principle the temperature range from room temperature up to 1400 °C. The deviation from the calculated function (3) shows the influence of other materials' parameters in addition to the density. An intensive investigation of the microstructure and the heat transportation mechanism are necessary to explain such differences in the diffusivity.

On compacted powder samples of the same material the results were completely different. The diffusivity versus pores ratio diagram (fig. 3) shows a more or less good agreement with the curve derived from the above formula but only at temperatures as high as 1400 °C. The diffusivity data at 300 °C are significantly lower. When these results were checked by repetition of the measurement, it was not possible to reproduce the data. The results of the second run of the measurement show a much (up to 6 times) higher diffusivity at low temperatures than the results of the first run. The difference decreases with increasing temperature and almost disappears at the maximum temperature of the first run (fig. 4). Taking the results from the second run at a test temperature of 300 °C and plotting them as a function of density you get a nearly identical curve as obtained with the high temperature data (fig. 3).

This result was rather surprising because the highest test temperatures were well below the sintering temperature. In order to verify this finding, extended measurements were carried out with tungsten specimens.

In a series of tests, compacted powder specimens of tungsten were measured up to a certain temperature in a first run. After cooling down the measurement was repeated up to just the same temperature. Therefore specimens were available for a post examination, which were never heated up above the given temperature. The high temperature results confirm the effect, which was previously found for the molybdenum alloy. An irreversible change of diffusivity occurred, the specimens kept their high diffusivity (fig.5). In order to detect the start of this change, the low temperature range was examined as well. Because of a disturbing influence of the lubricant, the investigation was started at 400 °C, a temperature at which the lubricant is completely removed by burning off. As an unexpected result this irreversible change could be observed even at temperatures as low as 600 °C (fig. 6).

Obviously an irreversible range in the material occurs caused by the heat treatment during the measurement. Since the heating rate, because of the quick measurement procedure. is high, it was of interest to look at the time dependence. Therefore the specimen was kept at 900 °C for half an hour and the diffusivity was measured after several periods (fig.7).

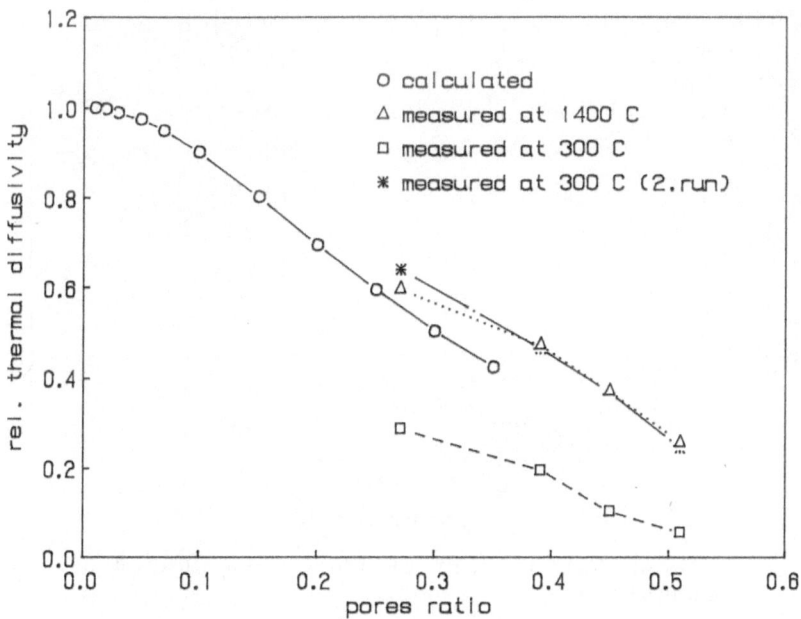

Fig. 3. INFLUENCE OF PORES ON THERMAL DIFFUSIVITY
(Molybdenum alloy)

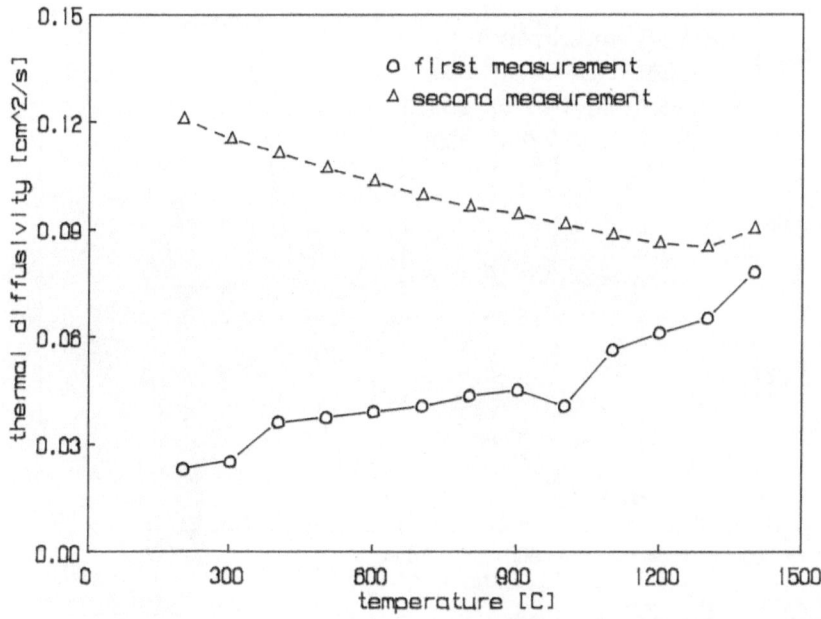

Fig. 4. THERMAL DIFFUSIVITY OF COMPACTED POWDER SPECIMENS
(Molybdenum alloy: 49 % dense)

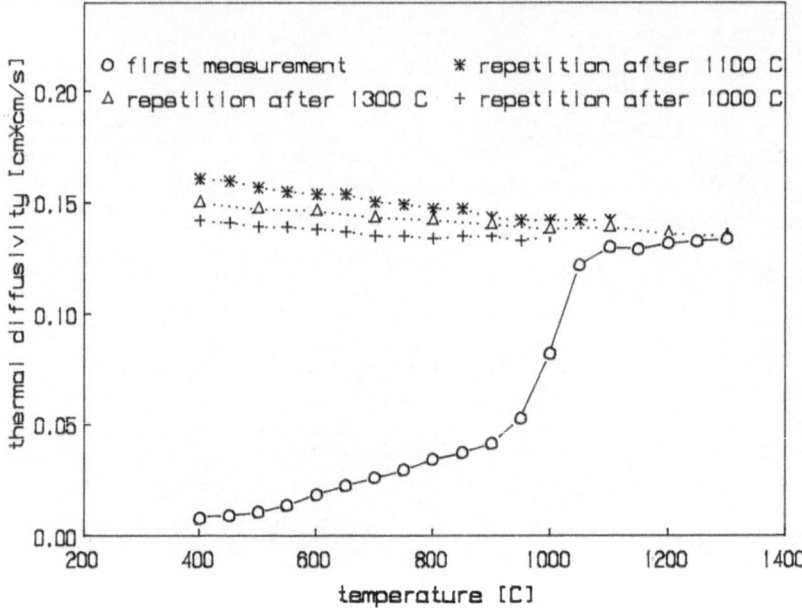

Fig. 5. THERMAL DIFFUSIVITY OF COMPACTED POWDER SPECIMENS
(tungsten)

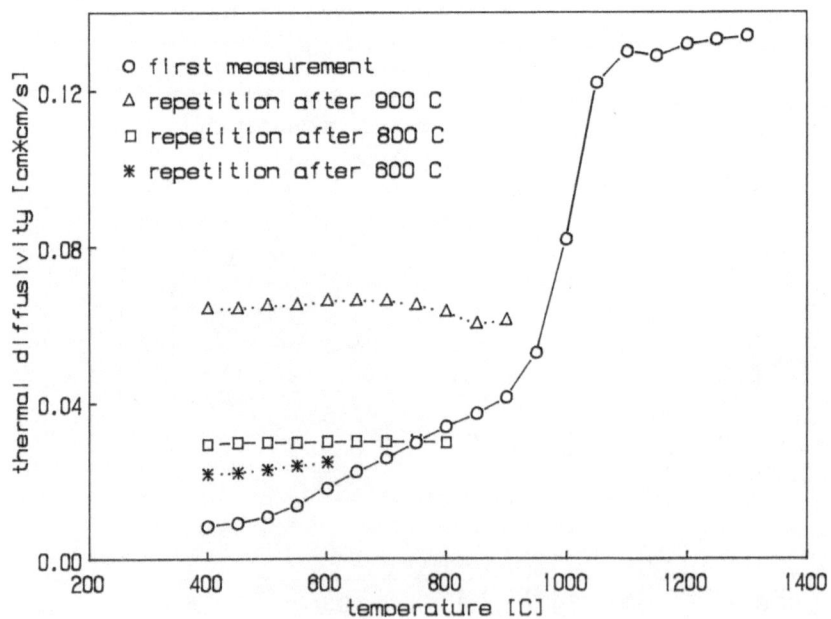

Fig. 6. THERMAL DIFFUSIVITY OF COMPACTED POWDER SPECIMENS
(tungsten)

The change in the material, which effects the thermal diffusivity, depends on the times and temperatures of the measurement procedure. Sometimes, when the cooling starts, the change in the material was not finished and in the second run the change continued. In that case the value of run 2 at the maximum temperature was higher than of run 1.

To confirm the effect with nonmetallic materials, the temperature dependence of the diffusivity was studied with cemented carbides too. Tungsten carbide based hard metals, materials usually used for cutting tools, were investigated from ambient temperature up to 1250 $^{\circ}$C. In principal, the same effect could be obtained (fig. 8). At 500 $^{\circ}$C the results of the first measurement could still be reproduced but at 700 $^{\circ}$C the significant increase occurred already.

To be sure that the effect was not an artifact caused by the Laser-Flash-Method the following experiment and estimation were performed.

One of the cemented carbide specimens was heated in the laser flash apparatus according the temperature programme of a measurement but never flashed by the laser. After this heat treatment the diffusivity was measured and the results compared with the data of another specimen of exactly the same material, which were obtained in a second measurement. The comparison shows almost identical diagrams (fig. 9). The estimation takes into account that the laser flash possibly sinter a layer on the surface of the specimen, which, because of its high diffusivity, leads to a higher overall diffusivity of the specimen. By the use of a model for the diffusivity of a two-layer composite, the thickness of a sintered layer necessary to increase the diffusivity to such an amount as experimentally obtained, was calculated. A layer of about 0.25 of the specimen thickness has to be sintered, which is not realistic. In addition the surface of the specimen was examined by means of a Scanning Electron Microscope. No evidence of sintering could be observed. An influence of the laser flash on the diffusivity increase can be excluded.

CONCLUSIONS

Because of the measurements of the thermal conductivity and diffusivity respectively and the characterisation work one can conclude:

- Models for two phase materials, in which pores represent the dispersed phase, can be applied to a certain extent to describe the dependence of the conductivity and diffusivity respectively of the density of sintered materials.

- Deviation from the calculated diffusivity - density diagram occurs with special treated sintered materials (e.g. after densification and recrystallisation) and in a very significant way with compacted powders.

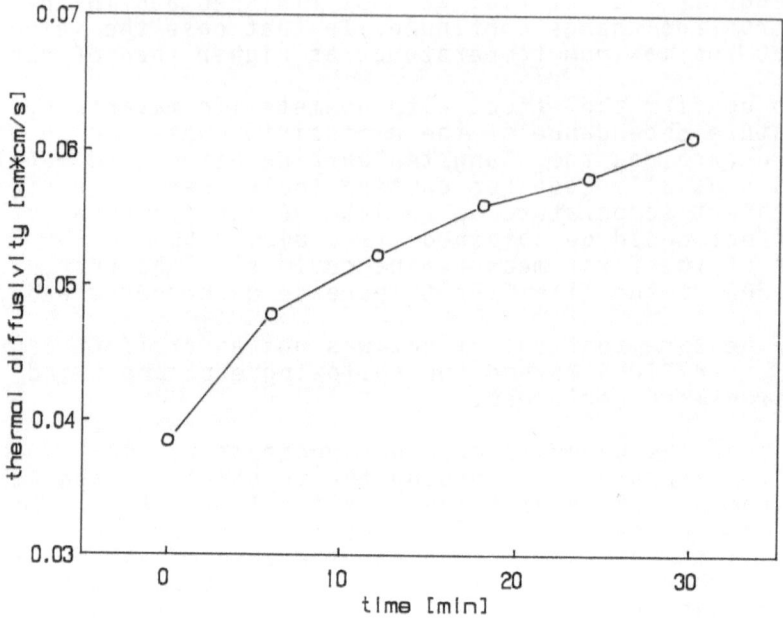

Fig. 7: CHANGE OF DIFFUSIVITY DURING HEATTREATMENT AT 900 °C

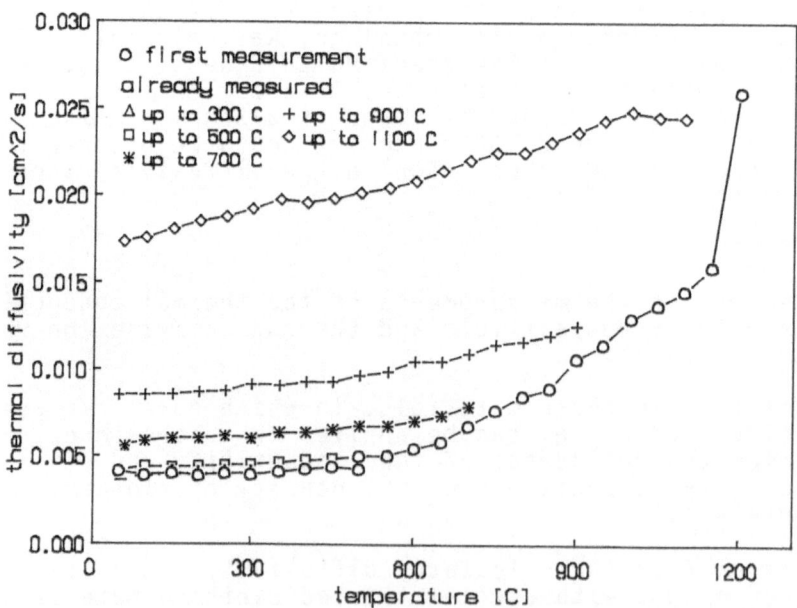

Fig. 8. THERMAL DIFFUSIVITY OF COMPACTED POWDER SPECIMENS
(cemented carbide)

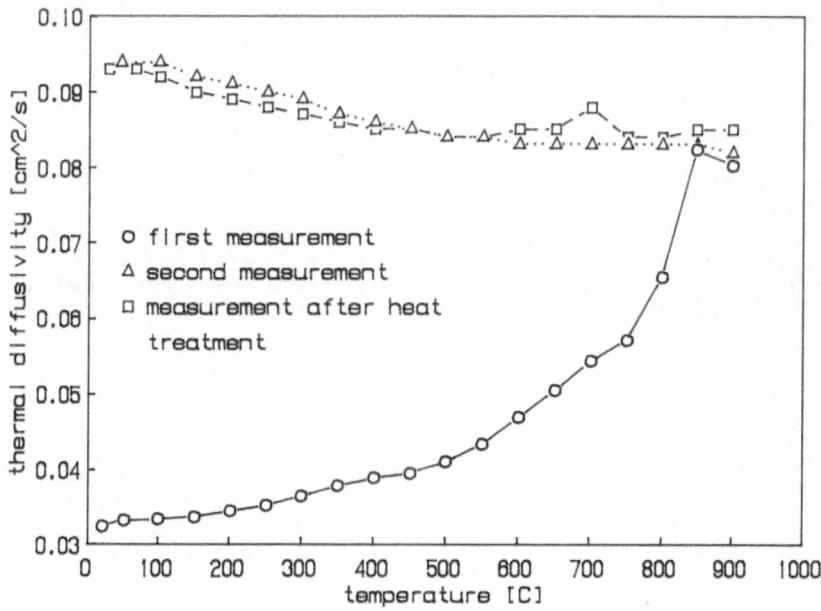

Fig. 9. THERMAL DIFFUSIVITY OF COMPACTED POWDER SPECIMENS
(cemented carbide)

- The irreversible change of diffusivity, which was obtained
  on different compacted powders after a short heat
  treatment far below the sintering temperature, cannot be
  explained by a change in density. At these relatively low
  temperatures compared to the sintering temperature the
  density remains nearly constant and also hardness and
  optical appearance do not change.

- Since it is not an artifact, which could be shown, one can
  assure that the diffusivity measurement is a very
  sensitive instrument to observe changes in the material,
  even such which could not be detected by measurement of
  the hardness or even electronmicroscopic examination.

- Calculations of temperature distributions in compacted
  powder parts during the sintering process have to take
  into account that the diffusivity of the material at every
  point of the part is a function of the temperature-time
  history.

ACKNOWLEDGEMENTS

The authors would like to thank Metallwerk Plansee/Austria,
Sandvik Hard Materials, Klosters Speed Steel and Craelius/
Sweden for the materials and their preparation.

References

1. W. Neumann, "Thermal Diffusivity of Cemented Carbides",
   Thermal Conductivity 18, Proc. of the 18th Int.Conf. on
   Thermal Conductivity, Plenum Press., p. 473

2. R.P. Tye, ASM publication 73-HT-47, American Society of
   Mechanical Engineers, New York

3. W. Neumann, "Thermal Conductivity Measurements on
   Granule-Based Composites", High Temp. - High Pressures
   13, P.687-694 (1981)

THE EXPERIMENTAL STUDY OF TRANSPORT AND
THERMODYNAMIC PROPERTIES OF NICKEL

K.D.Maglić, A.S.Dobrosavljević, and N.Lj.Perović

Boris Kidrič Institute of Nuclear Sciences,Vinča
Institute for Thermal Engineering and Energy
Research, Belgrade, Yugoslavia

ABSTRACT

Data on the thermal diffusivity, specific heat and electrical resistivity of high purity nickel are presented. The
thermal diffusivity results were obtained by the laser pulse
method in the 550-1500 K range, while the specific heat and
electrical resistivity were measured by the pulse heating technique in the 350-1500 K range. Thermal diffusivity measurements defined Curie point at 730 K. The measurement uncertainties were estimated at 3% for thermal diffusivity and specific heat and 1% for electrical resistivity. The discrete
values of thermal diffusivity and smooth tables of specific
heat and electrical resistivity are presented. The results
are analysed and compared with literature data.

INTRODUCTION

The study of thermophysical properties of ferromagnetic
materials and thermophysical characterization of the candida-
tes for thermophysical property standard reference materials
has been a continuing activity at the Boris Kidrič Institute
in the past seven years. The experimental study of 99.999%
pure nickel reported in this paper was a part of this program. It involved the study of its thermal diffusivity, specific heat and electrical resistivity in a wide temperature
range.

The thermal diffusivity of nickel has not been investigated to any great extent recently. The two recommended depen-
dencies, the one given in TPRC Series in 1973 (Touloukian et
al, 1973) and the eight years later CINDAS synthesis (Touloukian and Ho, 1981) differed amongst themselves, particularly
in the range above the Curie point, the difference reaching
10% at 1500 K. The laser pulse technique used at our Laboratory,
where the specimen temperature transients can be kept as small
as 1-2 K is a promising tool for the provision of additional
information on the magnetic transition and the high temperature ranges. On the other hand, the recent high temperature specific heat and electrical resistivity data (Cezairliyan and

Miiler, 1983) call for supplementary information on both properties in the lower temperature range. Our experimental apparatus employing essentially the same method adapted for the contact thermometry range was very convenient for the provision of new experimental data. Finally, the number of thermophysical property reference materials being so small and limited, effort that might have resulted in a contribution in this direction was justified on these grounds.

MEASUREMENTS

The thermal diffusivity was measured by the laser pulse technique. The specimens were 10.5 mm in diameter and 2.47 and 3.00 mm thick. The measurement system was linked to a computer for data acquisition and experiment control in real time. This included the filtering of the acquired data, their processing and computation of thermal diffusivity values at given temperatures. Comparison of the experimental data with the temperature response predicted by the mathematical model enabled the detection of any departures of the experimental conditions from those assumed in the mathematical model, as well as the introduction of appropriate corrections. The thermal diffusivity was computed from

$$a = K_x \frac{L^2}{\tau_x} \qquad (1)$$

where L is the specimen thickness, K_x is a constant corresponding to x per cent of the signal rise, and τ_x is the elapsed time from the initiation of the laser pulse until the rear surface temperature reaches x per cent of its maximum value. The thermal diffusivity measurements were limited to the PbS photoresistor response range, i.e. 550-1500 K.

The literature data on the thermal expansion of nickel (Touloukian and Ho, 1981) provided information for thermal expansion corrections of computed thermal diffusivity values.

The specific heat and electrical resistivity of a nickel wire 2.00 mm in diameter and 200 mm long were measured simultaneously, by the variant of the pulse heating method (Dobrosavljević and Maglić, 1985) based on the fast resistive heating of the speciment from room temperature to a predetermined temperature. Direct current pulses in the 200-600 A range enabled heating rates of 400 to 1000 KS^{-1}. Three 0.05 mm K-type thermocouples were welded intrinsically in the central portion of the specimen, at 10 mm separations. The central thermocouple was used for specimen temperature measurement. The other two thermocouples were used as the potential leads for voltage drop measurement and to monitor temperature uniformity in the measurement zone. The measurements were performed at 10^{-3} Pa. A computer system controlled the experiment and was used for real-time data acquisition and subsequent processing. Current, voltage drop across the measurement zone, and thermocouple emf data were collected during specimen heating, typically lasting 1 s, and during the initial cooling period. As the thermocouple was welded to the sample through which direct current flowed, the voltage drop across the thermocouple legs generated an error in the thermocouple output, which was

compensated for by a procedure described elsewhere (Dobrosavljević and Maglić, 1985). Several thousand data points per run
were collected, yielding 500 to 1000 specific heat and electrical resistivity values in the temperature range studied.
The specific heat is

$$C_p = \frac{(UI-P_r)}{n(dT/dt)} \qquad (2)$$

the electrical resistivity being computed from

$$\rho = \frac{U}{I} \frac{S}{l_e} \qquad (3)$$

where U is the voltage drop across the effective specimen length between the potential leads, l_e, I is the current, P_r is
the radiative power loss from the measurement zone, n is the
number of moles per measurement zone, dT/dt is the heating rate at the given temperature, and S is the specimen cross section.

RESULTS

The specimens for this study were purchased from Johnson
Mathey Chemicals Limited, UK. The certificate of analysis supplied by the manufacturer for the nickel rod from which the
specimens for thermal diffusivity measurement were machined
indicated that the material was 99.999% pure with the following
impurities in ppm by weight: Fe 1; Ag 1; Ca, Cr, Cu, Mg, each
less than 1. The certificate of analysis for nickel wire used
for the specific heat and electrical resistivity measurements
indicated all the elements detected: Al, Ca, Cu, Fe, Mg, Si,
Ag, each less than 1 ppm by weight. The method employed in the
determination of the impurity content was optical emission arc
spectrography using one or more of the following instruments:
(a) 3m Ebert graphing spectrograph, (b) Large, medium or small
prism spectrograph, (c) 1.5 m direct reading spectrometer.

All specimens were annealed at 1300 K for 1 hr and slowly
cooled. Specimens for thermal diffusivity measurement were annealed in the furnace of the laser pulse thermal diffusivity
apparatus, and the specimen for the specific heat and electrical resistivity measurements was annealed in the direct electrical heating apparatus. The annealing of the specimen for
specific heat and electrical resistivity measurement was not
repeated after each pulse.

Thermal Diffusivity

Thermal diffusivity was measured in the 550-1500 K range.
The lowest and the highest temperature were the result of the
limitations imposed by the optical transient temperature response detector and the low emissivity of nickel specimen. More than a hundred measurements were performed in this range,
which, after the averaging of individual measurements made at
the same reference temperature resulted in 41 thermal diffusivity values. They are listed in Table I of the Appendix together with the values corrected for the specimen thermal expansion
by way of $a_{corr}=a_{uncorr}(1+L/L_o)^2$. The values of the linear

thermal expansion, L/L_O, were taken from Touloukian and Ho, 1981. The computation of thermal diffusivity involved corrections for heat losses at elevated temperatures and for finite laser pulse duration.

Specific Heat

Four specific heat and electric resistivity experiments were performed on the same specimen. The information on the temperature ranges and the heating periods is summarized in Table II of the Appendix. The temperature and the heating ranges were varied for more detailed study of particular regions.

The shape of the specific heat function required interpretation of the experimental data with three cubic functions of the type

$$C_p = aT^3 + bT^2 + cT + d \tag{4}$$

in the three ranges:

     Range I   (350-620 K)
     Range II   (630-720 K)
     Range III (720-1500 K)

The coefficients of interpolation polynomials are listed in Table III of the Appendix. Figure 1 shows the deviation of the measured specific heat curves for the four runs from the smooth functions of interpolated polynomials in their respective ranges. Specific heat was computed with the atomic weight of nickel as 58.7 and expressed in $J\ mol^{-1}K^{-1}$. Discrete specific heat values are listed in Table IV of the Appendix.

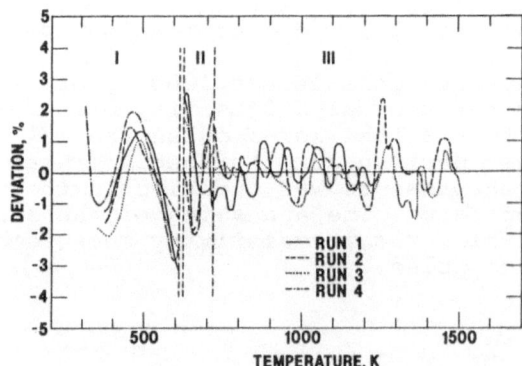

Fig.1. Deviation of specific heat
results for four runs of
nickel specimen from the
smooth functions

Electrical Resistivity

The experimental electrical resistivity function was interpreted by two quadratic functions of the type

$$\rho = aT^2 + bT + c \qquad (5)$$

in the ranges:

Range I   (350-600 K)
Range II (650-1500 K)

Coefficients are given in Appendix, Table III. Figure 2 shows the deviation of the four electrical resistivity measurements from the smooth functions in I and II respectively. The electrical resistivity values between the two polynomials have been computed in a number of discrete points as the arithmetic means of the data obtained in four runs. The electrical resistivity values correlated by Eq.(5) and given as discrete points in Table IV of the Appendix have not been corrected for thermal expansion.

Estimate of Errors

Error analysis of our thermal diffusivity results has been presented elsewhere (Perović et al, 1986), the maximum measurement uncertainty being estimated at 3%. The maximum uncertainty in the specific heat measurement was estimated at 3%, except in the vicinity of the magnetic transition (Dobrosavljević et al, in preparation). For the electrical resistivity measurements the maximum uncertainty was estimated at 1%.

The maximum departure of the polynomials from the experimental values is in the ±2% range for the specific heat and ±1% for the electrical resistivity, except in the 600-650 range, i.e., in the vicinity of the Curie point.

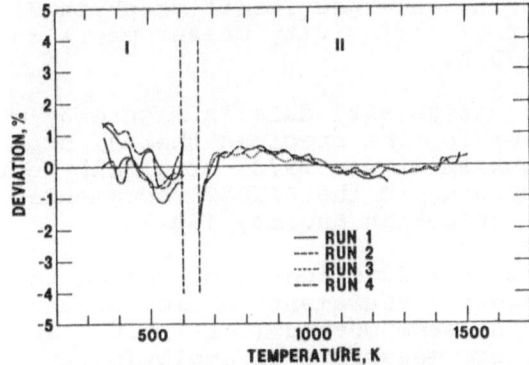

Fig.2. Deviation of electrical
resistivity results for
four runs of nickel
specimen from the smooth
functions

The thermal diffusivity measured in this work is presented graphically in Figure 3, together with two sets of data on the thermal diffusivity of nickel synthesized by TPRC (Touloukian et al, 1973) and CINDAS (Touloukian and Ho, 1981).

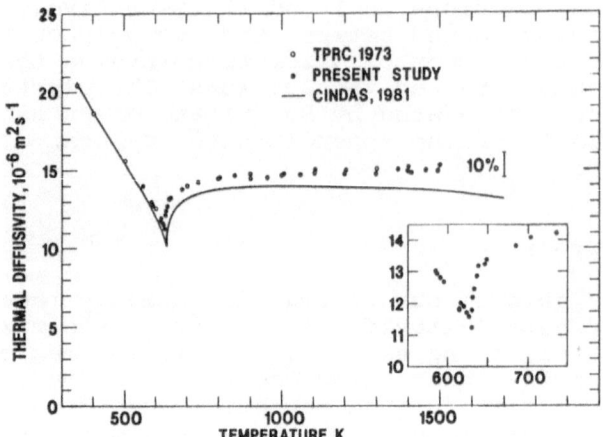

Fig.3. Thermal diffusivity of nickel:
present study and literature
reference data

The agreement with the TPRC recommended values is very good, mostly within 1% limits, and well within the uncertainty of 10% given for the former, or 3% for our measurements. The more recent, CINDAS recommended values lie lower, with a slight drop above 1100 K. The high temperature CINDAS values, however, seem to depend predominantly on the data obtained by the periodic heating technique (Zinov'ev, 1968) as the only information available above 1200 K.

As shown in the enlarged inset graph in Figure 3, and in Table I, our thermal diffusivity measurements identified the Curie Point at 630 K.

The thermal diffusivity data in Figure 3, were presented without correction for the specimen thermal expansion. The TPRC data (Touloukian et al. 1973) originally corrected, have been reduced according to the CINDAS recommended thermal expansion values (Touloukian and Ho, 1981).

The specific heat function is presented in Figure 4, and compared with recent measurements employing the high temperature variant of the same method (Cezairliyan and Miiller,1983), the high temperature measurements employing the modulation method (Glazkov, 1987), the results on specific heat of nickel in a wide temperature range by the adiabatic calorimetry (Novikov, 1981) and the recommended CINDAS data (Desai and Deshpande, 1984). The results of Cezairliyan and Miiller lie about 2% lower, with the same character in the high temperature region.

The results of Novikov conform very well with CINDAS recommended values, except above 1000 K, where they rise more steeply than our, Cezairliyan and Miiller and CINDAS data. Our data lie generally between 1% and 2% above the CINDAS recommended values. The specific heat obtained by Glazkov in his study of the formation of point defects in nickel at high temperatures, is slightly above our results, departing markedly above 1200K.

Our dynamic specific heat measurements located T_c at 624 K, as a mean of the four experiments.

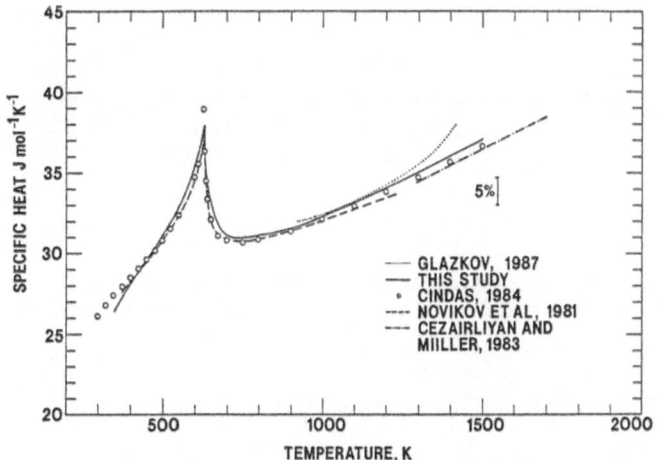

Fig.4. Specific heat of nickel: present study and data reported in the literature

The electrical resistivity results are compared with two summary recommended sets of data (Touloukian and Ho, 1981) and (Belskaya and Peletskii, 1981), and the high temperature data by (Cezairliyan and Miiller, 1984) in Figure 5. All the data fall within 3% margins, Cezairliyan and Miiller data for 99,98% pure nickel being the highest, and our results for 99,999% pure nickel the lowest. Two reference sets lie between them.

The study of transport and thermodynamic properties of nickel reported in this paper is a part of the program of study of ferromagnetic materials (Dobrosavljević et al,1985) and candidate materials for thermophysical property standard reference materials (Maglić et al, 1980).

The obtained results indicate that due to its stability and the shape of the temperature dependence nickel could be a convenient reference material for specific heat and electrical resistivity. Its low emissivity and the flat temperature function above the Curie point render it less attractive as the thermal diffusivity reference material.

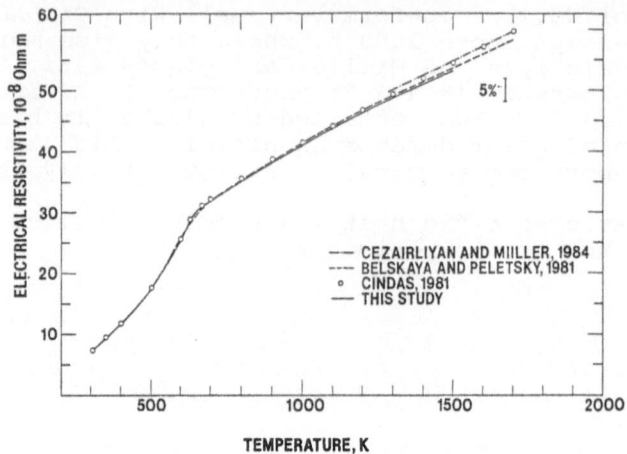

Fig.5. Electrical resistivity of nickel:
present study and data reported
in the literature

ACKNOWLEDGMENT

This study was supported in part by the U.S.-Yugoslav Joint Board on Scientific and Technical Cooperation.

APPENDIX

Table I. Experimental Thermal Diffusivity Values of Nickel

Temperature T,K	Thermal Diffusivity a, 10^{-6} (m^{2}s^{-1})		Temperature T,K	Thermal Diffusivity a, 10^{-6} (m^{2}s^{-1})	
	uncorr	corr.		uncorr	corr.
562	14.32	14.43	736	14.23	14.42
585	13.01	13.12	804	14.53	14.76
587	12.91	13.02	849	14.70	14.96
591	12.80	12.91	898	14.79	15.07
596	12.69	12.80	952	14.62	14.93
615	11.80	11.91	954	14.54	14.85
616	11.72	11.83	1003	14.70	15.04
617	12.00	12.12	1059	14.73	15.10
619	11.93	12.05	1084	15.01	15.41
624	11.69	11.81	1107	14.97	15.38
627	11.53	11.65	1193	14.96	15.42
630	11.19	11.30	1197	14.79	15.24
630	11.76	11.88	1295	14.98	15.50
631	12.19	12.31	1298	14.79	15.30
633	12.43	12.56	1369	15.06	15.63
636	12.86	13.00	1369	15.06	15.63
638	13.17	13.31	1397	14.92	15.51
646	13.22	13.39	1410	14.82	15.41
648	13.29	13.45	1455	15.01	15.64
684	13.79	13.94	1494	14.93	15.59
703	14.06	14.24			

Table II. Specific Heat and Electrical Resistivity Experiments

Run No.	Temperature range (K)	Heating period (ms)
Run 1	375-1239	1000
Run 2	380-1480	1300
Run 3	350-1540	1400
Run 4	310-710	1600

Table III. Coefficients of Interpolation Polynomials for Interpretation of Specific Heat and Electrical Resistivity Data of Nickel

Specific Heat, C_p (J mol^{-1}K^{-1})			
Coefficient	Range I	Range II	Range III
a	0.48683E-06	-0.18517E-05	-0.11321E-07
b	-0.64434E-03	0.42685E-02	0.42876E-04
c	0.31149	-0.32713E+01	-0.43936E-01
d	-0.24507E+02	0.86448E+03	0.44609E+02

Electrical Resistivity, (Ωm) (Uncorrected)		
Coefficient		
a	0.89682E-04	-0.73606E-05
b	-0.21653E-01	0.42676E-01
c	0.58699E+01	0.54985E+01

Table IV. Specific Heat and Electrical Resistivity of Nickel

Temperature (K)	Specific Heat (J mol^{-1}K^{-1})	Electrical Resistivity 10^{-6} (Ωm)
1	2	3
350	26.46	9.28
400	28.15	11.56
450	29.55	14.29
500	31.01	17.46
550	32.90	21.09
600	35.58	25.19
625	37.80	27.73
650	33.05	29.67
700	31.00	31.77
750	30.85	33.37
800	31.10	34.93
850	31.29	36.46
900	31.54	37.95
950	31.86	39.40
1000	32.23	40.81
1050	32.64	42.19
1100	33.09	43.54
1150	33.57	44.84
1200	34.06	46.11 (continued)

Table IV/Continued

1	2	3
1250	34.57	47.34
1300	35.08	48.54
1350	35.58	49.70
1400	36.07	50.82
1450	36.54	51.90
1500	36.97	52.95

REFERENCES

1. Bel´skaya E.A., Peletskii V.E., 1981, Elektrosoprotivlenie nikelja v oblasti temperatur 100-1700 K, Teplofizika visokih temperatur, Vol.19, No 3, pp.525.

2. Cezairliyan A. and Miiller A.P., 1983, Heat Capacity and Electrical Resistivity of Nickel in the Range 1300-1700 K Measured with a Pulse Heating Technique, Int. Journal of Thermophysics, Vol.4, No 4, pp.389.

3. Desai P.D. and Deshpande M.S., 1984, Thermodynamic Properties of Nickel, CINDAS, Report 80.

4. Dobrosavljević A.S. and Maglić K.D., 1985, in Measurement Techniques in Heat and Mass Transfer, Hemisphere Publishing Corp., Washington DC, pp.411-420.

5. Dobrosavljević A.S., Maglić K.D., Perović, Lj.N., 1985, Specific Heat Measurements of Ferromagnetic Materials by the Pulse-Heating Technique, High Temperatures-High Pressures, Vol.17, pp.591-598.

6. Dobrosavljević A.S., Maglić K.D., Perović Lj.N., Pulse Method for Specific Heat Measurement, in preparation.

7. Glazkov S.Y., 1987, Obrazovanie točečnih defektov i teplofizičeskie svoistva nikelja pri visokih temperaturah, Teplofizika visokih temperatur, Vol.25, No 1, pp.59.

8. Maglić K.D., Perović Lj.N., Životić P.Z., 1980, Thermal Diffusivity Measurements on Standard Reference Materials, High Temperatures-High Pressures, Vol.12, pp.555-560.

9. Novikov I.I., Roščupkin V.V., Mozgovoj A.G., Semaško N.A. 1981, Teploemkost nikelja i niobia v intervale temperatur 300-1300 K, Teplofizika visokih temperatur, Vol.19, No.5.

10. Perović Lj.N., Dobrosavljević A.S., Maglić K.D., 1986, The Laser Pulse Method for Thermal Diffusivity Measurement, IBK-ITE, Report No 566, Beograd.

11. Touloukian, Y.S., et al., 1973, Thermophysical Properties of Matter, Vol.13., IFI/Plenum, New York.

12. Touloukian Y.S. and Ho C.Y., 1981, Properties of Selected Ferrous Alloying Elements, McGrow-Hill/CINDAS, Vol.III-I.

13. Zinov´ev V.E., Koršunov,I.G., 1978, Obzori po teplofizičeskim svoistvam veščestv, ČI, IVI AN SSSR, Moskva.

ON MATERIAL CHANGES AND HEATING RATE DEPENDENT PROPERTIES

R. E. Taylor

Thermophysical Properties Research Laboratory
School of Mechanical Engineering
Purdue University
West Lafayette, IN 47906

ABSTRACT

The fact that material properties may be altered by temperature cycling has been known since antiquity and has long been employed usefully – especially in hardening of materials. The presence of such changes implies that the values obtained upon testing may depend on time–temperature history of the material. Thus, increased use of techniques capable of obtaining data on small samples over a large temperature range rapidly or of reaching and holding a prescribed temperature almost instantaneously opens the possibility of studying time–temperature changes on physical properties. In addition, the use of techniques employing high heating rates may make it possible either to observe or not observe the effects on thermophysical properties of certain processes. This is of increased interest to technology since high heating rates are being experienced in present day applications. Of particular interest to the thermophysics community is the meaning and usefulness of thermal diffusivity–conductivity data and their interrelationship for changing systems.

A number of examples of problems (opportunities) and observations encountered in thermophysical property testing of "unstable" systems are given.

INTRODUCTION

One candidate for the 4th Law of Thermophysics:"Everything is irreversibly altered upon heating to high temperatures". The definition of "high temperature" varies widely from discipline to discipline and even for the colleagues within a particular discipline. Nevertheless, the presence of such changes implies that the values obtained upon testing may depend on time–temperature history of the material. Thus, increased use of techniques capable of obtaining data on small samples over a large temperature range rapidly or of reaching and holding a prescribed temperature almost instantaneously opens the possibility of studying time–temperature changes on physical properties. In addition, the use of techniques employing high heating rates may make it possible either to observe or not observe the effects on thermophysical properties of certain processes. This is of increased interest to technology since high heating rates are being

experienced in present day applications. Of particular interest to the
thermophysics community is the meaning and usefulness of thermal diffu-
sivity-conductivity data and their interrelationship for changing systems.

The present paper draws upon the experience gained in examining the
thermophysical properties of a multitude of materials. The size of the
multitude can be envisioned from the fact that it involves about 900 reports
containing original data on the thermophysical properties of materials plus
20 graduate theses in the field. Unfortunately, most of the 900 reports
dealt with proprietary or ill-defined materials, so their usefulness is
limited. However, sharing the experiences gained may be of some benefit to
the thermophysics community.

GRAPHITIC MATERIALS

The changes in the properties of a carbon-carbon material pre-heated to
several temperatures below and equal to its nominal processing temperature
are shown in Fig. 1. The data were obtained from the pre-heated temperature
down. We see that the electrical resistivity (and hence thermal conductivity) and emissivity increased irreversibly when the sample was heated to
2240 K, even though this temperature is 260 K below its graphitization
temperature. Then upon heating to the processing temperature, larger

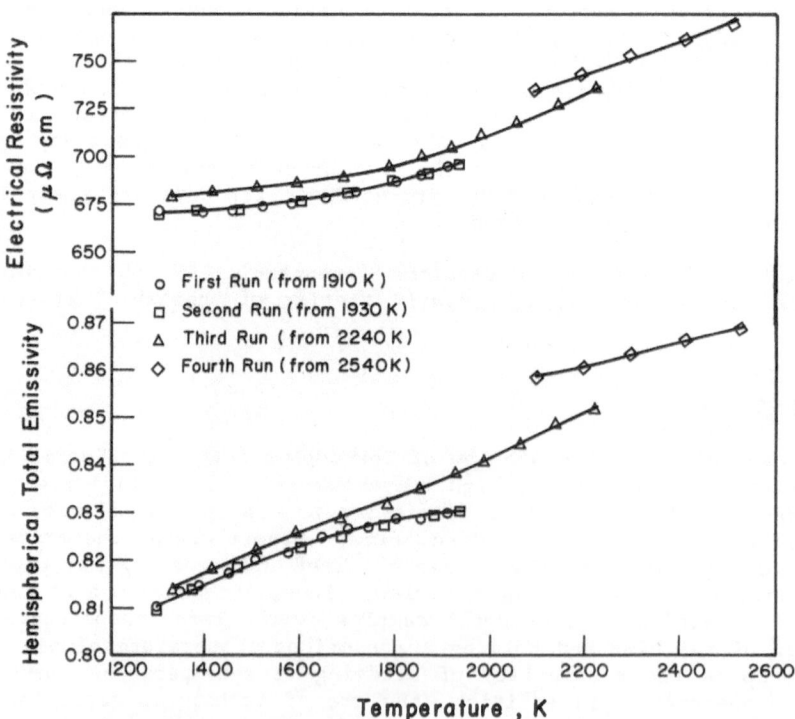

Fig. 1. Resistivity and Emissivity of a Carbon-Carbon Heated to
         Selected Temperatures.

irreversible changes occurred. This is typical of graphitic materials as the graphitization process is relatively slow and the materials have not completely stabilized.

Another example involving a graphitic material is shown in Fig. 2. In this example, thermal diffusivity was measured up to 1100 K using the standard laser flash technique. The diffusivity of a sister sample was measured using a newly developed modification of the laser flash technique (patent applied for) involving a directly heated (DH) sample. Data can be taken over a large temperature range very rapidly and it is possible to study degradation at temperature. All the data above 2200°C were obtained within fifteen minutes. The path of degradation is clearly evident.

CERAMIC MATERIALS

Since the plasma spray process involves high temperatures, one might believe that the properties of plasma-sprayed ceramics would be stable. That this is erroneous is illustrated in Fig. 3. Thermal diffusivity values for an as-sprayed ceramic, and sister samples heat treated for 36 hours at 1093°C (36-1093), 5 hours at 1371°C (5-1371) and 100 hours at 1371°C (100-1371) are shown. The increase in diffusivity values with increasing heat treatment is evident and the changes are substantial. Because the specific heat is essentially unchanged and the density changes are relatively small, conductivity value changes mirror the diffusivity value changes. Since the material is to be used in high temperature engines with a long operating life, these changes are important.

Incidentally, the results shown in Fig. 3 are just a small part of those obtained during a large program. During this program several sets of diffusivity results were obtained that did not fit the pattern. It was soon proven that these samples had been mislabeled by the heat-treat personnel. When the correct labels were used, the data all fit nicely. Thus we were

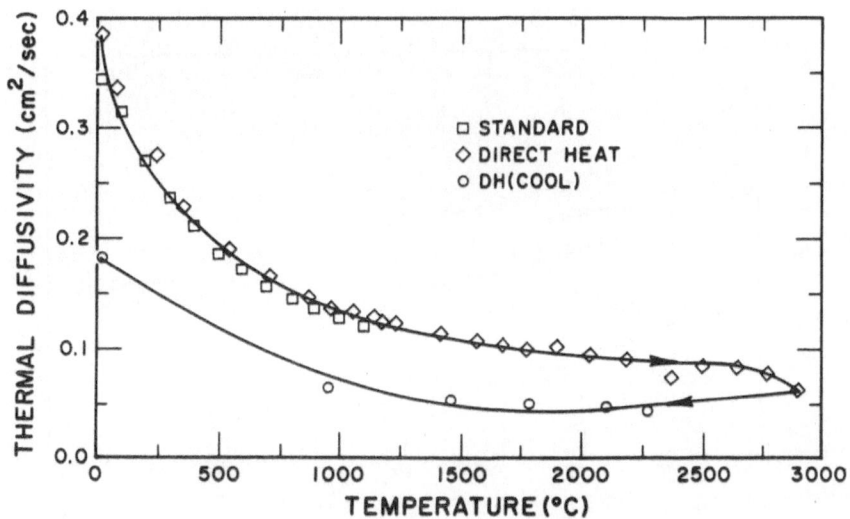

Fig. 2. Diffusivity Values for a Graphitic Material as it is Heated to High Temperatures.

able to show that by thermal diffusivity testing, we could determine the time-temperature history of the samples. We have also shown this with graphitic materials and with rocks.

CARBON-REINFORCED PHENOLICS

Thermal conductivity values were calculated from diffusivity and specific heat determinations of samples from a batch of a carbon-reinforced phenolic. Samples were heat-treated at 100, 200, 300, 400, 500, 600, 700, and 800°C. Measurements were made up to the heat-treat temperature and the results are shown in Fig. 4. The values for the 100 and 200°C samples were about the same. Conductivity values for the 300, 400, and 500°C heat-treat samples were progressively smaller, but then increased as the processing temperature increased above 500°C. The large changes caused by heat-treatment are clearly evident.

Because of the small size of the diffusivity samples and the fact that both sides are exposed to the environment (gas or vacuum), degradation as a function of time at temperature can be monitored.

ALLOYS

It is possible to rapidly heat a proprietary alloy to a sufficiently high temperature and observe property changes as a transformation occurs. This is shown in Figs. 5 and 6. In Fig. 5, the time-temperature history of Sample No. 3 during pulse heating (3-P) and during hold (3-Hold) is shown. The sample was heated from 275 to 800°C at 42°C/sec and then maintained at 800°C for 170 seconds. The electrical resistivity during the constant temperature period is shown in Fig. 6. The resistivity decreased rapidly at first and then approached a limiting value. When the experiment was performed on a sister sample to 750°C, no change in resistivity took place during the hold period. This is because 750°C is below the transformation temperature.

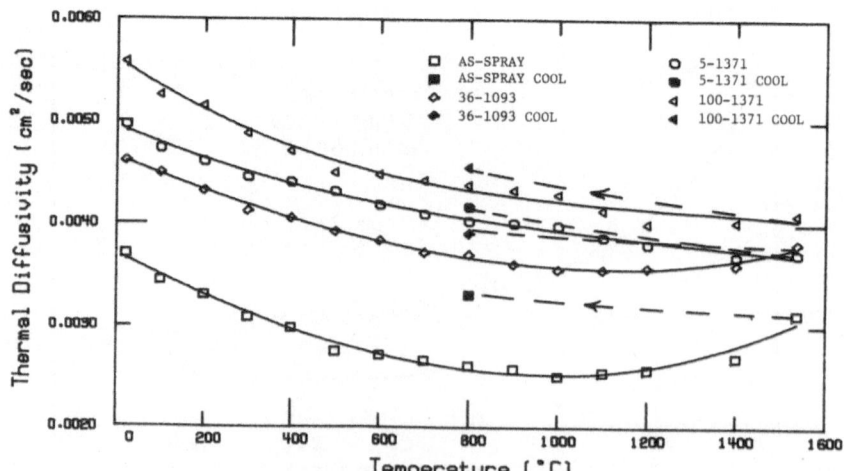

Fig. 3. Diffusivity of a Plasma-Sprayed Ceramic Subjected to Selected Heat Treatments.

An examination of the thermophysical properties of nickel-based alloys reveals an interesting situation. These materials have a complex metallurgy and exhibit both long range and short range order. Transformations, which occur over very long periods of time, take place at higher temperatures. Specific heat results obtained using the differential scanning calorimeter (DSC) and multiproperty (MP) apparatus are shown in Fig. 7. Values obtained using the multiproperty apparatus were obtained using heating rates from 3 to 26°C/sec. Also shown in Fig. 7 are specific heat values (CALC) calculated from measured conductivity and diffusivity values and the results

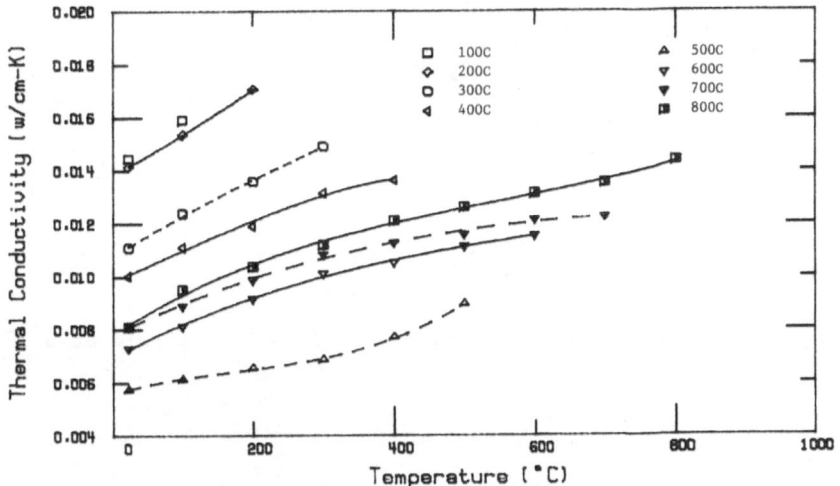

Fig. 4. Conductivity of a Carbon-Phenolic Heat Treated at Various Temperatures.

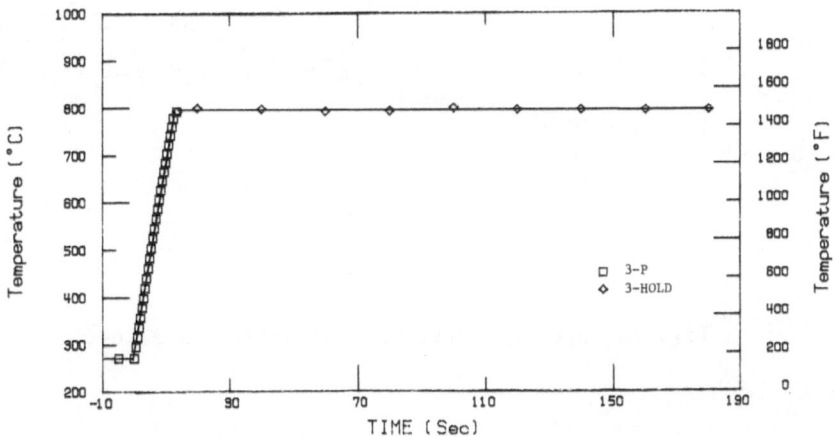

Fig. 5. Time-Temperature History of an Alloy Heated to 800°C.

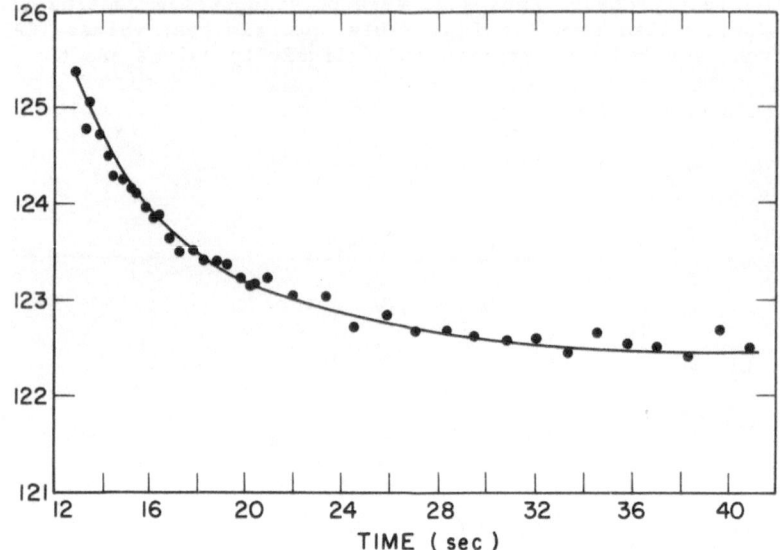

Fig. 6. Electrical Resistivity of an Alloy During Hold at 800°C.

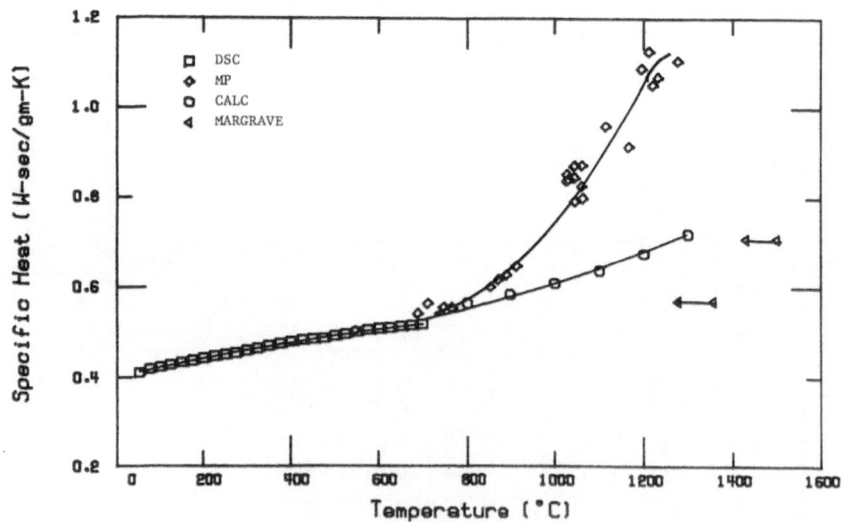

Fig. 7. Specific Heat of a Nickel-Based Alloy.

obtained by Margrave and Krishman [1]. Margrave and Krishman's results for the molten state look reasonable, but their results for the solid region are low due to quenching effects. The sharp rise in specific heat values at higher temperatures for nickel-based alloys have been observed [2] for more than twenty years.

Thermal conductivity values calculated from the diffusivity and specific heat results are shown in Fig. 8. The density and diffusivity values have been corrected for thermal expansion. These results indicate that the conductivity values increase rapidly above 800°C and then decrease sharply from the solid to the molten region. The conductivity values calculated from electrical resistivity results are shown as "ELECTRONIC" in Fig. 8. These results parallel the conductivity values calculated from density, diffusivity and specific heat up to 800°C, but do not show the sharp increase above 800°C. The thermal conductivity due to electrons approaches that calculated from the thermal diffusivity and specific heat data in the molten region.

Conductivity values (corrected for expansion) measured with the multi-property apparatus are included in Fig. 8. These results join nicely with the results calculated from diffusivity, density, and specific heat data below 800°C and in the molten region.

It is postulated that there are two limiting specific heat curves. Specific heat values, calculated from the conductivity, density, and diffusivity results are included in Fig. 7. These specific heat values are believed to be the ones associated with transient heat flow in this material. The specific heat values measured by pulse heating represent the other set. The possibility of two sets of specific heat values for this type of material has been suggested previously [2] and may be consistent with materials in which both short range and long range order exists.

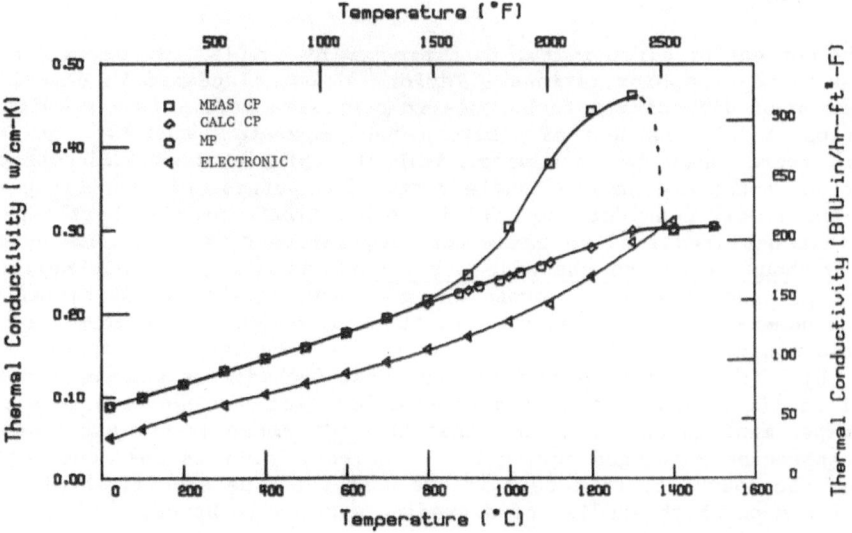

Fig. 8. Thermal Conductivity of a Nickel-Based Alloy.

The thermal conductivity in this material is largely, perhaps almost entirely above about 300°C, due to electrons. If L is about 30% greater than the classical value (L_0), the electronic contribution closely follows the recommended conductivity values between 200 and 1000°C. Above this temperature L probably approaches L_0 and becomes nearly equal to the classical value near the molten region.

It is concluded that the thermal conductivity increases smoothly from 0.0880 at 22°C to 0.300 W cm^{-1}K^{-1} at 1300°C and then remains nearly constant in the molten region - at least up to 1500°C.

COMMENTS ON DIFFUSIVITY/CONDUCTIVITY RELATIONSHIPS

The heat balance equation can be written as

$$\nabla \cdot \lambda \nabla T + \text{sources and sinks} = C_p \rho \ (dT/dt) \tag{1}$$

where λ is thermal conductivity, C_p is specific heat, and ρ is density. If there are no sources and sinks, then

$$\nabla \cdot \lambda \nabla T = C_p \rho \ (dT/dt) \tag{2}$$

If the thermal conductivity is independent of position (and temperature)

$$\lambda \nabla^2 T = C_p \rho \ (dT/dt) \tag{3}$$

Then

$$\nabla^2 T = \frac{C_p \rho}{\lambda} \ (dT/dt) = (1/\alpha) \ (dT/dt) \tag{4}$$

where the diffusivity, α, is defined as $\lambda/(C_p\rho)$.

For one dimensional heat flow

$$\alpha (d^2T/dx^2) = dT/dt \tag{5}$$

Thus the concept of diffusivity, as expressed by Eq. (5), involves a number of assumptions or approximations. Taylor [3] has discussed in detail the application of diffusivity techniques to composite materials, in which the assumption of λ independent of position and temperature must be considered. However, there seems to be a great deal of confusion over the role of sources and sinks and how to handle them. For purpose of clarifying this situation, I will consider the definition of "true" specific heat capacity to be that heat required to raise the temperature of a unit mass by one degree without otherwise changing its condition, i.e., without melting, vaporizing, desorbing gases, phase changes, etc. The heat associated with these phenomena will be considered in the sources and sinks term (Eq. 1). From this consideration it is seen that if one uses effective specific heat values which include such phenomena, one must include the sources and sinks term of Eq. (1). However, it may be possible that one can arrange a transient experiment in such a manner that the rate these events are occurring is not appreciably changed during the transient. In this case one need not consider the source or sink term if one uses the true specific heat. This is the basis on which studies of degrading systems is based.

Let us consider a specific example – the determination of the conductivity of moist soil. We could perform a standard thermal conductivity test involving steady-state conditions. If we should do so we would find that it would be difficult to establish equilibrium. When equilibrium is established, the moisture content would no longer be uniform; instead the moisture content would be less in the area of higher temperatures. Thus our values would depend upon the resulting moisture gradient and would not represent the original sample. On the other hand we might perform an experiment in such a fashion that a small temperature wave diffuses through the sample so fast that the moisture distribution is not affected. In such a case it would be possible to determine the diffusivity of the uniform sample. If we then determine the specific heat of this sample without removing water (true specific heat) we could calculate the conductivity of the original material. If we were to use an effective specific heat, however, we would obtain erroneous conductivity results.

REFERENCES

1. J. L. Margrave and S. Krishman, High Temperature Properties of Rene-N5, Private Communication (March 1987).
2. E. E. Stansbury, The Adiabatic Calorimeter, Third Conference on Thermal Conductivity, Vol. II, Gatlinburg, TN (October 1963).
3. R. E. Taylor, Thermal Diffusivity of Composites, High Temp. – High Pressures, 15, 299-309 (1983).

SESSION 3

HIGH-TEMPERATURE MATERIALS

THERMAL DIFFUSIVITY AND THERMAL CONDUCTIVITY

OF CARBON MATERIALS FOR TOKAMAK LIMITERS

E.P. Roth and M. Moss

Thermophysical Properties Division

Sandia National Laboratories
Albuquerque, NM 87185

ABSTRACT

A plasma limiter is used to control the plasma size in Tokamak
fusion research devices and serves as one of the main areas of plasma
wall interaction. Wall temperatures can reach as high as 3000 K resulting
in significant sputtering and wall erosion. Graphite and graphite
composites have been chosen as the materials which best meet the
requirements of this severe environment. The thermal conductivities of
these materials are critical to the modeling of the thermophysical
response of the limiters during operation. The thermal conductivities of
several candidate materials were measured. Three categories of materials
were investigated in this study: pyrolytic graphite and annealed
pyrolytic graphite; carbon/carbon fiber composites composed of two-
directionally woven fibers with a graphitized pitch matrix; carbon/carbon
fiber composites with a four-directional carbon fiber weave. The thermal
conductivities of the carbon/carbon fiber composites were determined as a
function of fiber orientation. Conductivities were measured using both
the laser flash diffusivity technique and the thermal comparative method.
The advantages and disadvantages of each of these methods will be
discussed and the data compared.

INTRODUCTION

Tokamak fusion research devices contain high-temperature, low-
density plasma with complex magnetic fields. However, the plasma does
interact with the walls of the reactor and can result in severe
thermomechanical stress on the wall materials. In particular, plasma
limiters, which are used to control plasma size and are the main area of
plasma-wall interactions, must withstand energy depositions of hundreds
of joules/cm^2 in tens of milliseconds during a plasma disruption and must
stand up to particle energies which can be as high as 200 keV [1]. In
addition, the wall materials must be restricted to low atomic number (Z)
elements so that surface erosion of the wall material does not poison the
plasma. The leading candidate materials for this application are
graphites and graphite composites.

Graphite materials are characterized by low atomic number, the
ability to withstand high temperatures, and relatively high thermal
conductivity. High thermal conduction, especially for transient
temperature excursions, is important in limiting the maximum surface
temperature which thus limits surface erosion. High thermal conduction in

these materials can be achieved either by taking advantage of the anisotropic nature of thermal conduction in pyrolytic graphites or by using composites of high thermal conductivity carbon fibers.

In this paper, we present data on several commercial materials which have been measured in our laboratory. Three classes of materials were investigated: pyrolytic graphites and annealed pyrolytic graphites; carbon/carbon fiber composites composed of two-directionally woven (2-D) carbon fiber weaves with a graphitized pitch matrix, and a more complex four-directionally woven (4-D) carbon fiber weave. The thermal diffusivities of these materials were measured using the laser flash method, which characterizes the transient thermal response of the materials as would be experienced in actual use. However, the use of this method in determining the effective homogeneous property of a highly inhomogeneous material poses several difficulties in data analysis and interpretation. In an attempt to address some of these problems, steady-state measurements were performed on the highly anisotropic 4-D fiber composites using the thermal comparative method.

FLASH DIFFUSIVITY METHOD

Flash diffusivity has now become a standard technique in many laboratories. The method first proposed by Parker et al. involves subjecting the front face of a small disk-shaped sample to a brief laser pulse and monitoring the temperature rise of the opposite face as a function of time [2]. For the limiting case of no heat loss from the sample, the diffusivity can be calculated from the measured time ($t_{1/2}$) to reach half of the peak temperature rise. The thermal diffusivity (α) is then given by:

$$\alpha = 0.139 \; \ell^2/t_{1/2} \; , \tag{1}$$

where ℓ is the sample thickness. The corresponding steady-state thermal conductivity (κ) is calculated using the simple relation:

$$\kappa(T) = \alpha(T)\rho(T)c_p(T) \; , \tag{2}$$

where $\rho(T)$ is the sample density, and $c_p(T)$ is the specific heat.

Accurate determination of thermal diffusivity requires making corrections for heat loss from the sample and for the effects of the finite laser pulse width. We have used data analysis programs developed to simultaneously correct for these effects [3]. The calculated diffusivity is determined from the measured half-rise-time ($t_{1/2}$), the pulse time to half-rise-time ratio, and the loss parameter determined from either the short-time shape of the leading edge or the long-time tail of the temperature-time curve. The results of these analyses usually agree within 5%.

Although the previous analysis works well for homogeneous samples, the model must be applied with caution to heterogeneous materials such as composites with highly preferred conduction paths. In these fiber-reinforced materials, the shape of the temperature rise curve is distorted in comparison with the theoretical homogeneous curve due to the early temperature rise of the high conduction fibers and radial heat flow originating from these fibers. However, previous work has shown that for materials with sufficient coupling between the thermally heterogeneous components and for sufficient sample thicknesses, the composite behaves as an effective homogeneous material [4,5,6]. It is not clear if the thermal conductivity obtained from this homogeneous diffusivity result should be in agreement with that obtained by steady-state methods.

The assumption of effective homogeneous behavior can be tested by performing a time-dependent analysis of the measured temperature rise curve. An effective time-dependent diffusivity can be determined at any point along the leading edge of the temperature rise curve to determine any deviations from the theoretical homogeneous curve. A time-dependent diffusivity can be calculated at pre-selected points of the temperature rise curve using the Cowan's modulus at each selected fraction of the peak temperature rise and the relation:

$$\alpha(t_x) = F(x)\ell^2/t_x \quad , \tag{3}$$

where t_x is the time to reach a fraction x of the peak temperature rise and F(x) is the corresponding Cowan's modulus. A homogeneous material will give the same diffusivity value for any point on the temperature rise curve, while the diffusivity for a heterogeneous, fiber-reinforced material will initially reflect that of the high-conductivity component and at later times take on a value between the equivalent homogeneous medium and the lower-conductivity matrix [4]. For sufficient sample thicknesses, the later time diffusivity should approach that of the equivalent homogeneous medium [4].

Equation 3 will give accurate diffusivity results only for the case where there are no heat losses and the finite pulse time effects can be ignored. However, this simple method can be used to determine if a particular sample is behaving as an effective homogeneous material. If that is the case, standard data analysis routines can be used which do correct for heat losses and pulse time effects.

EXPERIMENTAL EQUIPMENT AND METHODS

FLASH DIFFUSIVITY EQUIPMENT

The flash diffusivity apparatus consists of a pulsed ruby laser (λ = 6943Å) with maximum pulse energy of 35 joules and a pulse width of 1.35 ms, a high temperature furnace/vacuum system, and a spatially averaging InSb infrared detector [7,8]. The sample was held in the furnace using a low-thermal-contact tantalum holder with tantalum shields on both the front and back sample faces to eliminate laser flashby. The sample temperature rise as detected by the infrared detector was conditioned through a variable bandpass amplifier and biasing network and then recorded by a digital transient recorder. The output of the transient recorder was transferred to a Hewlett-Packard HP1000 minicomputer which then performed all data analysis [8,9].

System performance was checked using POCO AXM-5Q graphite over the experimental temperature range. The diffusivity obtained for this material was compared with that obtained for a round-robin study on the same type of material and compared with historical laboratory records for this sample [10]. Agreement was obtained to within ±5% of the accepted values.

STEADY-STATE EQUIPMENT

Steady-state thermal conductivity measurements were performed on the 4-D material using two experimental methods: a thermoconductometer for room temperature and a thermal comparative system for higher temperatures.

Thermal Comparative System

A Dynatech comparative thermal conductivity instrument, Model TCFCM, was used for measurements from room temperature to 700°C [8,11-15]. The comparative method of measuring thermal conductivity uses two identical reference standards of known and stable conductivity placed in intimate contact with the top and bottom of a sample of equal cross section. The standards are chosen to match the expected conductivity of the sample as closely as possible. Heaters are placed on the top and bottom of the stack, and insulation and guard heaters are used to reduce radial temperature gradients. A determination of thermal conductivity requires the measurement of temperature gradients along the flux direction in the sample and standards by means of thermocouples. The heat fluxes in the sample and the standards are made as nearly equal as possible. The flux in the sample is considered to be the average of the fluxes in the standards.

All comparative measurements were made in argon gas with a reference standard of Vacumet Consumet high-purity iron [16]. Overall accuracy of this method is considered to be ±5%.

Thermoconductometer

A Colora thermoconductometer was used to perform rapid (≈ 10 min) room temperature measurements [8,11,17]. The instrument employs two liquids with different boiling points to heat and cool the specimen. If calibrated samples are measured with a selected liquid pair, the thermal resistance of any sample can be read from a calibration curve which plots resistance vs. time with an estimated accuracy of ±10%.

Specific Heat

The specific heats of representative samples for each type of material were measured with a Perkin-Elmer DSC2 differential scanning calorimeter (DSC) from room temperature to 700°C [8,18]. NBS sapphire standards were used to verify system calibration to within ±2%. The measured specific heat values were compared with literature values recommended for graphite [19]. Maximum deviation from the recommended curve for all of the materials was 3%. In order to maintain a common specific heat reference curve for all samples, the literature values for specific heat were used in all thermal conductivity calculations.

SAMPLE MATERIAL

Three carbon-based material types were measured in this study: pyrolytic graphites, two-directionally woven carbon/carbon fiber weaves, and a four-directionally woven carbon/carbon fiber weave. The materials were obtained from various U.S. and European manufacturers and in most cases the details of the sample preparation methods are not available.

Pyrolytic Graphite

Pyrolytic graphite samples were obtained from B.F. Goodrich and Pfizer [20,21]. Samples were provided both in the as-deposited condition and after having undergone annealing at 3000°C for 1 hour. A separate B.F. Goodrich sample was measured which had been annealed for an extended but unknown time. The thermal diffusivities for all samples were measured parallel to the deposition plane which is the high conductivity direction in these highly anisotropic samples. The diffusivity samples were discs with diameters of 0.127 cm and thicknesses ranging from 0.3 cm to 1.0 cm, where the greater thicknesses were used for the high-conductivity annealed samples.

Two-Directionally Woven Carbon/Carbon Fiber Weaves

Four 2-D samples were obtained from European manufacturers: two samples from Carbone Lorraine Corp. (designations AO5 and A223G), one sample from SIGRI, and one sample from Schunk & Ebe [22-24]. The materials were orthogonally woven composites and had a range of densities and fiber dimensions. The measured physical parameters for these samples are included in the figures. The source materials and manufacturing methods for these samples are not known. Three samples were cut from each material: two samples representing orthogonal heat flow directions in the plane of the

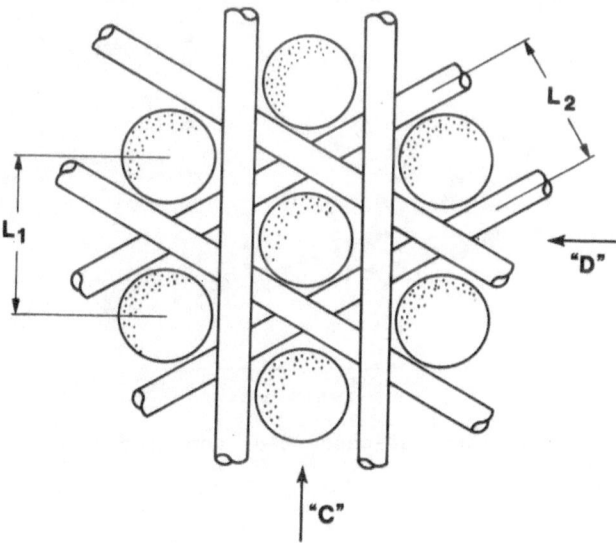

Fig.1   Schematic of four-directionally woven carbon/carbon composite.
         L1= .32 cm; L2= .28 cm; "C,D" are heat flux directions in plane.

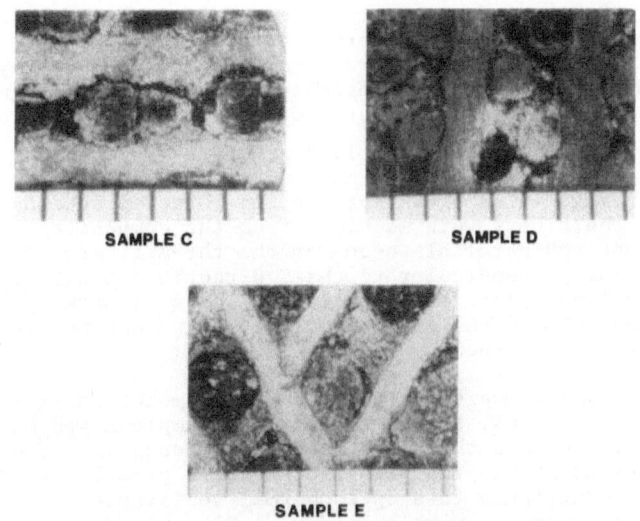

Fig.2   Photomicrographs of 4-D carbon/carbon composites for three heat
         flux directions. Scale is 0.1 cm.

fiber weave and one sample with heat flow perpendicular to the fiber plane.
Diffusivity sample thicknesses varied from 0.15 cm to 0.31 cm which were
much greater than the fiber thicknesses (≈0.02 cm). The samples were of
sufficient diameter that a representative pattern of fibers was measured.

Four-Directionally Woven Carbon/Carbon Fiber Weaves

    The 4-D carbon/carbon fiber materials were manufactured by Fiber
Materials, Inc. [25]. These materials consisted of three in-plane fibers
(diameter ≈ 0.1 cm) woven at 60 degrees relative orientation. Center-to-

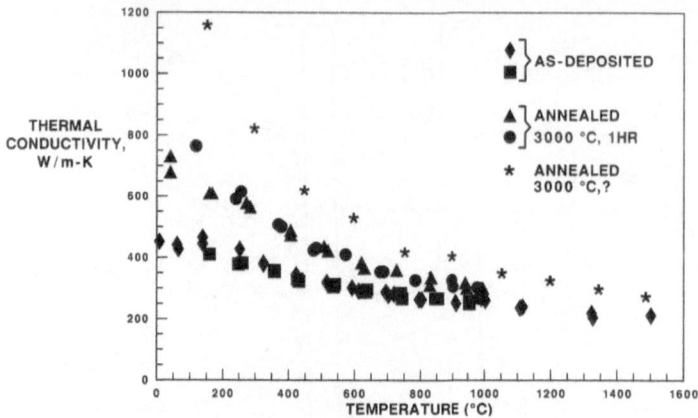

Fig. 3   Pyrolytic graphites: as-deposited; annealed 3000°C, 1 hr; extended
anneal 3000°C.

center fiber separation in the plane was 0.28 cm with a plane spacing of
0.32 cm. A fourth fiber was woven perpendicular to the plane with a diameter
of approximately 0.2 cm and a center-to-center distance of 0.32 cm. Figure 1
shows a schematic of the weave pattern. Nominal fiber volume fraction was
64%. These materials were manufactured from a pitch precursor yarn and
densified through several stages consisting of pressure impregnation with
pitch, high pressure carbonization at 600-700°C followed by graphitization
near 2500°C [26]. Final density was 1.95 g/cm^3.

     As shown in Fig. 1, samples representing three directions of heat flux
with respect to the fibers were machined from the composites: C direction -
in the fiber plane and parallel to one of the three fibers; D direction - in
the fiber plane and perpendicular to the C direction; E direction -
perpendicular to fiber plane and parallel to large diameter fibers. Figure 2
shows photomicrographs of the sample surfaces in which the fiber
orientations are readily seen.

     Comparative samples were discs 5.1 cm in diameter and 1.5 to 2.0 cm
thick. Heat flux was normal to the flat faces. Sample E was machined as one
piece from the bulk composite while C and D were each made from three pieces
cemented together such that the plane joints were parallel to the heat flux
and thus did not contribute to sample thermal resistance.
Thermoconductometer samples were cylinders 1.27 cm in diameter and 1.98 cm
in length. The diffusivity samples were single-piece discs with 1.27 cm
diameters. Several thicknesses were measured for each orientation from 0.4
cm to 2.0 cm.

RESULTS AND DISCUSSION

Pyrolytic Graphites

     Figure 3 shows the calculated thermal conductivity data for the
pyrolytic graphite samples. No significant difference was seen between the
B.F Goodrich and Pfizer materials either before or after annealing. The
thermal conductivity of the as-deposited samples decreased slightly with
increasing temperature and was fairly constant at the higher temperature
range. The annealed samples showed a marked increase in conductivity over
the as-deposited samples at low temperatures, increasing from approximately

450 W/m-K to 700 W/m-K at 50°C. However, the conductivities of the as-deposited and the annealed samples converged near 1000°C. The conductivity of the extended-annealed B.F. Goodrich sample was much greater than seen for the one-hour annealed materials. The conductivity at 200°C was almost 1200 W/m-K which is approximately three times greater than that for copper at the same temperature. However, the conductivity had essentially converged with that of the other pyrolytic samples by 1500°C.

The measured properties of these pyrolytic graphite samples are in good agreement with the known behavior of these types of materials. Oriented pyrolytic graphites are highly anisotropic heat conductors [27]. These materials consist of grains with most of the carbon atoms arranged in planar hexagonal arrays linked together with strong covalent bonds and weak van der Waals forces between the layers. These grains primarily align their basal planes parallel to the deposition surface. However, very high annealing temperatures for extended times are required to achieve a high degree of ordering in the bulk material. Heat conduction in the plane can be hundreds of times greater than perpendicular to the plane and is primarily determined by grain size and umklapp scattering of phonons [28]. At higher temperatures, umklapp scattering dominates the phonon mean free path and the thermal conductivities of samples with different grain sizes will converge.

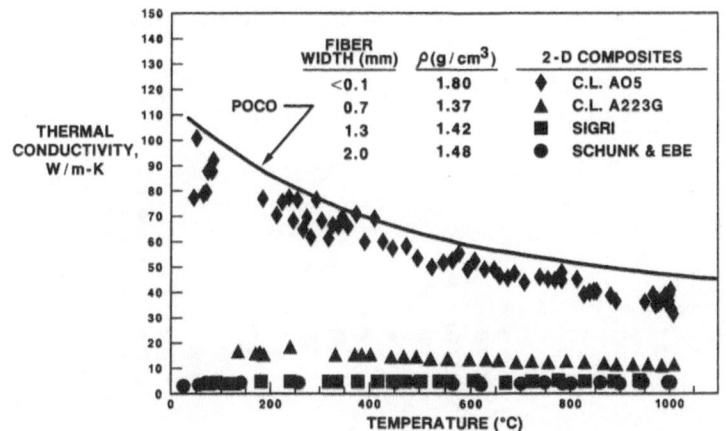

Fig. 4   2-D carbon/carbon composites. Four European samples measured perpendicular to fiber plane. Solid line: POCO graphite reference.

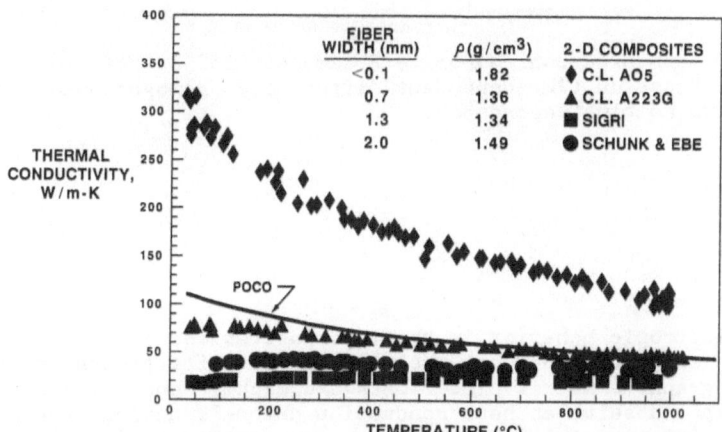

Fig. 5   2-D carbon/carbon composites. Four European samples measured parallel to fiber plane. Solid line: POCO graphite reference.

A combination of grain growth and grain orientation would account for our observed increase in conductivity, although x-ray diffraction analysis would be required to identify the details of the microstructural changes.

2-D Carbon/Carbon Composites

Figures 4 and 5 show the data for the thermal conductivity of the 2-D composites perpendicular and parallel to the fiber planes respectively. The thermal conductivity of POCO AXM-5Q graphite is shown as a common reference. No difference was seen between the two orthogonal directions in the plane and, for clarity, both sets of data are not shown. Comparison of Figs. 4 and 5 shows that the heat conduction perpendicular to the fiber planes was significantly less than for conduction in the planes. This anisotropic heat conduction ratio increased with increasing fiber dimensions, varying from approximately 8 for the coarse-fiber Schunk & Ebe material to only 3 for the fine-fiber C.L. A05. The conduction in the coarse-fiber materials was fairly constant over the measured temperature range, while the conduction of the fine-fiber C.L. A05 exhibited a temperature dependence similar to that of POCO graphite. The data show that the conduction increased with decreasing fiber size and increasing density.

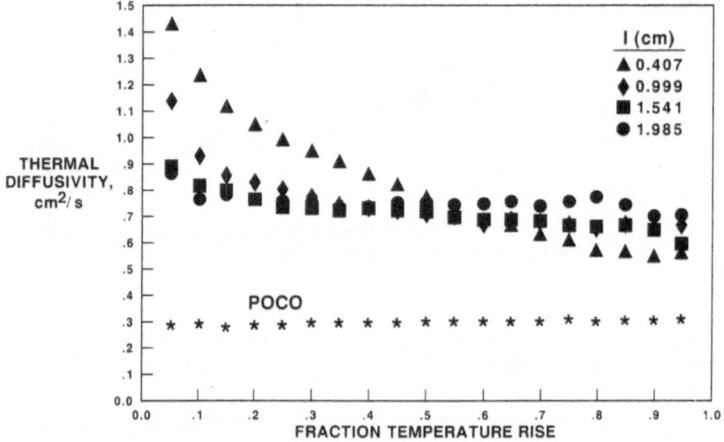

Fig. 6   4-D carbon/carbon composite E samples, 150°C. Effect of sample thickness on time-dependent diffusivity. Homogeneous POCO graphite shows no time dependence.

This anisotropic behavior in these composites results from the difference in conduction between the high-conductivity fibers and the lower conductivity graphite matrix. The fibers are highly conductive parallel to the fiber axis and serve as heat conduction channels through the composite [5]. The C.L. A05 material has a high density similar to most fully-densified carbon/carbon composites and shows a temperature dependence similar to that of graphite indicating that this material can be modeled as a closely connected two component system. Heat conduction in the plane can be modeled as parallel regions of high and low conduction, while heat conduction perpendicular to the plane is modeled as a layered series of high and low conductivity material. The remaining three materials had significantly lower densities (1.4-1.5 g/cm^3) and thus must have included a significant amount of open porosity or separation between the fibers and matrix. A thermal model of these materials would then necessarily include high contact resistance between the fibers and matrix and possibly between the contacting fibers themselves which would account for the lower overall conductivity and the lack of significant temperature dependence.

4-D Carbon/Carbon Composites

The thermal diffusivities of the 4-D carbon/carbon materials were measured from 150-950°C. Three thicknesses each of the C and D samples and four thicknesses of the E samples were measured. Typical time, temperature traces for each material over the measured temperature range were analyzed using Eq. 3 for distortions of the temperature rise function due to inhomogeneous heat flow. The thermal conductivity was calculated for each sample to show the effects of any distortions and then compared with the values obtained from the steady-state thermal comparative system.

Figure 6 shows the thermal diffusivities as a function of fractional temperature rise for the four E samples at the lowest measured temperature. This measurement represents the most inhomogeneous heat flow condition: heat flow parallel to the large diameter, high-conductivity fibers at a temperature where the fibers have the greatest difference in diffusivity from the matrix. The figure shows a significant distortion in the temperature rise curve, especially for the thinnest samples. At early times, the diffusivity is determined primarily by the high-diffusivity fibers, while at later times the diffusivity is closer to that of the equivalent homogeneous medium. As sample thickness increases, the transient thermal wave has time to achieve equilibration between the fibers and matrix by means of radial heat flow. The data show that except for the 0.4 cm sample, the samples had achieved a homogeneous temperature rise profile by 0.3 of the maximum temperature. Also included in Fig. 6 is the time-dependent diffusivity of a homogeneous POCO graphite sample which shows that the diffusivity is indeed constant over the entire temperature rise profile for a homogeneous transient wave.

Figure 7 shows a comparison of time-dependent diffusivity determined for the three heat flux directions in the low-temperature range. The minimum thicknesses of all three samples ($\approx$0.4 cm) were measured and each showed a distorted temperature rise function. However, the E samples showed the greatest distortion compared with the C and D samples. The effect of a larger sample thickness is shown in Fig. 8. The data show that an effective homogeneous temperature rise curve was achieved after only a temperature rise fraction of 0.15. Figure 9 shows the effects of increasing temperature on the time-dependent diffusivity. The early time diffusivity for the E samples decreased significantly as temperature was increased from 150°C to 800°C. This decrease could result from decreased contact resistance between the fibers and matrix at higher temperatures, but more likely results from the rapid decrease in diffusivity of the fibers as was seen for the pyrolytic graphites.

The corrected diffusivities based on the the half-rise times were used to calculate the thermal conductivities of each sample. Figures 10-12 show the conductivities for each heat flow direction in the 4-D composites. The data show that for the C and D samples, good agreement was obtained in the conductivities between thicknesses even though some distortion was seen in the temperature rise curves for the thin samples. However, the diffusivity for the 0.4 cm thickness of the E samples was approximately 10-15% higher than for the three thicker E samples over the measured temperature range. Thus, the severe distortion observed for the thin samples due to inhomogeneous heat flux leads to artificially high diffusivity values.

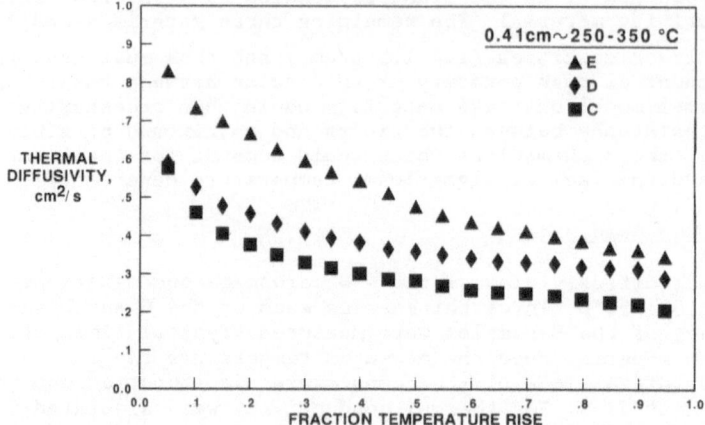

Fig. 7　Comparison of time-dependent diffusivity for three heat flux directions at low-temperature range.

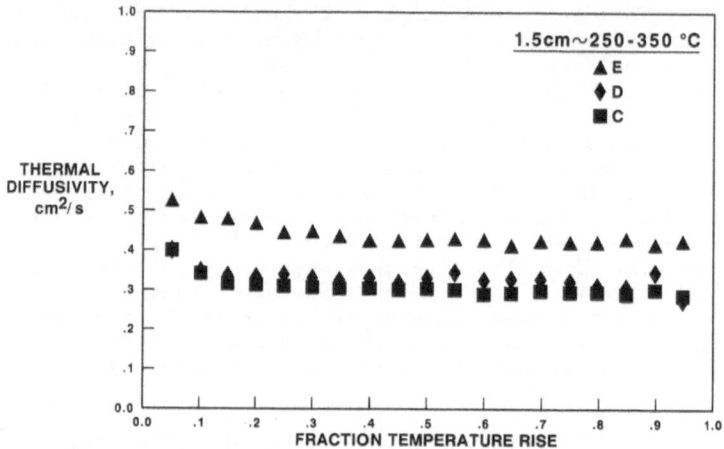

Fig. 8　Comparison of time-dependent diffusivity for three heat flux directions for thick (1.5 cm) samples.

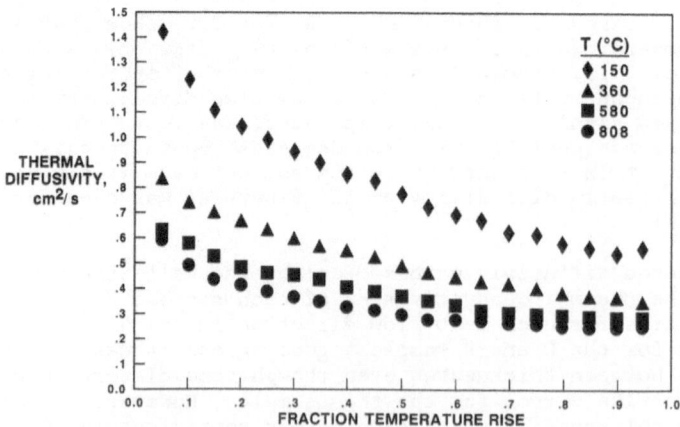

Fig. 9　Effect of temperature on time-dependent diffusivity for E samples.

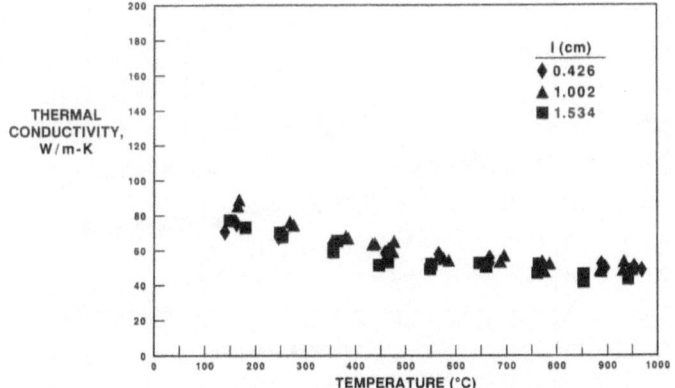

Fig.10   Thermal conductivity as a function of thickness for heat flux direction C.

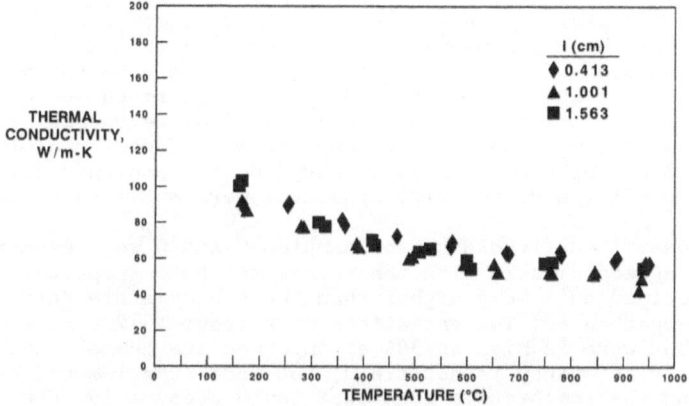

Fig.11   Thermal conductivity as a function of thickness for heat flux direction D.

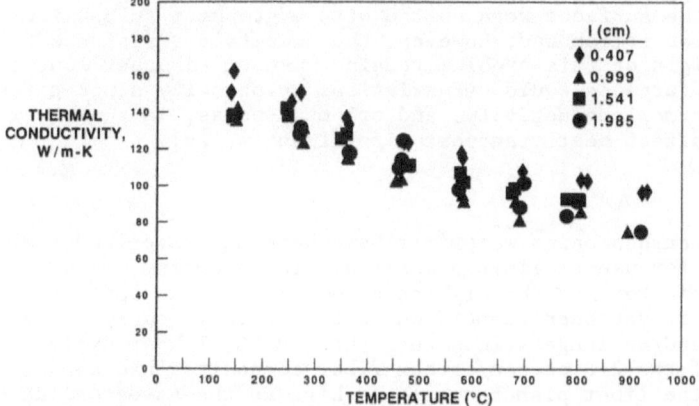

Fig.12   Thermal conductivity as a function of thickness for heat flux direction E.

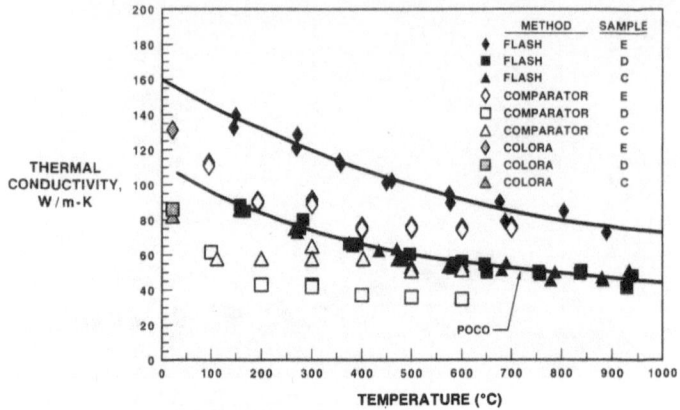

Fig.13   Comparison of thermal conductivity for 4-D composites derived from
transient and steady-state methods.

Thermal conductivity data for the 1 cm thick samples were chosen as
representative of the homogeneous limit for each heat flux direction. Figure
13 shows a comparison of these data with the results from the steady-state
methods. Values near 30°C were measured with the thermoconductometer while
the remaining steady-state data were obtained with the thermal comparative
system. All comparative samples were measured as a function of increasing
temperature while sample C was also measured during cooling. All data
obtained from the flash diffusivity apparatus were obtained during heating.

The diffusivity-derived data for samples C and D were essentially
identical and agreed closely with the values for POCO graphite. However, the
diffusivity derived data were higher than the steady-state data except at
the highest temperatures. The variations were about 15-20% at room
temperature, but were as high as 30% at intermediate temperatures. The
origin of this discrepancy is not clear, but some experimental problems were
incurred during the comparative runs that could account for a portion of the
difference. Abnormally large temperature differences appeared across the
sample-standard interfaces in all comparative runs. These interfacial
temperature gradients could produce radial temperature gradients which would
be a source of error depending on thermocouple placement. On the assumption
that surface pits due to machining accounted for the high interface
resistance, the surfaces were coated with a graphite suspension and dried,
and the samples remeasured. However, the interface resistance remained and,
thus, the origin of this problem remains unknown. Another source of sample-
to-sample differences would be variations in porosity since pores strongly
lower the thermal conductivity, and oriented pores, in analogy with oriented
fibers, can affect heat transport significantly [29].

SUMMARY

Several carbon based materials have been characterized that meet the
requirements for use as limiter materials in Tokamaks. The oriented
pyrolytic graphites had the highest conductivity. The optimum annealing
process has not yet been identified, but certainly requires higher
temperature and/or longer times than the 3000°C, 1 hour cycle investigated
here. Some of the 2-D carbon/carbon fiber composites had conductivities
parallel to the fiber planes nearly as high as the as-deposited pyrolytic
graphites. The small fiber, high-density composites had conductivities
significantly higher than the other 2-D composites. These materials also
have superior mechanical properties which make them good candidate limiter
materials. The 4-D carbon/carbon composite had a thermal conductivity

parallel to the major fiber axis higher than all but the highest 2-D material at low temperatures, and equal to the highest 2-D material at high temperatures.

The question of the relation of homogeneous heat wave propagation in a coarse, fiber-reinforced material to the steady-state conduction still remains. In contrast to previous work, we have observed that the conductivity derived from the transient method is equal to or higher than a steady-state measurement even in the homogeneous limit, but the uncertainty in the measurements did not allow a clear comparison between the two methods [4,5,6].

Acknowledgements

This work supported by the U.S. Department of Energy under Contract DE-AC04-76-DP00789. We thank G.M. Haseman, P.E. Quesenberry, and W.E. Fowler for assistance in the laboratory.

References

1. R.W. Conn, "Relation of Surface Interactions to First-Wall and In-Vessel (IVC) Design and Materials Performance in Fusion Devices", J. Nucl. Mater., 103 & 104, 7 (1981).

2. W.J. Parker, R.J. Jenkins, et al, "A Flash Method of Determining Thermal Diffusivity, Heat Capacity, and Thermal Conductivity", U.S. Navy Technical Report USNRDL-TR-424, May, 1960.

3. J.A. Koski, "Improved Data Reduction Methods for Laser Pulse Diffusivity Determination with the use of Minicomputers", Proceedings of the Eighth Symposium on Thermophysical Properties, Vol. II: Thermophysical Properties of Solids and of Selected Fluids for Energy Technology, ASME, New York, (1982), pp. 94-103.

4. D.L. Balageas and A.M. Luc, "Transient Thermal Behavior of Directional Reinforced Composites: Applicability Limits of Homogeneous Property Model", AIAA J., 24, 109 (1986).

5. A. Whittaker, R. Taylor, and H. Tawil, "Thermal Diffusivity of Some Fine-Weave Carbon/Carbon-Fiber Composites", High Temperatures-High Pressures, 17, 225 (1985).

6. R.L. Shoemaker, "Limitations of the Pulse Diffusivity Method as Applied to Composite Materials", High Temperatures-High Pressures, 18, 645 (1986).

7. Infrared detector from Properties Research Laboratory , Box 2224, East Lafayette, Indiana 47906.

8. Reference to a particular product or company implies neither a recommendation nor an endorsement by Sandia National Laboratories, nor a lack of suitable substitutes.

9. Hewlett-Packard Co., Cupertino, CA 95014.

10. E. Fitzer, "Results of the Cooperative Measurements on Heat Transport Properties up to 2800 K", AGARD-R-606, Technical Editing and Reproduction Ltd., Harford House, 7-9 Charlotte St., London WIP 1H, 1973.

11. The model TCFCM comparative thermal conductivity instrument is manufactured by Dynatech R/D Co., Cambridge, MA. The Colora

thermoconductometer is manufactured by Colora Messtechnik GMBH, Lorch/Wurttemberg, FRG, and is sold by Dynatech.

12. M. Moss, J.A. Koski, and G.M. Haseman, "Measurement of Thermal Conductivity by the Comparative Method", Report No. SAND82-0109, Sandia National Laboratories, Albuquerque, NM (1982).[*]

13. J.N. Sweet, M. Moss, and C.E. Sisson, "The Use of Numerical Heat Transfer Techniques to Analyze Thermal Comparator Conductivity Measurements", Thermal Conductivity 18, T. Ashworth and D.R. Smith eds., Plenum, New York (1985).

14. J.N. Sweet, E.P. Roth, M. Moss, G.M. Haseman, and A.J. Anaya, "Comparative Thermal Conductivity Measurements at Sandia National Laboratories", Report No. SAND86-0840, Sandia National Laboratories, Albuquerque, NM (1986).[*]

15. J.N. Sweet, "Establishments of Accuracy Limits and Standards for Comparative Thermal Conductivity Measurements, Int. J. Thermophysics, 7, 743 (1986).

16. Carpenter Technology Corp., Reading, PA.

17. L.C. Beavis and M. Moss, "Thermally Conductive Silicone Based Materials for Attaching Concentrator Solar Cells to Heat Sinks", Symposium Series, No. 245, Vol. 81, Proc. 23rd AIChE/ASME National Heat Transfer Conf., Denver, CO, 1985 Amer. Inst. Chem. Eng., New York (1985).

18. Perkin-Elmer Corp., Norwalk, CT.

19. A.T.D. Butland and R.J. Madison, J. Nucl. Mater., 49, 45 (1973).

20. B.F. Goodrich, Aerospace and Defense Division, Super-Temp Operations, 11120 South Norwalk Blvd., Santa Fe Spring, CA 90670.

21. Pfizer, Inc., 640 N. 13th St., Easton, PA 18042-1497.

22. Carbone Lorraine, Carbone USA Corp., 400 Myrtle Ave., Boonton, NJ 87005.

23. SIGRI GmbH, Post Box 11 60, 8901 Meitingen, FRG.

24. Schunk Kohlenstofftechnik GmbH, Post Box 64 20, 6300 Geissen, FRG.

25. Fiber Materials Inc., Biddeford Industrial Park, Biddeford, ME 04005.

26. J.B. Smith, R.L. Burns, and L.L. Lander, "Low Cost/High Performance Carbon-Carbon Nozzles", Tech. Report: RK-CK-83-2 for Propulsion Directorate, U.S. Army Missile Laboratory, Redstone Arsenal, AL 35898 (1982).

27. B.T. Kelly, "The Thermal Conductivity of Graphite", Chem. Phys. Carbon, 5, 119 (1969).

28. R. Taylor, "The Thermal Conductivity of Pyrolytic Graphite", Phil. Mag., 13, 157 (1966).

29. M. Moss and G. Haseman, "A Proposed Model for the Thermal Conductivity of Dry and Water-Saturated Tuff", Materials Research Society Symp. Proc. 26, Elsevier, New York (1984), p. 967.

---

*Available from NTIS, U.S. Department of Commerce, 5285 Port Royal Road, Springfield, VA 22161

THERMAL PROPERTIES OF MULLITE-CORDIERITE COMPOSITES

Stephen C. Beecher*, Ryan E. Giedd**, and David G. Onn[†]

Applied Thermal Physics Laboratory (ATPL)
Department of Physics and Astronomy
University of Delaware
Newark, DE 19716

Richard M. Anderson, John B. Wachtman

Center for Ceramics Research
Rutgers University
Piscataway, NJ 08854

ABSTRACT

The thermal conductivity and specific heat of mullite-cordierite composites have been measured in the temperature range 90 K to 420 K. The sound velocity, as determined from a fit to the specific heat data, agrees well with velocities determined from mechanical measurements. The temperature and composition dependence of the phonon mean free path is established. The thermal conductivity as a function of composition passes through a minimum for the mullite-rich compositions due to a minimum in the upper limit of the phonon mean free path. Solid solution formation is a possible cause.

INTRODUCTION

Mullite-cordierite composites represent an important technological class of materials for use in the electronics packaging industry. Previous studies of these composite materials have been primarily concerned with their synthesis, dielectric and mechanical properties.[1-4] These materials have a lower dielectric constant ($\sim$5-8) than alumina ($\sim$9), at present a commonly used substrate material.[2-4] However, the thermal conductivity of alumina,[5] ($\sim$33 W/m.K), is much higher than that of either mullite or cordierite ($\sim$4-7 W/m.K).[6,8]

The thermal conductivity of mullite, cordierite and their composites is expected to be low due to the complexity of the unit cell.[7] Literature values for cordierite ($2MgO \cdot 2A\ell_2O_3 \cdot 5SiO_2$) are 4.0 and 4.2 W/m.K while for

---
*Supported by the Center for Ceramics Research, Rutgers University.
**Supported by Delaware Research Partnership and E.I. Du Pont de Nemours and Co. Inc.
[†]Fellow, Center for Advanced Study, University of Delaware.

mullite $(3Al_2O_3 \cdot 2SiO_2)$ they range from 4.0 to 6.7 W/m.K.[6,8] The factors affecting the thermal conductivity of mullite-cordierite composites have not previously been studied systematically.

EXPERIMENTAL PROCEDURE

The mullite-cordierite composites were synthesized from powders of > 99% stated purity. They were hot pressed and sintered at temperatures below their melting temperatures to produce highly dense ceramic composites. Details of their synthesis and characterization can be found elsewhere.[3-4] The notation used to describe the compositions throughout this paper is weight % mullite / weight % cordierite e.g. 90/10 etc.

Table 1. Measured bulk density of our mullite-cordierite composites

Sample	Density (g/cm^3)
10/90	2.59
25/75	2.65
50/50	2.82
75/25	2.98
90/10	3.08

The bulk densities for these materials, measured by the Archimedes technique, are shown in Table 1.

The thermal properties of these materials were measured using an electronic flash system.[9] This system measures the thermal diffusivity (α) and the specific heat (C_p) of the samples and uses the following relation to determine the thermal conductivity (κ).

$$\kappa = \rho \alpha C_p \qquad (1)$$

where ρ is the measured bulk density of the material. This recently developed technique has been calibrated by measurements on sapphire which agree within experimental error with NBS tabulated values.

The thermal conductivity was also determined at room temperature using a single probe thermal comparator.[10] The values obtained were consistently higher than those from electronic flash.[4] However, the probe technique is known to be more responsive to surface than to bulk thermal conductivity. We are continuing to study the source of the differences obtained by our two techniques.

RESULTS AND DISCUSSION

We present here our results for the specific heat (C_p) and thermal diffusivity (α) experimentally measured using the electronic flash. From the latter we obtain the temperature dependence of the phonon mean free path (l). The derived thermal conductivity (κ) (Eq. 1) is then discussed in terms of the temperature dependence of l.

Specific Heat

The room temperature specific heat for each composition studied is shown in Figure 1. 100% cordierite and mullite samples with identical processing were not available to us so that the end point values of

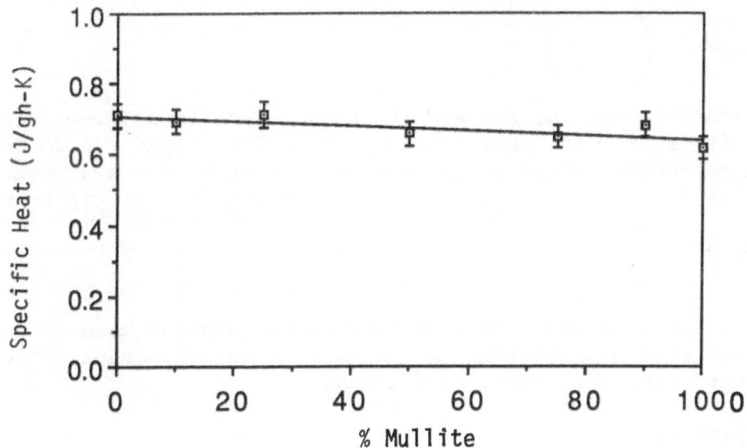

FIGURE 1. Specific heat against composition at room temperature. Room temperature specific heat end points are from reference (6). The solid line is a least squares fit to all points.

0.71 J/g.K and 0.62 J/g.K respectively were taken from the literature.[6] Since no error bars were cited we assign them the same uncertainty as in our own measurements. From Figure 1 we see that, using a least square fit to all seven points, C_p at room temperature follows a simple rule of mixtures within experimental error. Our data for C_p between 90 K and 420 K adheres to the same rule although end-point values are not available for these temperatures.

The specific heat as a function of temperature for each composition is shown in Figure 2. Error bars were omitted for clarity but may be visualised by reference to Figure 1. We have used the Debye function for crystalline solids[11] to model the specific heat and to obtain values of the Debye temperature(Θ_D):

$$C_v = 9rNk_B \left(\frac{T}{\Theta_D}\right)^3 \int_0^{\Theta_D/T} \frac{x^4 e^x dx}{(e^x - 1)^2} \qquad (2)$$

where r is the number of atoms per molecule, N is the number of molecules, and k_B is Boltzmann's constant. In the absence of known measurements of the elastic bulk modulus for these materials, we assume $C_p = C_v$ over our temperature range recognizing that the resulting values of the Debye temperature may be slightly high.

By fitting the specific heat data to Eq. (2) between 200 K and 400 K we have been able to determine the Debye temperature for these materials. These values are listed in Table II. Using this Debye temperature to determine the specific heat over the temperature range measured, it was found that below about 200 K the Debye function predicted a much lower C_p than was measured. The source of this excess specific heat, which may be due to low-lying optical modes, will require measurements to lower temperatures in the future.

From the Debye temperature we can determine the average speed of the acoustic phonons since:[12]

$$V_{th} = \frac{2\pi\Theta_D k_B}{h} \left(6\pi^2 \frac{N}{V}\right)^{-\frac{1}{3}} \tag{3}$$

where h is Planck's constant and N/V the number density of atoms. The resulting velocities are also shown in Table II. For comparison we list in Table II the mechanical sound velocities V_m, inferred from measurements of Young's modulus (E) and the mass density (ρ) of similar materials[2] by using:

$$V_m = \left(\frac{E}{\rho}\right)^{.5} \tag{4}$$

The sound velocities obtained from thermal and mechanical measurements are remarkably consistent considering that they were determined on different samples prepared under differing conditions.

Thermal Diffusivity

The thermal diffusivity, which we measure directly by the electronic flash technique, provides more direct information on the temperature dependence of the phonon mean free path than does the thermal conductivity. In fact the temperature dependence of the thermal diffusivity is identical to that of the phonon mean free path. This can be seen from a kinetic approximation for the thermal conductivity of a phonon gas:[12]

$$\kappa = \frac{1}{3} (C \cdot v \cdot \ell) \tag{5}$$

where C is the specific heat per unit volume, v is the velocity of sound and ℓ is the phonon mean free path which, when combined with Eq. (1) above gives the thermal diffusivity as:

$$\alpha = \frac{1}{3} (v\ell) \tag{6}$$

v is almost temperature independent and ℓ varies as 1/T due to dominant Umklapp scattering at temperatures above the temperature where κ has its maximum value.[13] As a result a log-log graph of α against T will have a slope of -1.

The thermal diffusivity of our specimens as a function of temperature is shown in Figure 3 where log-log axes are used. The data has a slope of -1 over most of the temperature range. (Note that in the figures as drawn the two log axes are not exactly commensurate.) For the 90/10 and 10/90 composites the $\ell \sim 1/T$ dependence continues to below 90 K. In contrast the 25/75, 50/50 and 75/25 compositions show a distinct departure from the straight line behavior at a composition dependent temperature.

For the 50/50 and 75/25 compositions the thermal diffusivity becomes almost constant below the departure from 1/T behavior implying that the phonon mean free path has reached its maximum value presumably due to impurity scattering. This in turn suggests that for these compositions some solid solution formation has occurred during the synthesis. Indirect evidence for the possible formation of solid solutions in the mullite-cordierite system has previously been obtained from mechanical[2] rather than thermal properties. However, the mechanical measurements on our composites do not support this evidence.[3,4]

The phonon mean free paths at 90 K for all of the samples can be calculated from Eq. (6). They are shown against composition in Figure 4. A minimum in the phonon mean free path occurs near the 75/25 composition.

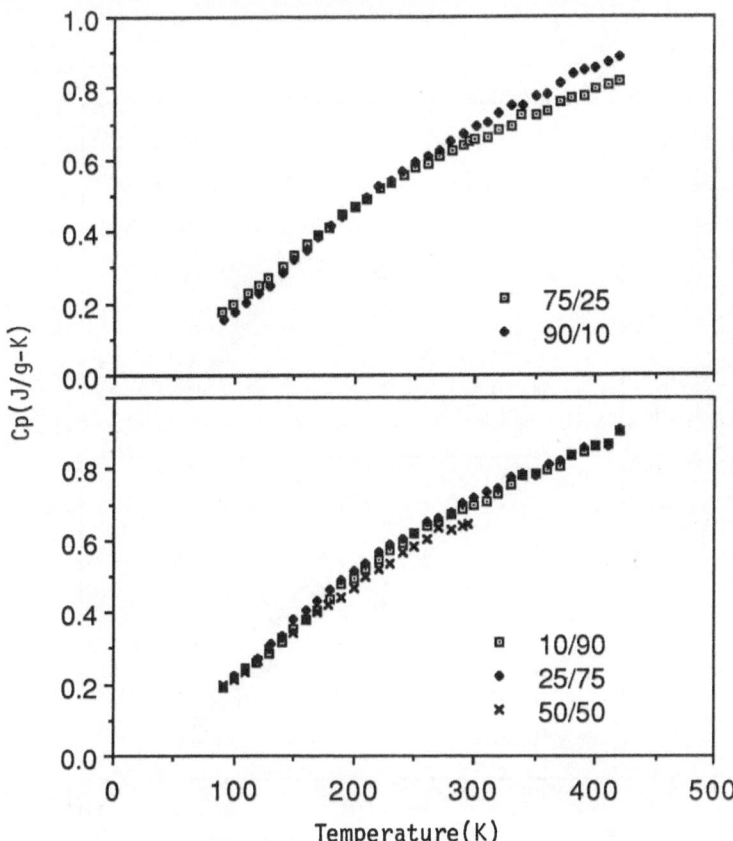

FIGURE 2. Specific heat against temperature for mullite/cordierite composites. Error bars were omitted for clarity but are typically 5%.

TABLE II. Debye temperatures and sound velocities by thermal and mechanical measurement (see text).

Sample	Θ_D(K)	Vth (10^5 cm/sec)	Vm (10^5 cm/sec)
10/90	1040	8.4	7.5
25/75	1040	8.4	7.7
50/50	1060	8.2	8.8
75/25	1140	8.8	8.0
90/10	1070	8.1	8.0

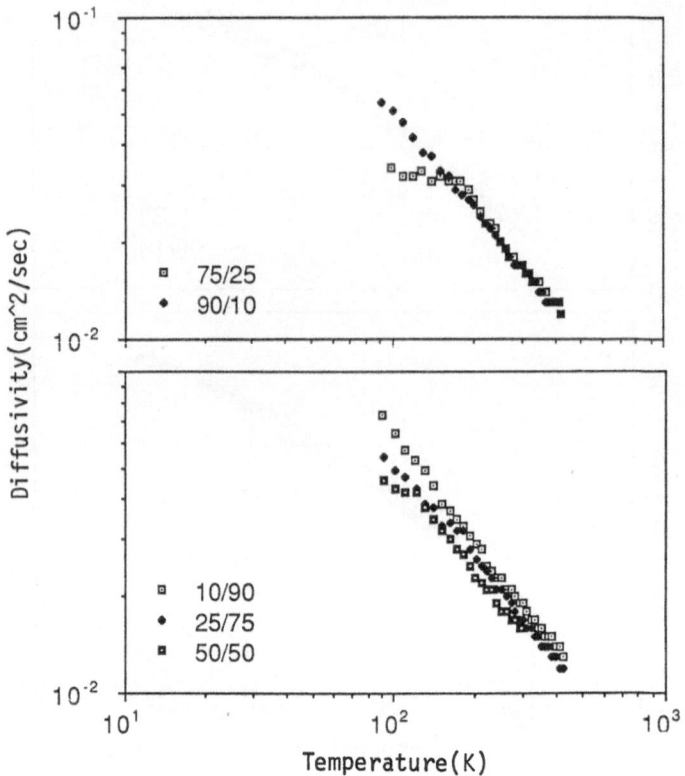

FIGURE 3. Thermal diffusivity as a function of temperature. Note that the vertical and horizontal log scales are not commensurate. The slope of the 10/90 composition is -1.

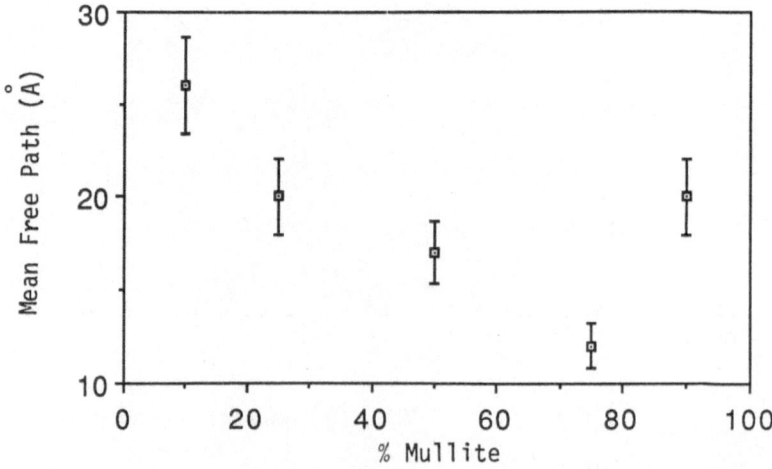

FIGURE 4. Phonon mean free path at 90 K as a function of composition.

The phonon mean free path can be expected to show a minimum near the mullite-rich end of the composite range, where mullite solid solutions may be formed. The short mean free paths in these materials preclude the possibility of grain-boundary scattering and are presumably due to impurity or defect scattering

Thermal Conductivity

The thermal conductivity as a function of composition at two different temperatures is shown in Figure 5. The thermal conductivity for both 100% mullite and 100% cordierite is taken to be 4 (W/m.K) in agreement with ref. (6).

For a two phase system one would expect a simple rule of mixing for these materials.[13] The room temperature thermal conductivities show a distinct concave behavior as a function of composition. This concave behavior appears at all temperatures but is more pronounced at lower temperatures. It is a reflection of the composition and temperature dependence of the phonon mean free path discussed in the previous section. Such concave compositional trends in thermal conductivity are suggestive of the formation of a solid solution in composite systems.[13]

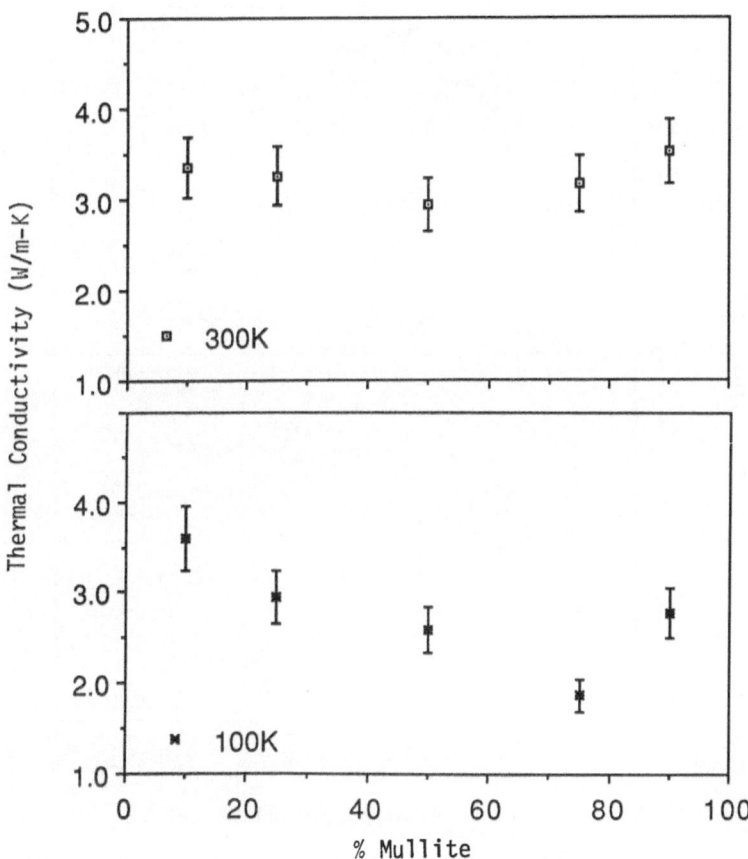

FIGURE 5. Thermal conductivity against composition at 100 K and 300 K.

For crystalline materials the thermal conductivity is usually proportional to the specific heat at temperatures below the peak where the phonon mean free path is temperature independent. It is proportional to the phonon mean free path at temperatures far above the peak where the specific heat is almost constant. The peak in the thermal conductivity is due to the dominance of the contribution of the specific heat over the 1/T dependence of the mean free path with decreasing temperature (see Eq. 5).

The thermal conductivity as a function of temperature for these materials is shown in Figure 6. Again error bars were omitted for clarity, but there is an uncertainty of 10% in each measurement. In contrast to the temperature dependence of the thermal diffusivity, the thermal conductivity has a maximum for all of the samples within our temperature range. For the 10/90 composition the very gradual low temperature decrease in the thermal conductivity below the peak is attributable to a balance between the rapid decrease in the specific heat and steadily increasing phonon mean free path (Figures 2, 3 and Eq. 5). For the 75/25 composition the more abrupt decline is additionally due to the achievement of the maximum phonon mean free path as described in the previous section. For the 25/75, 50/50 and the 90/10 compositions the maximum mean free path is approached but not yet achieved at 90 K leading to an intermediate trend.

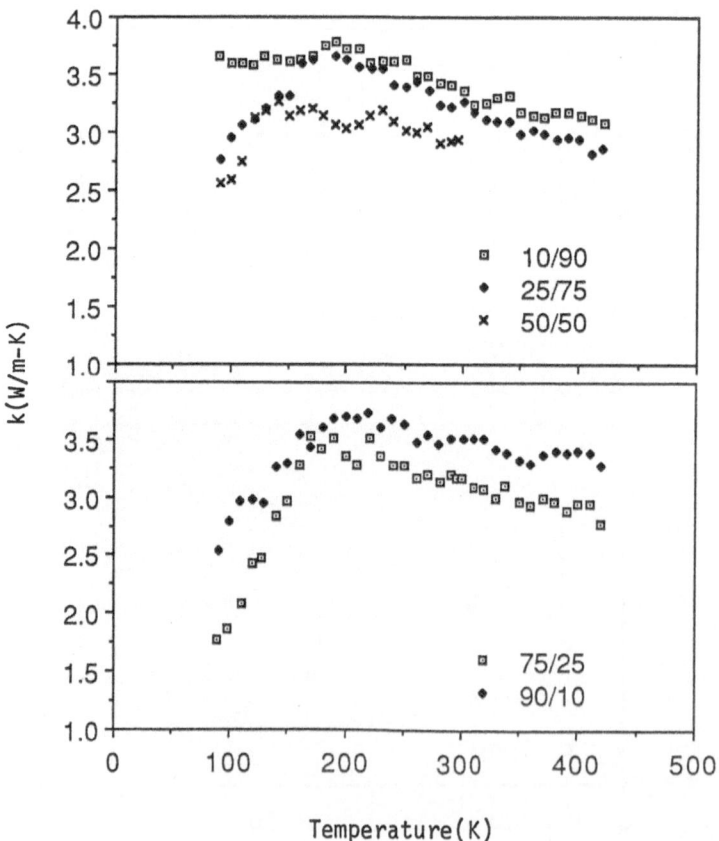

FIGURE 6.   Thermal conductivity against temperature.   Error bars were
omitted for clarity.   The typical error is 10%.

SUMMARY AND CONCLUSIONS

We have measured the thermal diffusivity and specific heat of a series of mullite-cordierite composites in the temperature range of 90 K to 420 K and from them obtained the composition and temperature dependence of the thermal conductivity. Above 200 K the specific heat can be described by the Debye model and yields sound velocity values that agree very well with those obtained from Young's modulus. Contributions to the thermal conductivity from the specific heat and the phonon mean free path (which is proportional to the thermal diffusivity) are separable. This permits us to observe the trends of the phonon mean free path as a function of temperature and composition. These trends suggest the possible formation of mullite-solid solutions in these materials.

In order to identify if there is a minimum in the compositional dependence of the phonon mean free path, we conclude that it is essential to make measurements at temperatures near or lower than the temperature where the maximum occurs in the thermal conductivity. Further studies of this potentially useful composite system are in progress.

REFERENCES

1. M. P. Davis and W. C. Hackler, Correlation Between Physical Properties and Thermochemical Reactions in a Mullite-Cordierite Composition, Bul. Am. Cer. Soc., Vol. 40, 6:362 (1961).
2. B. H. Mussler and M. W. Shafer, Preparation and Properties of Mullite-Cordierite Composites, Bul. Am. Cer. Soc. Vol. 63, 5:705-710, 714 (1984).
3. R. M. Anderson, Cordierite-Mullite Composites: A Study of Their Mechanical, Thermal and Dielectric Properties, Ph.D. Thesis, Rutgers University (1987).
4. R. M. Anderson, R. Gerhardt, J. B. Wachtman, D. G. Onn, and S. C. Beecher, Proc. International Symposium "Ceramic Substrates and Packages", Denver, CO, Oct. 1987 (in press).
5. H. C. Graham and N. M. Tallan, Polycrystalline Insulators, in: "Physics of Electronic Ceramics," L. L. Hench and D. B. Dove, ed., M. Dekker, New York (1982).
6. R. Kamo and W. Bryzik, Cummins/Tacom Advanced Adiabatic Engine, Ceram. Eng. Sci. Pro., 5(5-6):312 (1984).
7. R. Berman, "Thermal Conduction in Solids," Clarendon Press, Oxford (1978).
8. M. G. Woods, W. F. Mandler and T. L. Scofield, Designing Ceramic Insulated Components for the Adiabatic Engine, Am. Ceram. Soc. Bul., 64(2):287 (1985).
9. R. E. Giedd and D. G. Onn, Electronic Flash: A Rapid Method for Measuring the Thermal Conductivity and Specific Heat of Dielectric Materials, to be published in these proceedings.
10. R. B. Dinwiddie, A. J. Whittaker and D. G. Onn, A Rapid Screening Thermal Conductivity Comparator, Talk Summaries 1987 Electronics Division American Ceramics Society Annual Meeting.
11. E. S. R. Gopal, "Specific Heats at Low Temperatures," Plenum Press, New York (1966).
12. C. Kittel, "Introduction to Solid State Physics," 6th Edition, Wiley, New York (1986).
13. W. D. Kingery, H. K. Bowen and D. R. Uhlman, "Introduction to Ceramics," 2nd Edition, Wiley, New York (1976).

EFFECT OF PROCESSING CONDITIONS ON THERMAL CONDUCTION IN ALUMINA[†]

Ralph Dinwiddie, Andrew Whittaker, and David G. Onn[*]

Applied Thermal Physics Laboratory
Department of Physics and Astronomy
University of Delaware
Newark, DE 19716

ABSTRACT

Thermal conductivity versus temperature data, in the temperature range 77 to 500 K, is presented for a series of seven alpha alumina samples sintered at temperatures varying from 1273 K to 1873 K. Peaks in thermal conductivity are observed for the five samples sintered at < 1773 K. A shift in the magnitude and location of the peaks and an increase in the overall magnitude of thermal conductivity over the entire temperature range are observed with increasing sintering temperature. It is concluded that the temperature dependence of thermal conductivity of these materials is determined by two contributing factors:

a) An increase in density as the sintering temperature is increased which leads to the removal of internal thermal barriers and an increase in the effective cross sectional areas of the samples.
b) An increase in the phonon mean free path for samples sintered at higher temperatures. This may be due in part to an increase in the intrinsic thermal conductivity of the grains.

INTRODUCTION

Alumina is used widely in the electronics packaging industry as a substrate material for mounting electronic components.[1] Efficient thermal management in this application plays a critical role in ensuring system reliability, and it is the moderately high thermal conductivity ($\sim$40 W/m.K for single crystal[2]) afforded by alumina, coupled with good electrical insulation properties that make it so attractive to manufacturers.

Thermal conduction in alumina is dominated by phonon transport and the moderately high values observed are largely a result of the strong covalent bonding between the atoms and the relatively low atomic mass of the structure. If the phonons are assumed to behave as a gas[3] then the kinetic

[†]Research supported by Delaware Research Partnership and E. I. Du Pont de Nemours and Co. Inc.
[*]Fellow, Center for Advanced Study, University of Delaware.

theory may be applied to give

$$\kappa = 1/3 \, \rho.C_v.v.l \tag{1}$$

where κ = thermal conductivity, $\rho.C_v$ = heat capacity per unit volume, v = average sound velocity and l = effective mean free path of phonons. Although values of C_v are formally required in equation 1, values of C_p are more frequently encountered in the literature since C_p is much more readily determined experimentally. For the materials under investigation C_v is approximately equal to C_p over the entire temperature range.

Thermal conductivity is both temperature and structure dependent. Phonon propagation is severely limited by anharmonicity within the lattice. At low temperatures this may be introduced by internal boundaries. At slightly higher temperatures scattering from various sites including point defects such as vacancies or localized mass differences due to the occupation of vacant lattice sites by impurity atoms may occur and eventually direct phonon-phonon interactions (Umklapp processes[4]) begin to dominate.

Consequently an understanding of the temperature dependent thermal conductivity of alumina and the relationship to the processing history and the resulting structure is of both practical and scientific importance.

This report presents the results of an investigation into the temperature and structure dependence of the thermal conductivity of a series of seven alumina samples sintered at temperatures ranging from 1273 to 1873 K (1000 C to 1600 C).

SAMPLE PREPARATION

The samples were manufactured from Sumitomo AKP-15 alumina powder with a narrow submicron size distribution. The starting powder was mixed with 50 micron diameter polyethylene spheres, which tend to promote uniform densification during sintering, and bound with palm oil. The bound medium was moulded into bar shaped samples of a suitable size (2 x 1/4 x 1/4 inches) for the thermal conductivity measurements. The moulded bars were then heated to 462 K and pressed to form the green compact. This palm oil was dissolved using Freon TF vapor, leaving open channels of porosity which later permits the egress of the volatilised polyethylene. All the samples were then heated to 1273 K, at a rate of 1 K/min, to remove the polyethylene. After cooling to room temperature the seven samples were reheated to their final sintering temperatures which ranged from 1273 K to 1873 K at 100 degree intervals. Each sample was maintained at its sintering temperature for 1 hour. Densities in the range 60 to 96% of the theoretical density of alumina were achieved (figure 1).

Proton induced X-ray emission studies[5] carried out on the samples indicated the presence of impurity elements in relatively low concentrations:

Mo 527 ppm, Fe 30.8 ppm, Cu 11.6 ppm

Since the chemical synthesis of the starting powder yields a very high purity product it seems likely that these trace impurities were introduced into the samples during the course of processing.

EXPERIMENTAL

The thermal conductivity of the samples was measured in the temperature

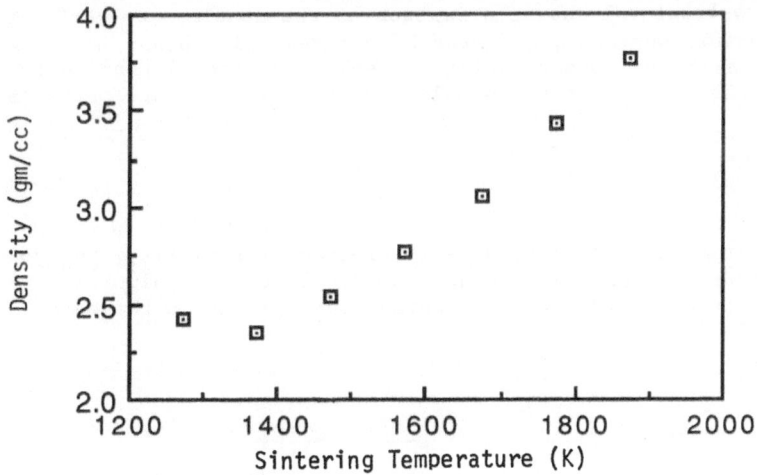

FIGURE 1. Sample density as a function of sintering temperature.

range 77 to 500 K using a microcomputer[6] controlled guarded longitudinal system in a uniform furnace.[7] This apparatus employs both active and passive radiation shielding to reduce radiative power loss from the sample.

The experimental arrangement consists of a rod or bar sample with a cold sink and heater mounted on opposite ends. Two thermocouples are mounted a known distance apart in the central portion of the sample to monitor the temperature gradient and the thermal conductivity is obtained directly from the relation

$$\dot{Q} = - \kappa.A.dT/dx \tag{2}$$

where $\dot{Q}$ is the heat flow rate supplied to the sample, dT/dx is the resulting temperature gradient, A is the cross sectional area of the sample and κ is the thermal conductivity.

Laubitz[8] determined the absolute accuracy of the technique to be better than 2% up to a temperature of 400 K. Calculations performed in this laboratory indicate that the determinate error increases to 5% at 500 K. At higher temperatures it becomes increasingly difficult to adequately compensate for the effect of radiative heat loss. For this reason an upper limit of 500 K has been set on the measurements.

Several thermal conductivity standards (e.g. Armco iron, 710 glass) were measured to establish the accuracy of the system and this was confirmed by cross correlation experiments performed on the electronic flash[9] and laser flash[10] techniques also located in this laboratory.

Specimen bulk densities were calculated from measurements of the mass in air and the geometrical volume. The accuracy of the density measurement was assessed to be 3%. Mercury porosimetry measurements were also made on the samples. These data are not included in this report due to the presence of closed porosity.

A microstructural characterisation of the samples was performed by viewing fracture surfaces, effected by a three point bend, in a scanning electron microscope. Quantitative stereological techniques[11] were applied to micrographs of each of the samples to determine average particle and grain sizes.

RESULTS

The thermal conductivity data is plotted, for clarity, in two separate figures (Figure 2a and 2b). Values for single crystal alumina[3] are included for comparison. Two important features are apparent in the data. As the

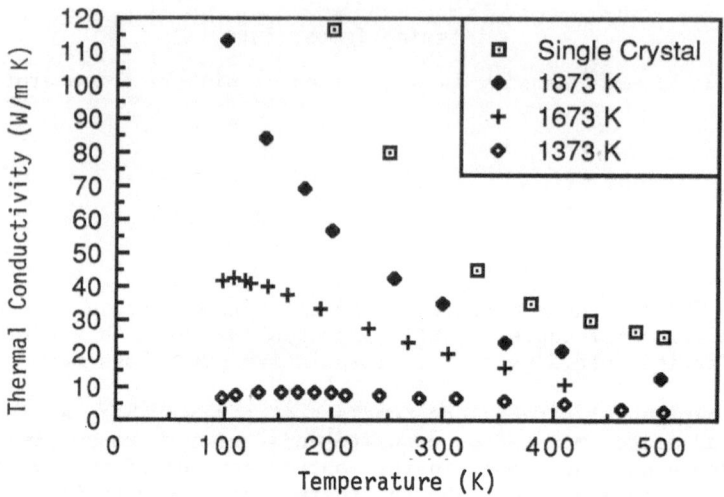

FIGURE 2a. Temperature dependence of thermal conductivity. Single crystal data (Berman 1951) is supplied for comparison.

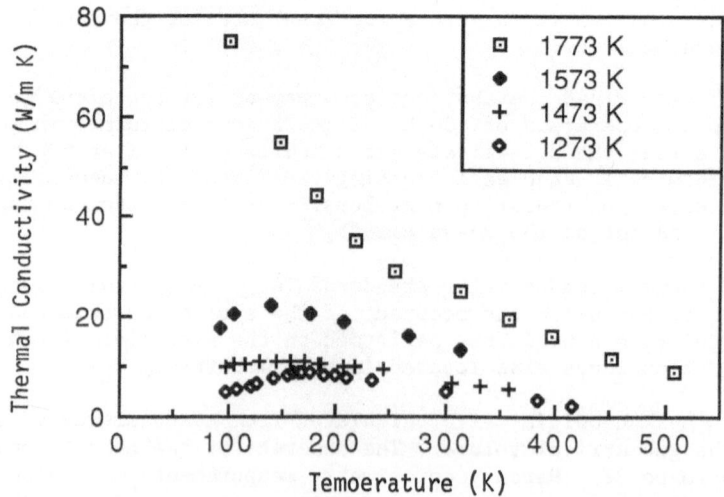

FIGURE 2b. Temperature dependence of thermal conductivity.

sintering temperature of the sample is increased, an increase in thermal conductivity, over the entire temperature range, is observed. Room temperature thermal conductivity data for each of the seven sintering temperatures is given in Table 1. The increase is demonstrated more clearly in Figure 3, where thermal conductivity at 200 K and 300 K is graphed as a function of sintering temperature. The second interesting feature is the behavior of the peak in the data. A peak in thermal conductivity was detected for the five samples sintered below 1773 K. As the sintering temperature increases the magnitude of the peak increases and the temperature at which this peak occurs decreases.

Scanning electron photomicrographs of the samples appear in Figure 4a through 4g. All photomicrographs are of fracture surfaces magnified 10,000 times. The mean linear intercept method[11] was used to determine the average grain size in each sample. It is evident from the photomicrographs in

TABLE 1. Variation of sample characteristics with sintering temperature.

Sintering Temperature (K)	Percent Porosity (%)	Thermal Conductivity At 300 K (W/M K)	Average Grain Size (μm)
1273	39.0	5.40	0.40
1373	40.6	6.50	0.39
1473	35.9	7.50	0.40
1573	30.0	14.6	0.45
1673	22.6	22.9	0.50
1773	13.2	26.7	0.63
1873	4.70	35.0	1.40

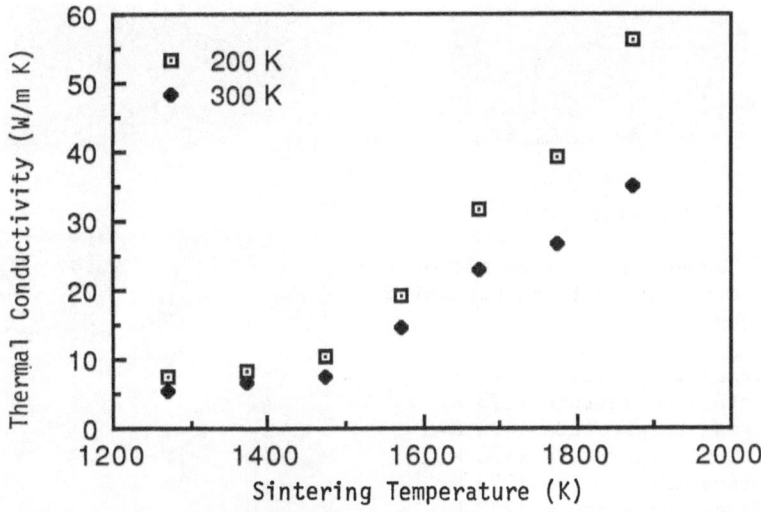

FIGURE 3. Thermal conductivity at 200 K and 300 K as a function of sintering temperature.

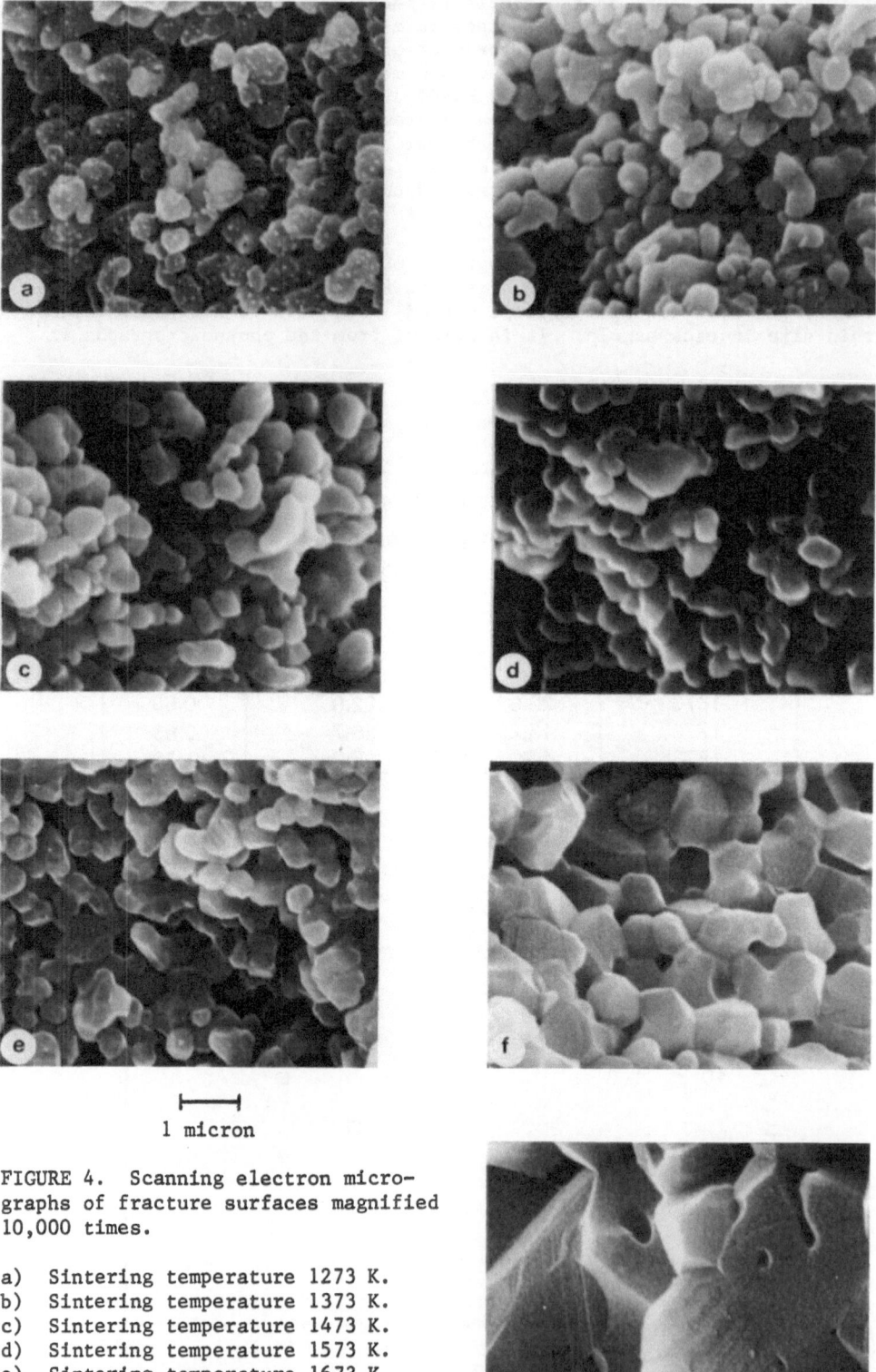

1 micron

FIGURE 4. Scanning electron micrographs of fracture surfaces magnified
10,000 times.

a) Sintering temperature 1273 K.
b) Sintering temperature 1373 K.
c) Sintering temperature 1473 K.
d) Sintering temperature 1573 K.
e) Sintering temperature 1673 K.
f) Sintering temperature 1773 K.
g) Sintering temperature 1873 K.

Figures 4a, 4b and 4c, that virtually no grain growth takes place in samples sintered below 1573 K. Figures 4d, 4e and 4f demonstrate a steady increase in grain size with increasing sintering temperature. There is a large increase in grain size between the sample sintered at 1773 K (Figure 4f) and the sample sintered at 1873 K (Figure 4g). The variation of grain size with sintering temperature is depicted in Figure 5.

The photomicrographs also demonstrate the morphological development taking place during sintering. In Figures 4a, 4b and 4c the grains appear as an agglomeration of irregularly shaped particles. In Figure 4d we begin to see clear evidence of sintering. Many of the particles have begun to form necks with adjacent particles. These particles also appear irregular in shape. Faceting of the grains first appears in the sample sintered at 1673 K (Figure 4e). The faceting clearly becomes more prominent in Figures 4f and 4g with the grains appearing as equiaxed polyhedra.

The effective mean free path of phonons in the specimens was derived from Equation (1) using known values of specific heat,[12] single crystal speed of sound[13] and measured thermal conductivity data. The maximum mean free path occurs at the same temperature as the peak in thermal conductivity. This value of mean free path is associated with the correlation length of structural artifacts acting as scattering centers. At higher temperatures the mean free path is governed by Umklapp scattering. The maximum mean free path of the five samples in which a peak in thermal conductivity was observed, is plotted as a function of sintering temperature in Figure 6. The maximum mean free path increases with sintering temperature much faster than the grain growth (Figure 5). This maximum in mean free path is much smaller than the average grain size.

DISCUSSION

The observed increase in thermal conductivity with sintering temperature can be explained qualitatively in terms of the accompanying increase

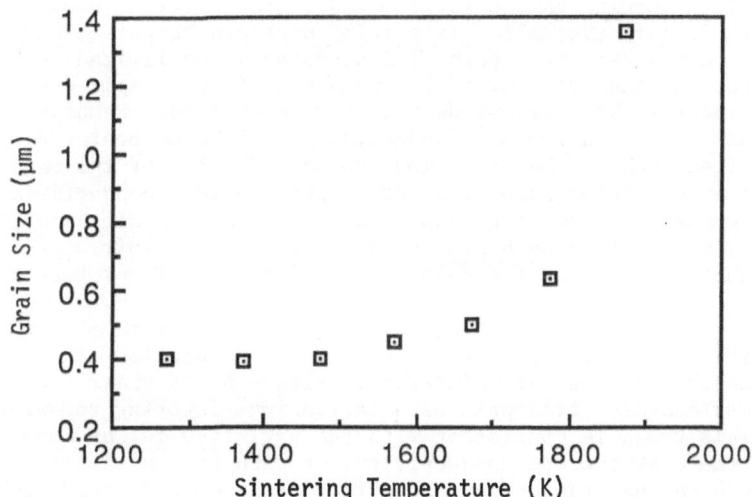

FIGURE 5. Average grain size as a function of temperature.

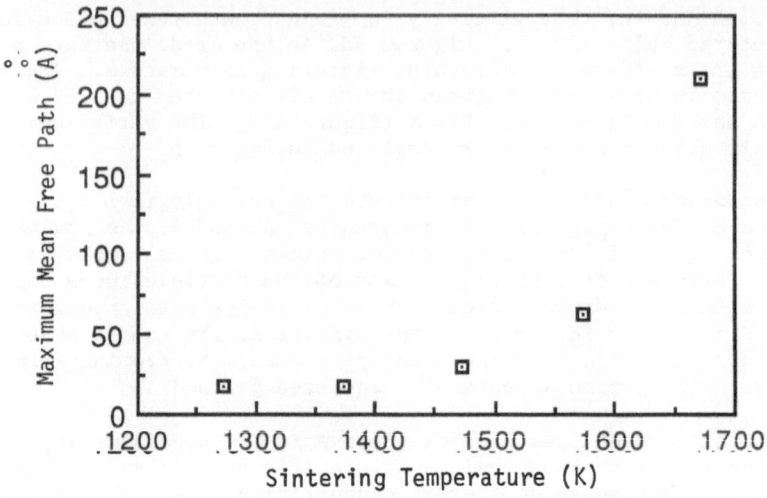

FIGURE 6. Maximum mean free path as a function of grain size.

in bulk density, from 60 to 96% of the theoretical value. The presence of
porosity will tend to reduce the cross sectional area available for heat
flow and may also act as a thermal barrier depending on the geometry and
orientation of the porosity. The density increase with increasing sintering temperature depicted in Figure 1 is closely paralleled by the increase
in thermal conductivity as a function of sintering temperature plotted in
Figure 3. The microstructural development evident in Figures 4 a to 4g
lends a more graphic substantiation to this hypothesis.

An explanation of the variation in the magnitude of the peak thermal
conductivity and the temperature at which it occurs can be advanced by considering the variation of the phonon mean free path with temperature. In
a pure single crystal the phonon mean free path follows a decreasing exponential function of temperature due to the occurrence of Umklapp scattering.
Approaching lower temperatures the mean free path will continue to increase
sharply until it becomes physically limited by the external boundaries of
the crystal. In polycrystalline materials containing significant levels of
porosity the phonon mean free path will necessarily be limited to length
scales far shorter than the physical boundaries of the sample due to the
presence of internal boundaries, defects and impurities. Equation (1)
indicates that a peak in thermal conductivity will be encountered when the
phonon mean free path reaches a maximum value. The longer the maximum mean
free path of a particular sample, a larger peak thermal conductivity and a
lower peak temperature are to be expected. Conversely a sample with a
shorter maximum mean free path can be anticipated to exhibit a peak in
thermal conductivity of lower magnitude occuring at a higher temperature.

The maximum mean free paths calculated for the five samples whose thermal conductivity exhibited a peak are plotted as a function of the temperature at which the samples were sintered in Figure 6. A clear trend of
increasing maximum mean free path with increasing sintering temperature is
observed. This trend is consistent with the variation in the magnitude and
location of the peak thermal conductivity for each of the samples. It is
important to note that the maximum mean free paths were derived from the
raw experimental thermal conductivity data and as such are not corrected for
the presence of porosity in the samples. This correction will certainly

have the effect of increasing the values obtained for the lower density samples since the phonons are physically limited to transit within regions of condensed matter. The trend of the data presented in Figure 6 will not however be altered. It seems reasonable to assume that the two most dense samples will exhibit peaks in thermal conductivity at lower temperatures and that they will have longer phonon mean free paths.

The observation that minimal grain growth occurs at temperatures < 1573 K (Figures 4a to 4c) may be apportioned to the fact that the grain boundaries are effectively pinned by the appreciable number of pores present. At higher temperatures neck formation between adjacent grains begins (Figure 4d) and at higher temperatures still obvious grain growth is evident (Figures 4e to 4g). This is reflected in the graph of grain size versus sintering temperature shown in Figure 5.

A comparison of Figures 5 and 6 indicates that the maximum mean free path tends to increase at a greater rate than the grain size as a function of the sintering temperature. This is demonstrated more clearly in the plot of maximum mean free path versus grain size shown in Figure 7. As a result of this, since the maximum mean free path never exceeds the average grain size for any of the samples exhibiting a peak in thermal conductivity, one might postulate that the intrinsic thermal conductivity of the individual grains is enhanced by increased sintering temperatures. This is certainly a plausible assertion if it is assumed that the particles of the starting powder contain a significant concentration of structural defects and impurities whose deleterious influence on thermal conductivity would be removed by the annealing effect of elevated sintering temperatures. However such an assertion must be tempered by the fact that the ultimate thermal conductivity of the samples under investigation is probably governed by a very complex interaction of variables. These include the presence of porosity in varying concentrations (the normalization of which has largely been ignored due to the complexity of accommodating the potential variation in pore geometry, which is as yet uncharacterized), the development of necks between adjacent particles which will tend to increase the effective contact area

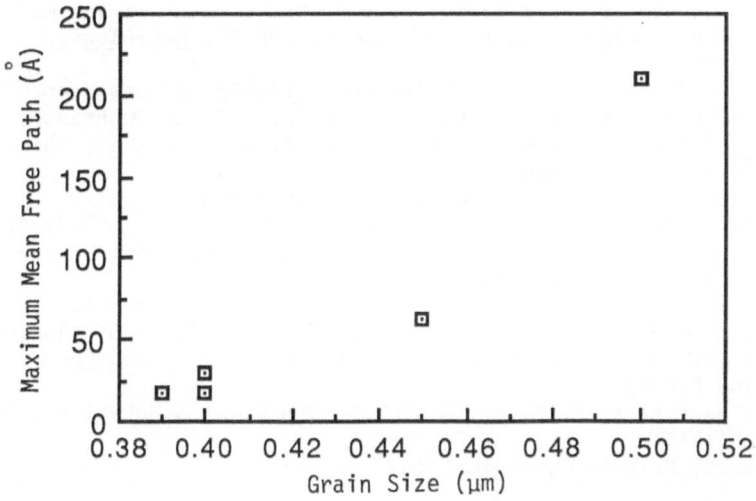

FIGURE 7. Maximum mean free path as a function of grain size.

of the particles while not necessarily leading to any significant increase in the bulk density,[14] and the state of structural perfection of the starting powder compared to that of the grains in the well sintered samples. These questions are currently being addressed in this laboratory and the results will be published in due course.

CONCLUSION

The effect of increased sintering temperature on the structural development of monolithic alumina is clearly demonstrated. There are two contributing factors that determine the temperature dependence of thermal conductivity of these materials.

a) The increase in density from 60 to 96% of the theoretical value, which results from increasing sintering temperature, is accompanied by an increase in thermal conductivity over the whole temperature range. This is attributed to the removal of internal thermal barriers and an increase in the effect cross sectional areas of the samples.

b) The shift in the magnitude and location of the peak in thermal conductivity is apportioned to an increase in the phonon mean free path for samples sintered at higher temperatures. This may be due in part to an increase in the intrinsic thermal conductivity of the grains.

REFERENCES

1. H. J. Sanders, High-Tech Ceramics, Chem. & Eng. News, 27-40, July 9 1984.
2. R. Berman, (1951) reported in "Introduction to Ceramics," 2nd Edition, by W. D. Kingery, H. K. Bowen, D. R. Uhlman, John Wiley & Sons, New York (1960).
3. N. W. Ashcroft, N. D. Mermin, "Solid State Physics," Saunders College, Philadelphia (1976).
4. C. Kittel, "Introduction to Solid State Physics," 6th ed., John Wiley & Sons Inc., New York (1986).
5. C. M. Fou, Private Communication.
6. Commodore Business Machines, West Chester, PA.
7. R. B. Dinwiddie, An Apparatus for the Measurement of Thermal Conductivity in the Temperature Range 80 to 550 K, M. S. Thesis, University of Delaware, 1986.
8. M. J. Laubitz, D. L. McElroy, Precise Measurement of Thermal Conductivity at High Temperatures (100 - 1200 K), Metrologia, Vol. 7 1:1 (1971).
9. R. E. Giedd, David G. Onn, "Electronic Flash: A Novel Approach to Thermal Property Measurement," Proceedings of the Twentieth International Thermal Conductivity Conference, October 1987, this volume, Plenum Press, New York, in press.
10. A. J. Whittaker and D. G. Onn, to be published.
11. Standard Methods for Determining Average Grain Size, "E112-82 Annual Book of ASTM Standards, Part 2, Metallography: Nondestructive Testing," 135 (1982).
12. E. C. Kerr, H. L. Johnstron, N. C. Hallet, Low Temperature Heat Capacities of Inorganic Solids, III. Heat Capacity of Aluminum Oxide (Synthetic Sapphire) from 19 to 300 K, J. Phys. Chem., 72:4740 (1950).
13. T. F. Heuter, R. H. Bolt, "Sonics," John Wiley & Sons Inc., New York, (1955).
14. K. Ewsuk, Private Communication.

ACKNOWLEDGEMENTS

The authors wish to thank Charles Booth and Larry Harrison of the Dupont Company for processing the samples used in this study.

ROLE OF STRUCTURE AND COMPOSITION IN THE

HEAT CONDUCTION BEHAVIOR OF SILICON CARBIDE

D.P.H. Hasselman

Department of Materials Engineering
Virginia Polytechnic Institute and State University
Blacksburg, Virginia 24061 USA

ABSTRACT

A review is presented of experimental data for the thermal
conductivity and diffusivity of silicon carbide obtained by the author and
his co-workers over the last few years. It is shown that additives in the
form of sintering aids, structural imperfections as those found in
amorphous silicon carbide fibers and stacking faults can lower the thermal
conductivity or diffusivity of silicon carbide significantly. In contrast,
significant improvements in the heat conduction behavior of silicon carbide
can be achieved by grain boundary additives which act as impurity traps and
by additional heat treatments which cause impurity precipitation and
reduction in crystal lattice imperfections. However, for all silicon
carbides investigated, the thermal conductivity is well below literature
values for high-purity single crystals.

INTRODUCTION

Silicon carbide, in view of its chemical stability and oxidation
resistance, high wear resistance and excellent mechanical behavior at
elevated temperatures, represents an excellent candidate material for
load-bearing purposes under corrosive or erosive conditions at high
temperatures. In general, components or structures for high-temperature
service will be subjected to transient or steady-state heat flow. For this
reason, the heat conduction properties such as thermal conductivity and
diffusivity are also expected to play a significant role in meeting
engineering performance as governed by specific design criteria. For such
components as heat exchanger tubes, the thermal conductivity should be as
high as possible. High thermal conductivity and diffusivity also are
essential in order to reduce the possibility of thermal shock failure to
which brittle structural materials are particularly susceptible. In
contrast, components for service under conditions for which heat losses
should be kept to a minimum, the thermal conductivity should be as low as
possible. It is the purpose of this paper to discuss the variables and
present experimental data for the heat conduction behavior of single-phase
silicon carbide and silicon carbide based composites.

BACKGROUND

The thermal conductivity (K) of a material for uniaxial heat flow in the x-direction is defined by [1]:

$$q = -K dT/dx \qquad (1)$$

where q is the heat flux per unit area and unit time, T is the temperature and dT/dx is the temperature gradient.

The thermal diffusivity of a solid can be defined in terms of the transient heat conduction equation [1]:

$$K \partial^2 T/\partial x^2 = c\rho \partial T/\partial x \qquad (2)$$

where K, ρ and c represent the thermal conductivity, density and specific heat, resp. The thermal diffusivity (κ) is defined by the ratio:

$$\kappa = K/\rho c \qquad (3)$$

The thermal diffusivity which represents the "diffusion constant" for the propagation of a spatially non-uniform temperature is a more fundamental property of a material than the thermal conductivity, which represents the product of the thermal diffusivity and the volumetric heat capacity.

The specific heat (c) of a material is defined by:

$$c = dQ/dT \qquad (4)$$

where Q is the internal energy per unit mass. The quantity $C = c\rho$ is the volumetric heat capacity.

In general, the conduction of heat in solids is controlled by variables at the atomic level and at the microstructural or continuum level.

At the atomic level, the conduction of heat in single-phase materials occurs by three primary mechanisms, namely phonon, electron [2, 3] and photon transport [4]. Photon transport is effective primarily at elevated temperatures. Theoretical calculations of the thermal conductivity for any of these three mechanisms relies on the general equation:

$$K = 1/3 C v \ell \qquad (5)$$

where C is the volumetric heat capacity, v is the carrier velocity and ℓ is the "mean-free-path" between collisions of the phonon or electron or the distance of propagation of a photon between emission and re-absorption. In eq. 5, the quantity $1/3 v \ell$ represents the thermal diffusivity. For this reason, theoretical estimates for the thermal conductivity rely on separate derivations of the specific heat, the carrier velocity and the mean-free path. From the perspective of the primary objective of this paper, the specific heat and carrier velocity for any given material at any temperature are intrinsic properties not strongly affected by extrinsic variables. Any major changes in the thermal conductivity and diffusivity can be brought about by corresponding changes in the mean-free-path only.

The mean-free-path is controlled by intrinsic variables such as other phonons and thermally induced vacancies as well as extrinsic variables such

142

as alloying elements, impurities, vacancies, crystal defects, elastic and optical discontinuities, etc., which can lead to phonon and electron scattering or the absorption of photons.

The total mean-free-path, ℓ_{total} in terms of the intrinsic and extrinsic mean-free-path, ℓ_{int} and ℓ_{ext}, resp., is:

$$1/\ell_{total} = 1/\ell_{int} + 1/\ell_{ext} \qquad (6)$$

which indicates that ℓ_{total} in a material in which the mean-free-path is affected by extrinsic variables is always less than ℓ_{int}. Because of this the important conclusion can be reached that the presence of structural imperfections or foreign atoms in a crystal lattice will always tend to lower the thermal conductivity. In other words, the thermal conductivity of a material can never be raised above the value controlled by the intrinsic variables only.

The purpose of this paper is to present experimental data obtained by the author and his co-workers which illustrates a number of the above compositional and structural effects on the heat conduction behavior of single-phase silicon carbide and silicon carbide-based composites.

RESULTS AND DISCUSSION

Fig. 1 shows typical experimental data for the thermal diffusivity for four different types of dense polycrystalline structural silicon carbide with a high and normal content of boron and carbon sintering aids, a silicon carbide with a BeO grain boundary phase and a silicon carbide subjected to a heat-treatment subsequent to sintering, referred to as "crystallized" silicon carbide. These three sets of data represent the range of the lowest to the highest values for the thermal diffusivity of polycrystalline dense silicon carbide obtained by the writer and his co-workers. These differences in thermal diffusivity are illustrative of the effect of impurities or additives on heat transport, the thermal diffusivity controlled primarily by phonon scattering due to phonon-phonon interactions and scattering at foreign solid-solution atoms. For this reason, the silicon carbide with the high content of sintering aid should exhibit a lower value for the thermal diffusivity than the silicon carbide with the normal amount of sintering aid. This difference is thought to be due primarily to the differences in boron content in solid solution in the silicon carbide matrix. It appears almost ironic that the very presence of the sintering aid, essential for densification also is highly effective in lowering the thermal conductivity. This is particularly unfortunate for the synthesis of dense polycrystalline silicon carbide which for its intended design application should exhibit a thermal conductivity as high as possible. Fortunately, as illustrated by the data of fig.4, there appear to be at least two methods by which the thermal diffusivity and conductivity of polycrystalline ceramics can be enhanced.

The first method consists of subjecting the polycrystalline silicon carbide to "recrystallization," i.e., an annealing treatment subsequent to densification. As shown by the experimental data of fig. 1, this can lead to significant increases in thermal diffusivity. Although further research is required to establish the primary reason for this increase, it is thought at this time that the "recrystallization" treatment promotes the precipitation of the boron atoms in the form of boron nitride or boron carbide [11, 12]. This will lower the density of phonon scatterers and thereby increase the thermal diffusivity. In this respect it is of interest to note that silicon carbide sintered in argon, generally exhibits values

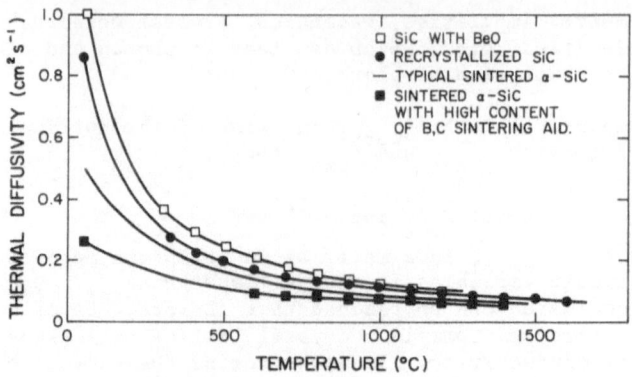

Fig.1. Thermal diffusivity of various structural
silicon carbides as a function of temperature.

for thermal conductivity much higher than the same silicon carbide sintered
in nitrogen [13, 14]. At the same time, however, the electrical
conductivity of silicon carbide sintered in Ar can be much higher than the
corresponding value for silicon carbide in nitrogen. The reasons for these
effects is that the argon atoms are much too large to go into solid
solution in the silicon carbide. This, however, is not the case for the
nitrogen atoms, which introduces additional phonon scatterers and a
corresponding decrease in thermal conductivity. At the same time, the
nitrogen atoms which contribute donor electrons to the silicon carbide,
compensates for the other impurities which act as acceptors and thereby
create electron holes. The electrons and holes compensate each other,
which lowers the electrical conductivity.

The second method for enhancing the thermal diffusivity and
conductivity of silicon carbide consists of adding a grain boundary phase
which exhibits very low mutual solid solubility with silicon carbide but
which exhibits a high solubility for the impurities found in silicon
carbide. By this method, during grain growth the impurities are swept
along by the grain boundary phase, analogous to the zone-refining effect,
thereby purifying the grains and associated increase in the thermal
diffusivity and conductivity of the silicon carbide grains.

The above mechanism is thought to be responsible for the improvement
in the thermal conductivity of polycrystalline silicon carbide with a few
weight percentage of BeO added as a grain boundary phase [13]. For such a
SiC-BeO composite the thermal conductivity at room temperature has been
reported to be as high as 270 W/m$^\circ$K with a corresponding value of
diffusivity of approximately 1.2 cm^{2}s^{-1}. This value is well above the
value of 0.82 for the recrystallized silicon carbide at room temperature
as shown in fig. 1. The SiC with BeO added consists of electrically
resistive grain boundaries and grains with an electrical conductivity
higher than the original silicon carbide. Of interest to note is that by
the addition of the BeO, the electrical conductivity of the silicon carbide
changes from n-type to p-type. This indicates that the BeO removes the
n-type donor impurity atoms preferentially over the p-type. This, in turn,
reduces the extent the n- and p-type impurities compensate for each other's
contribution to the electrical conductivity. Although the addition of the
BeO has led to an improvement in the thermal conductivity of impure silicon
carbide, the values attained are still well below the potential value for
silicon carbide, as judged by the data for the high-purity single crystal
reported by Slack [14] with a value at room temperature near 490 W/m$^\circ$ K with
a correspondent value for the thermal diffusivity in excess of 2 cm^{2}s^{-1}.

Silicon carbide exists also in a non-crystalline amorphous state, in the form of fibers made by the pyrolysis of organic fibers at moderate temperatures. These fibers also contain applicable amounts of other elements such as oxygen and carbon. In general, dielectric amorphous materials at low and moderate temperatures at which radiation effects are neglected, exhibit very low thermal conductivity and diffusivity primarily because of very small values of the phonon mean-free-path. Foreign elements in an amorphous material are expected to decrease the phonon mean-free-path further. For this reason, amorphous silicon carbide fibers are expected to exhibit values for the thermal conductivity and diffusivity comparable to those found for multi-component glassy materials and well below the values for crystalline silicon carbides, such as those shown in fig. 1.

Because of the lack of ready availability of an appropriate technique for measurements of the thermal diffusivity and conductivity of these dielectric fibers, Brennan, Bentsen and Hasselman [15] obtained these values for the amorphous silicon carbide fibers indirectly by means of the composite technique. By this method, the required values for the fibers were obtained indirectly, be measuring the corresponding values for an lithia-alumino-silicate glass-ceramic with and without a reinforcement of the silicon carbide fibers. From these data the values for the thermal conductivity and diffusivity were then calculated with the aid of composite theory for heat flow parallel to uniaxial fibers and perpendicular to the fibers given by the expression derived by Raleigh [5].

The data obtained for the thermal conductivity and diffusivity of the amorphous silicon carbide fibers are shown in fig. 2. The slight degree of indicative of the existence of a preferred orientation in the original polymeric fiber. The near independence of the thermal diffusivity on temperature compared to crystalline silicon carbide is indicative of a

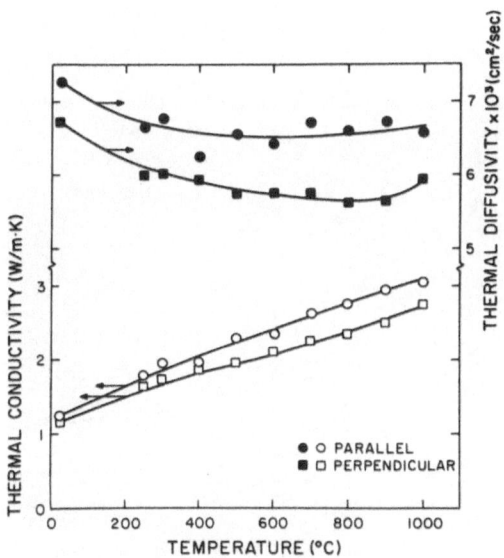

Fig. 2. Thermal diffusivity and conductivity of Nicalon silicon carbide fibers parallel and perpendicular to fiber axis [15].

phonon mean-free-path controlled by the close spacing of the lattice
defects rather than by phonon–phonon interactions. For this reason, the
positive temperature dependence of the thermal conductivity is governed
primarily by the positive temperature dependence of the specific heat.

Crystallization of amorphous materials as the direct result of reduced
phonon scattering at lattice defects is expected to lead to an increase in
the thermal diffusivity and conductivity, as for instance illustrated by
data for the thermal diffusivity of a glass-ceramic subjected to a range of
heat treatments [16]. The heating of the amorphous silicon carbide to
temperatures in excess of 1000°C leads to crystallization accompanied by a
change in composition by the loss of nitrogen and oxygen [17]. As
indicated by experimental data for CVD-SiC reinforced with the amorphous
silicon carbide fibers shown in fig. 3, the crystallization of such fibers
can result in significant increases in the thermal diffusivity of these
composites when heated and cooled to temperatures of 1400 and 1800°C [18].
Especially for this latter temperature this increase at room temperature
amounts to a factor well in excess of two. The latter data were obtained

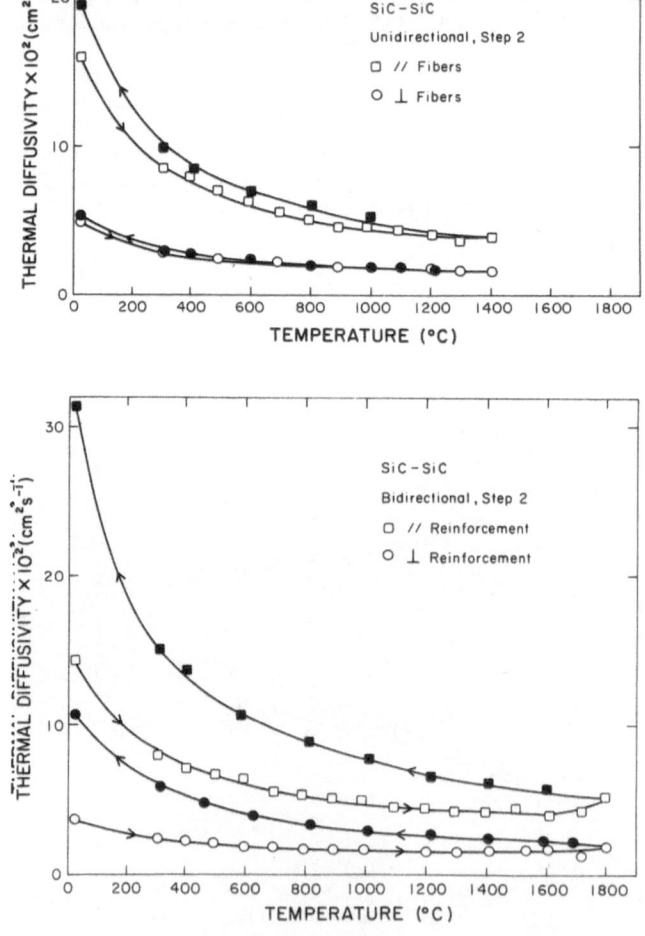

Fig.3. Effect of thermal cycling to 1400 and 1800° on the
thermal diffusivity of SiC-fiber reinforced CVD-SiC
matrix.

over a period of approximately eight hours. Because crystallization and compositional changes are time-dependent it is anticipated that heating cycles of longer duration would have resulted in even greater relative increases.

Crystalline silicon carbide also exists in the form of whiskers, made either by the pyrolysis of rice-hulls (RH) or by the vapor-liquid-solid (VLS) process [19]. The thermal conductivity and diffusivity of such whiskers, determined from the corresponding data of whisker reinforced ceramic matrix composites, described earlier, showed some interesting differences in behavior. Fig. 4 shows the experimental data for the thermal diffusivity at room temperature for composite samples of RH-SiC whisker-reinforced polycrystalline aluminum oxide as a function of whisker content [20].

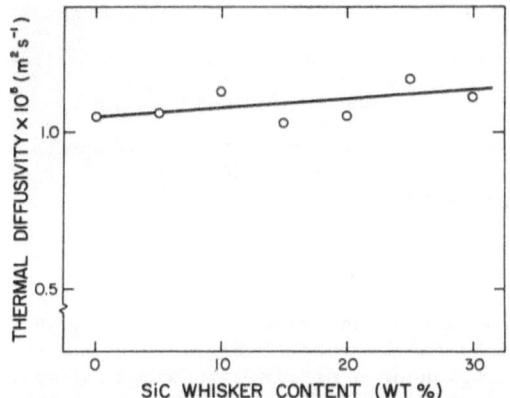

Fig.4. Effect of RH-SiC whisker content on the thermal diffusivity of an alumina matrix at room temperature (ref.20)

These data clearly indicate that the presence of the whiskers had only a slight positive effect on the thermal diffusivity. This indicates that the corresponding value for the RH-whiskers is only slightly higher than the thermal diffusivity of the alumina matrix of approximately 0.1 $cm^2 s^{-1}$. No effect of orientation with respect to the hot-pressing direction was found either. Fig. 5 shows the values of the thermal conductivity calculated from the experimental data of the thermal diffusivity given in fig. 4 and the measured values of the specific heat and density of the composite samples. At room temperature the thermal conductivity is independent of whisker content. This implies that the thermal conductivity of the RH-SiC whiskers is identical to the corresponding value for the alumina matrix of approximately 32 W/m$^{\circ}$K. This is the lowest value determined the thermal conductivity of dense crystalline silicon carbide. This low value of thermal conductivity is thought to be the direct result of the very high density of stacking-faults in these RH-SiC whiskers coupled with a relatively high oxygen content, both of which act as effective phonon scatterers. Included in fig. 5 also, are the calculated values for the thermal conductivity of the composites at 300 and 600°C. These data show that at these higher temperatures, in contrast to room temperature, the presence of the silicon carbide whiskers increases the value of the thermal conductivity significantly. This increase is related to the differences in the relative temperature dependence of the thermal conductivity of the alumina matrix and the RH-SiC whiskers, which in the latter material is suppressed due to the already short phonon

147

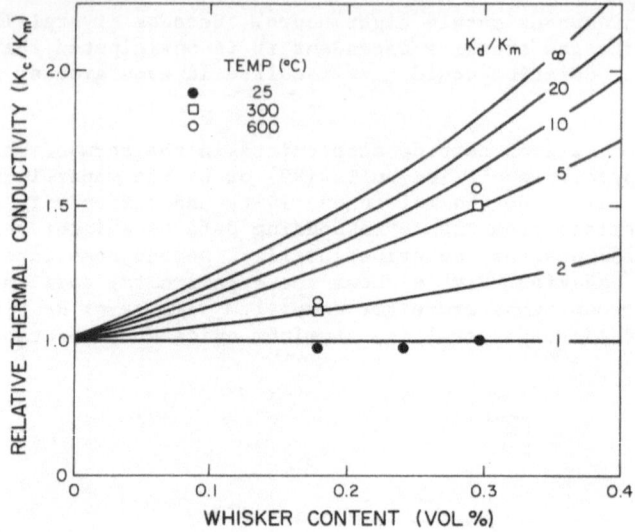

Fig.5. Relative thermal conductivity of rice-hull SiC whisker
reinforced alumina at 25, 300 and 600°C (ref.20)
($K_m \stackrel{\sim}{-} 30$ W/m°K).

mean-free-path due to phonon scattering at the stacking-faults and oxygen
impurities. Superposed on the data of fig. 5, is the theoretical behavior
for a composite consisting of a matrix with dilute dispersions in the form
of parallel circular cylindrical inclusions for heat flow perpendicular to
the cylinder axes for a range of values for the ratio of the thermal
conductivity (K_d) of the dispersions to that of the matrix (K_m). Although
approximate, the choice of this model is justified as during hot-pressing
the whiskers tend to align themselves preferentially perpendicular to the
hot-pressing direction. Comparison of the theoretical curves with the
experimental data suggest that at 300 and 600°C, the thermal conductivity
of the RH-SiC whiskers is some two to five times higher than for the
aluminum oxide.

Fig. 6 shows the relative dependence of the thermal conductivity at
room temperature of polycrystalline mullite (alumino-silicate) ceramic
reinforced by either RH or V.L.S.-SiC whiskers, as a function of whisker
content [21]. Both types of whiskers cause a significant increase in
thermal conductivity, but this effect is far more pronounced for the
V.L.S.-whiskers than for the RH-whiskers. This implies that the thermal
conductivity of the V.L.S.-whiskers is much higher than for the
RH-whiskers. This is to be expected as the V.L.S.-whiskers have much lower
densities of lattice-defects and much higher purity than the RH-whiskers.
Nevertheless, the level of impurities of the V.L.S. whiskers is
sufficiently high that the electrical resistivity is sufficiently low that
they can be self-heated by the application of an electrical field [22].
This implies that the thermal conductivity of the V.L.S.-whiskers is still
expected to be below the value for pure single crystals. The theoretical
curves for composite behavior identical to those of fig. 5, included in
fig. 6 permit estimating the thermal conductivity of the RH and V.L.S.-SiC
whiskers relative to the mullite matrix. The RH whiskers appear to have a
thermal conductivity approximately five to six times that of the mullite
matrix. For the latter, the actual value is approximately 5.5 W/m°K, which
suggest that the thermal conductivity of the RH-whiskers is approximately

148

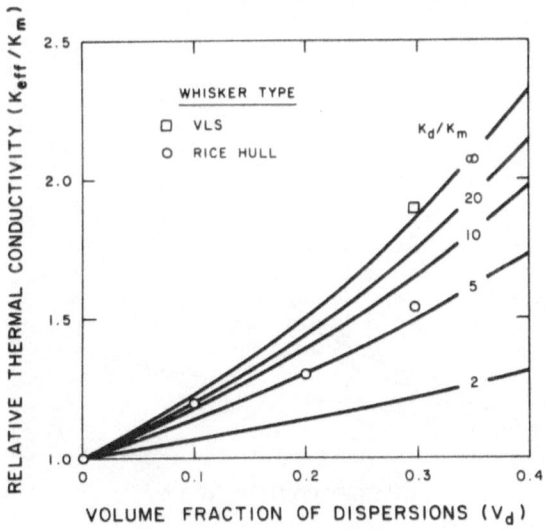

Fig.6. Relative thermal conductivity of SiC whisker-
reinforced mullite with $K_m \simeq 5.5$ W/m$^\circ$K (ref.21).

30 W/m$^\circ$K. This value is very close to the value inferred from the data for
RH—whisker reinforced alumina, shown in fig. 5. Because of the very close
spacing of the curves for $K_p/K_m > 10$, actual estimate of the value of K_p/K_m
for the V.L.S.—whiskers cannot be made. However, it can be concluded that
the thermal conductivity of the V.L.S.—whiskers exceeds the value for the
mullite by well over an order of magnitude. A more accurate estimate
requires a matrix phase with a thermal conductivity which is comparable in
value to that of the V.L.S.—whiskers.

For such a composite system consisting of V.L.S.—SiC whisker
reinforced silicon nitride, fig. 7 shows the data for the thermal
conductivity as a function of whisker content [23]. These data show a
downward curvature with increasing whisker content in contrast to the
upward curvature expected from composite theory. It is thought that these
data are affected by another variable which tends to decrease the thermal
conductivity with increasing whisker content. Nevertheless, comparison of
the experimental data with the theoretical curves suggests that the thermal
conductivity of the V.L.S.—whiskers is some five to ten times higher than
the value for the silicon nitride of $\sim$ 48 W/m$^\circ$K. Accordingly, the thermal
conductivity of the V.L.S.—whiskers is expected to range from about 240 to
480 W/m$^\circ$K. This latter value is very close to the value obtained by Slack
[14] for a high-purity single crystal. It is reasonably safe to assume
that the actual value for the V.L.S.—SiC whiskers will be near the average
value of 360 W/m$^\circ$K. As an aside, data for the thermal conductivity
ascertained indirectly from composite data should be treated with some
caution, especially for those composites which exhibit microcracking or the
existence of a thermal barrier resistance at the interface which are known
to lower the effective thermal conductivity significantly [24, 25]. If
these effects are ignored, a value of the thermal conductivity of any one
of the components inferred from composite data will be underestimated.

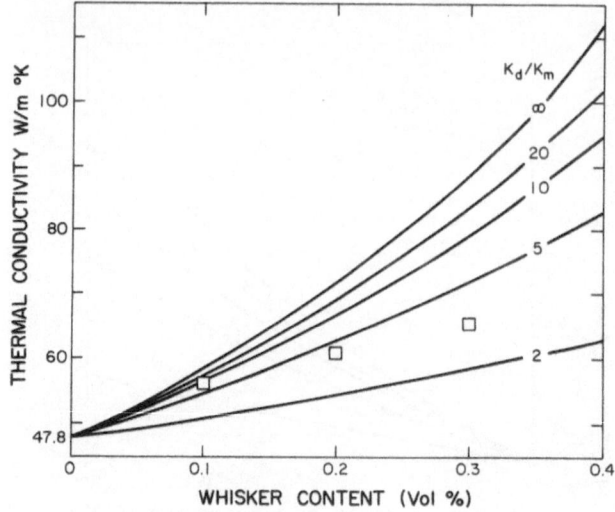

Fig.7. Thermal conductivity at room temperature for VLS
SiC—whisker reinforced silicon nitride (ref.23).

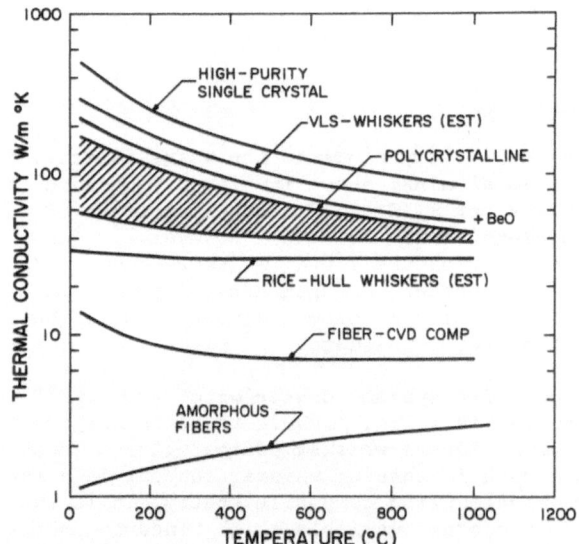

Fig.8. Thermal conductivity of various silicon carbides
as a function of temperature.

Fig. 8 shows the collective data for thermal conductivity from room
crystal of silicon carbide obtained by Slack [14] are included. Of
interest to note is that at room temperature the highest and lowest values
for the thermal conductivity differ by some two and a half orders of
magnitude. Because of the negative temperature dependence of the thermal
conductivity of the single crystal and the positive temperature dependence
for the amorphous fibers at cryogenic temperatures at which the single
crystal exhibits its highest values, the total range may well approach four
orders of magnitude.

In summary, because of the significant effect of impurities, additives and lattice imperfections, the thermal diffusivity and conductivity of silicon carbide can vary significantly. This should be of practical interest as it permits the materials technologist with a wide range of choice of materials with varying thermal conductivity to meet specific design requirements.

ACKNOWLEDGEMENTS

The author is indebted to the Standard Oil Company for providing a sample of the SiC-BeO and for permission to publish the data for the recrystallized silicon carbide and to the Journal of Material Science for permission to publish some of the figures.

REFERENCES

1. H.S. Carslaw and J.C. Jaeger, Conduction of Heat in Solids, 2nd Ed. Clarendon Press, Oxford (1960).
2. C. Kittel, Introduction to Solid State Physics, 2nd Ed. John Wiley, New York (1962).
3. R. Berman, Thermal Conduction in Solids. Clarendon Press, Oxford (1976).
4. W.D. Kingery, H.K. Bowen, D.R. Uhlmann, Introduction to Ceramics, 2nd Ed. John Wiley, New York (1976).
5. L. Rayleigh, Phil. Mag. 34 (1892) 481.
6. J.C. Maxwell, A. Treatise on Electricity and Magnetism, 1. 3rd Ed. Oxford University Press (1904).
7. D.A.C. Bruggeman, Annalen Physik 24 (1935) 636.
8. A.E. Powers, Conductivity in Aggregates, Knolls Atomic Power Laboratory Report-2145, General Electric Corp. (1961).
9. Z. Hashin, J. Comp. Mat. 2 (1968) 284.
10. S.C. Cheng and R.I. Vachan, Int. J. Heat and Mass Transfer, 12 (1969) 249.
11. C.H. McMurtry, Private Communication.
12. R.A. Youngman and R.C. Enck, Paper 65-BP-87, Presented at 89th Annual Meeting of the American Ceramic Society, Pittsburgh, PA, April 26-30 (1987).
13. M. Ura and O. Asai, Japan Fine Ceramics Association, F.C. Report vol. 1, no.4 (1984).
14. G.A. Slack, J. Appl. Phys. 35 (1964) 3460.
15. J.J. Brennan, L.D. Bentsen and D.P.H. Hasselman, J. Mat. Sc. 17 (1982) 2337.
16. K. Chyung, G.E. Youngblood, D.P.H. Hasselman, J. Am. Ceram. Soc. 61 (1978) 530.
17. T. Mah, N.L. Hecht, D.E. McCullem, J.R. Hoenigman, H.M. Kim, A.P. Katz, H.A. Lippsitt, J. Mat. Sci., 19 (1984) 1191.
18. H. Tawil, L.D. Bentsen, S. Baskaran, D.P.H. Hasselman, J. Mat Sc., 20 (1985) 3201.
19. J.V. Milewski, F.D. Gac, J.J. Petrovic, S.R. Skaggs, J. Mat. Sc., 20 (1985).
20. L.F. Johnson, D.P.H. Hasselman, J.F. Rhodes (in preparation).
21. L.M. Russell, L.F. Johnson, D.P.H. Hasselman, R. Ruh, J. Am. Ceramic Soc. (in press)
22. J.J. Petrovic, Private Communications.
23. L.M. Russell, L.F. Johnson, D.P.H. Hasselman, J.J. Petrovic (in preparation).

24. D.P.H. Hasselman, J. Comp. Mat., 12 (1978), 403.
25. D.P.H. Hasselman and L.F. Johnson, J. Comp. Mat. 12 (1987) 508.

THERMAL CONDUCTIVITY MEASUREMENTS OF CERAMIC

NUCLEAR FUELS BY LASER FLASH METHOD

A. K. Sengupta and C. Ganguly

Radiometallurgy Division
Bhabha Atomic Research Centre
Bombay 400 085, INDIA

ABSTRACT

The design and in-pile performance of nuclear fuels to a great extent are dependent on their thermal conductivity. The thermal conductivities of UO_2, $UO_2 - 4\%$ PuO_2, (UPu)C and (UPu)N fuels have been determined in the temperature range of 1000 - 1800 K from measurements of thermal diffusivity by the laser heat pulse technique.

TiO_2 dopant (0.05 - 0.1 w/o) was found to increase the thermal conductivity of oxide pellets prepared by the powder route. The sol-gel microsphere pelletisation (SGMP) process led to oxide pellets having higher thermal conductivity compared to those prepared by the powder route.

The thermal conductivity of $(Pu_{0.7}U_{0.3})C$ and $(Pu_{0.7}U_{0.3})N$ has been evaluated for the first time.

The present paper highlights the role of titania dopant and the SGMP process in improving the thermal conductivity of the oxide fuels and the effect of plutonium content in the thermal conductivity of (UPu)C and (UPu)N fuels.

INTRODUCTION

The thermal conductivity of a ceramic nuclear fuel is of great technical importance as it determines its surface and centre temperatures during irradiation and its maximum permissible rate of linear heating.

The high performance thermal reactor fuels under consideration are large grain UO_2 and $(UPu)O_2$ pellets having a uniformly distributed pore structure. Large grain (30 - 40 μm) oxide pellets have been successfully fabricated by the addition of titania dopant[1]. The sol-gel microsphere pelletisation (SGMP) process has been reported[2] to produce high density oxide pellets containing uniformly distributed spherical closed pores in the ideal diameter range of 2 - 5 μm. Mixed uranium plutonium monocarbide and mononitride are advanced fuels for liquid metal cooled fast breeder reactors. These pellets are fabricated by carbothermic reduction of oxide followed by cold pelletisation and sintering[3].

The thermal conductivity of the oxide, carbide and nitride fuels at different temperature depends on their stoichiometry, grain size and porosity. In the present study, thermal conductivity of the following fuels was determined from measurements of thermal diffusivity up to 1800 K using the 'laser heat pulse' technique:

- UO_2 and UO_2 - 4 % PuO_2
- TiO_2 doped large grain UO_2 and UO_2 - 4 % PuO_2
- UO_2 prepared by the 'sol-gel microsphere pelletisation (SGMP)' process
- $(U_{0.8}Pu_{0.2})C$ and $(U_{0.3}Pu_{0.7})C$
- $(U_{0.8}Pu_{0.2})N$ and $(U_{0.3}Pu_{0.7})N$

EXPERIMENTAL

Sample Preparation

UO_2 and UO_2 - 4 % PuO_2 pellets were prepared from ammonium diuranate (ADU) derived UO_2 and plutonium oxalate derived PuO_2 powders by the conventional 'cold-pelletisation ($\sim$ 300 MPa) and sintering' (1900 K, 8 hours, Ar + 8 % H_2) process. A small amount of TiO_2 (0.05 w/o and 0.1 w/o) was added in some batches of UO_2 and UO_2 - 4 % PuO_2 powders. UO_2 pellets were also prepared by the non-conventional 'sol-gel microsphere pelletisation (SGMP)' route. The gel microspheres were prepared by the 'external gelation of uranium (EGU)' process and subjected to controlled calcination in order to get microspheres of desirable specific surface area, crushing strength and oxygen to uranium ratio. The microspheres were directly pelletised ($\sim$ 300 MPa) and sintered at 1873 K for 4 hours in Ar + 8 % H_2 atmosphere to get high density UO_2 pellets containing uniformly distributed spherical pores.

The mixed uranium plutonium monocarbide (MC) and mononitride (MN) were prepared by carbothermic reduction of tabletted UO_2, PuO_2 and graphite powder mixture in vacuum and flowing N_2, respectively. The MC and MN clinkers were then crushed, milled, pelletised (at 300 MPa) and sintered at 1773 - 1973 K in Ar + 8 % H_2 atmosphere.

The sintered pellet samples for measurement of thermal diffusivity were 1 - 2 mm in thickness and approximately 10 mm in diameter.

Characterisation

The characteristics of the sintered pellets in terms of density and composition are summarised in Tables 1 & 2.

The microstructures of conventional UO_2 pellet and TiO_2 doped large grain UO_2 pellets prepared by the powder route are shown in Fig. 1. The UO_2 pellets prepared by the SGMP process had uniformly distributed and nearly spherical pores in the diameter range of 2 – 5 microns as shown in Fig. 2.

Table 1. Characteristics of UO_2 and UO_2-4% PuO_2 sintered pellet samples

Sl. No.	Sample	TiO_2 (w/o)	Density (%T.D.)	Grain Size (μm)	O/M (M=U or U+Pu)
1.	UO_2	Nil	92%	12	2 ± .002
2.	UO_2	0.05	89%	20	"
3.	UO_2	0.01	93%	45	"
4.	UO_2^*	Nil	91%	15	"
5.	UO_2+4% PuO_2	Nil	91%	10	1.997
6.	UO_2+4% PuO_2	.05	94%	35	1.998

* Prepared from SGMP route.

Table 2. Characteristics of uranium plutonium mixed carbide and nitride sintered pellet samples

Sl. No.	$\frac{Pu}{U+Pu}$	Carbon (w/o)	Oxygen (w/o)	N (w/o)	M_2C_3 or M_2N_3 (w/o)	Percentage Theoretical Density (% T. D.)
1.	0.70	4.86	0.60	0.10	15	91
2.	0.20	4.75	0.15	0.05	9	88.5
3.	0.70	0.44	0.50	4.88	Nil	88
4.	0.20	0.45	0.28	4.8	Nil	86.5

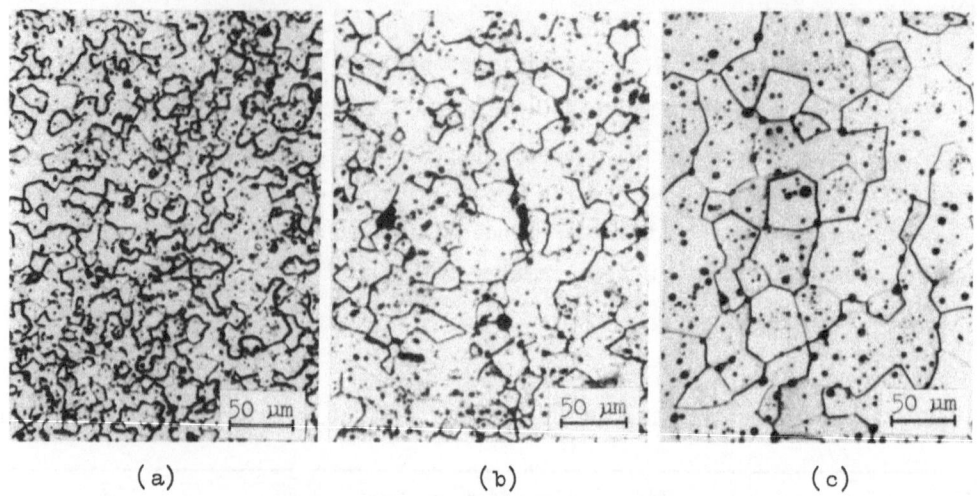

(a)   (b)   (c)

Fig. 1. Microstructures of doped and undoped UO_2 sintered pellets.
(a) UO_2 (grain size: 12 μm); (b) UO_2 + 0.05 % TiO_2 (grain
size: 25 μm); (c) UO_2 + 0.01 % TiO_2 (grain size: 45 μm)

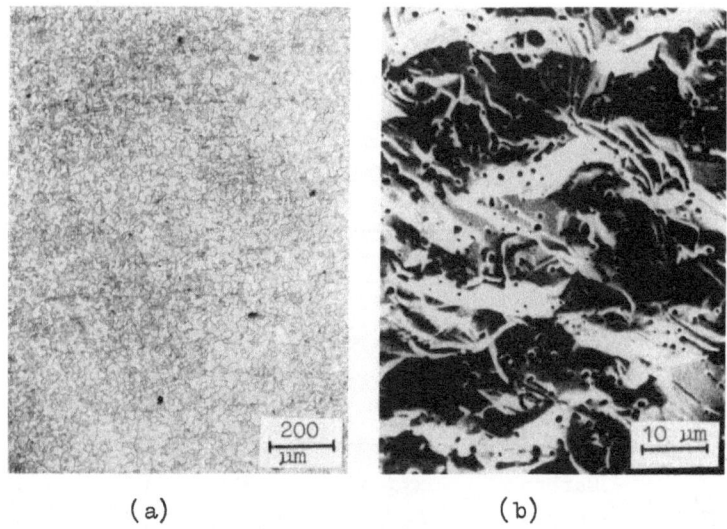

(a)   (b)

Fig. 2. (a) Microstructures of UO_2 sintered pellets
prepared by SGMP process
(b) Scanning Electron Micrograph of sintered UO_2
pellets showing uniformly distributed
spherical pores (2 - 5 μm)

Measurement Procedures

The thermal diffusivity 'a' was measured experimentally by the
transient flash method of Parker et al[4]. A square wave, pulsed ruby

laser of energy 6 joules/pulse was used as the flash source for irradiating one surface of the thermally insulated sample. The rise in temperature on the other surface was measured by In–Sb IR detector. The thermal diffusivity 'a' in cm^2/s was then calculated from the relation

$$a = \frac{1.37\ L^2}{\pi^2 \cdot t_{\frac{1}{2}}} \qquad\qquad \cdots\cdots (1)$$

where 'L' is the corrected thickness of the sample at the measurement temperature T and $t_{\frac{1}{2}}$ is the time in seconds taken for half the maximum rise in temperature of the back surface of the sample. The diffusivity value was corrected for radiation heat loss whenever necessary using Cape and Lehman's[5] method.

The specific heat data of UO_2,[6] PuO_2,[7] UC, PuC,[8] UN and PuN[9] were used for calculation of the specific heat of the mixed oxide, carbide and nitride, assuming Kope's law which states that the specific heat of an ideal solid solution is proportional to the mole fraction of the solutes.

The porosity corrections were made according to the following relations:

$$i)\ k_{measured} = k_{theoretical}\ (1 - \alpha \cdot P)\quad 0 \leqslant P \leqslant 0.1 \qquad \cdots\cdots (2)$$

$$ii)\ k_{measured} = k_{theoretical}\ \frac{1 - P}{1 + \beta \cdot P}\quad 0.1 \leqslant P \leqslant 0.2 \qquad \cdots\cdots (3)$$

where 'P' is the pore fraction of the pellet sample and 'α' and 'β' are constants whose values depend on pore size, shape and orientation. In the present investigation the pore shape was assumed to be spherical. The Loeb[10] equation with $\alpha = 2.5$ was used for oxides while the Maxwell Eucken[11] equation with $\beta = 1$ was utilised for the carbides and nitrides.

The thermal diffusivity measurements were carried out from 700 K to 1800 K under vacuum (1.33×10^{-3} Pa). For samples containing high (66 %) plutonium relatively lower vacuum (1.33 Pa) was maintained to minimise 'Pu' loss by volatilisation. To ensure that there was no significant Pu loss at higher temperature, a few measurements were also carried out during the cooling cycle of the sample. The thermal diffusivity data obtained during the heating and cooling cycles were in good agreement.

RESULTS AND DISCUSSIONS

Oxides

The thermal conductivity of conventional UO_2 pellets, titania doped

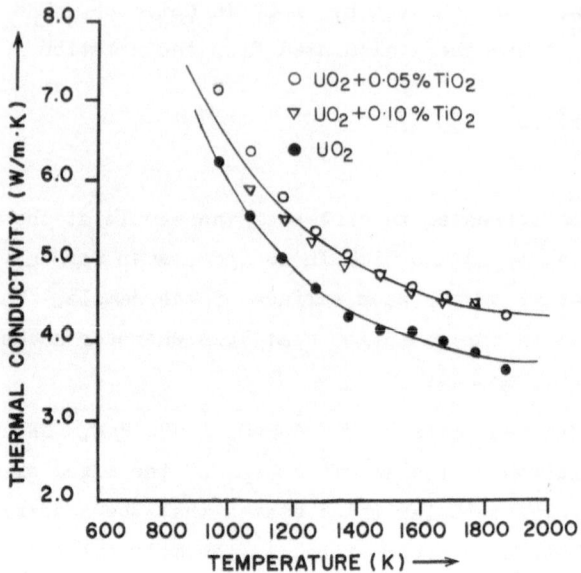

Fig. 3(a). Thermal conductivity of titania doped
and undoped UO_2 sintered pellets
(corrected to Theoretical Density).

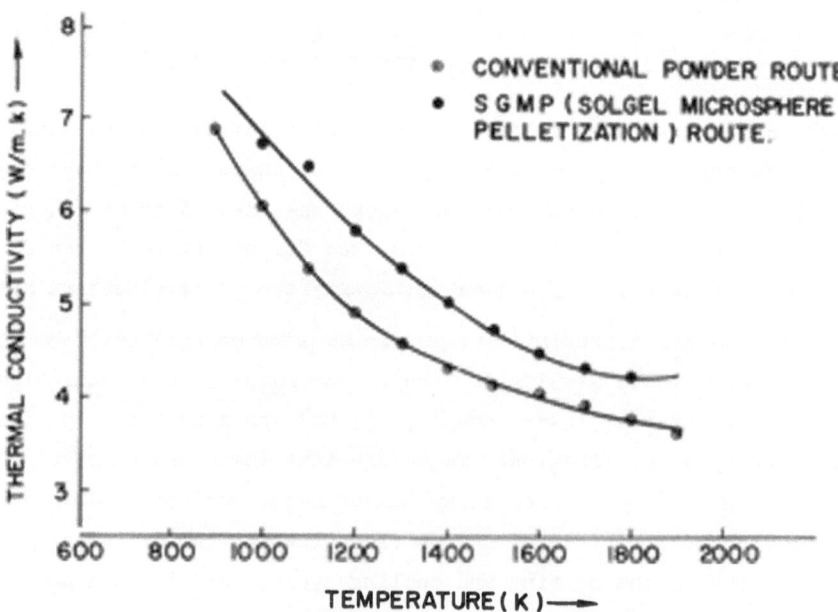

Fig. 3(b). Thermal conductivity of sintered UO_2 pellets
(corrected to Theoretical Density)

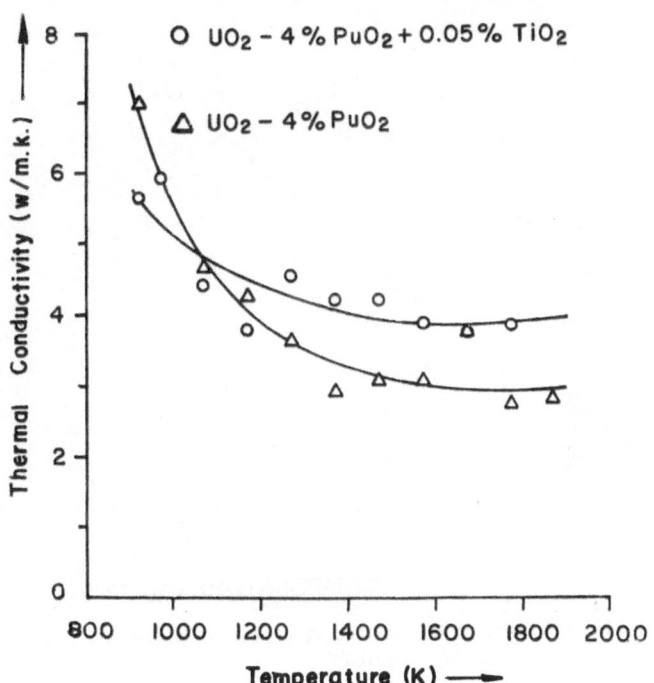

Fig. 4. Thermal conductivity of titania doped and
undoped sintered UO_2–4% PuO_2 pellets
(corrected to Theoretical Density)

UO_2 and UO_2 prepared by the SGMP route are given in Figs. 3(a) & 3(b).
The thermal conductivity of TiO_2 doped and undoped UO_2–4% PuO_2 is shown
in Fig. 4.

The thermal conductivity of TiO_2 doped UO_2 and UO_2+4% PuO_2 are about
10–15% higher than the undoped ones. The higher values could be due to
the presence of Ti^{3+} contributing to electronic heat conduction. Further,
titania enhances the grain size (2–3 times) and has a spherodization
effect on the pore structure [Fig. 1(a), (b) and (c)]. These microstructural changes would reduce the thermal resistance to heat flux and
in turn increase the thermal conductivity.

The UO_2 prepared by the SGMP route had a uniformly distributed
spherical pore structure [Fig. 2(a)] which would be responsible for
improving the thermal conductivity values by about 10 – 15%.

Mixed Carbide

The thermal conductivity of uranium and plutonium rich mixed carbide
at different temperatures is shown in Fig. 5. The values for $(U_{0.8}Pu_{0.2})C$
are in good agreement with that reported by Lewis & Kerrisk[12]. The thermal

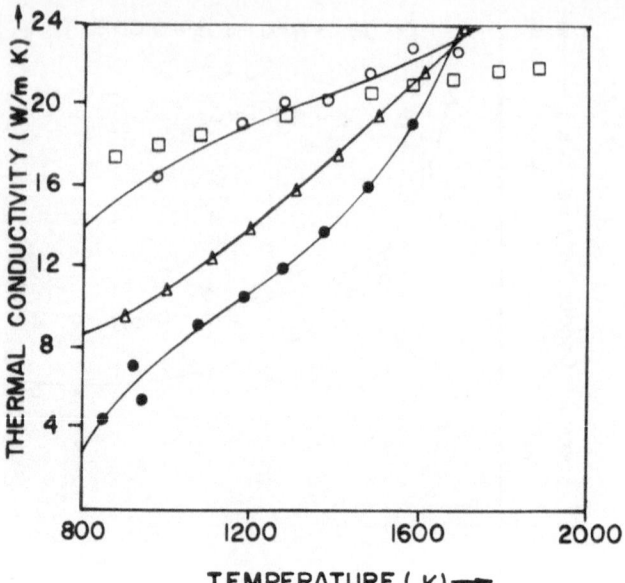

Fig.5. Thermal conductivity of uranium plutonium
mixed carbide
o : $(U_{0.8}Pu_{0.2})C$; △: $(U_{0.3}Pu_{0.7})C$;
● : PuC_{1-x}(Ref.13); ▫: $(U_{0.8}Pu_{0.2})C$(Ref.12)

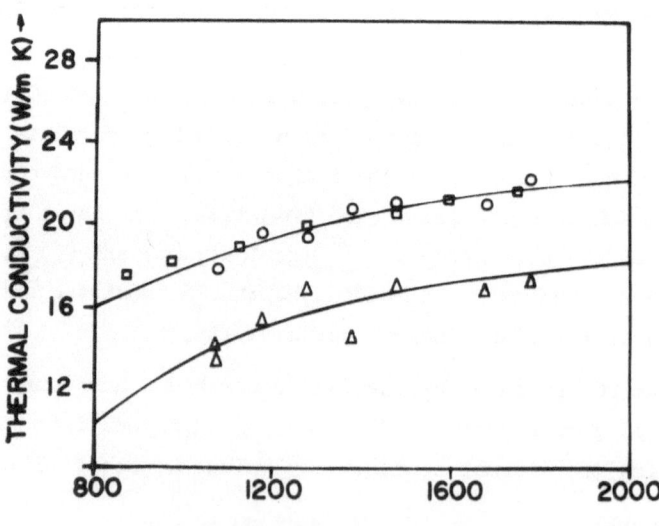

Fig. 6. Thermal conductivity of uranium plutonium
mixed nitride (corrected to 100% T.D.)
o : $(U_{0.8}Pu_{0.2})N$; △: $(U_{0.3}Pu_{0.7})N$;
▫ : $(U_{0.8}Pu_{0.2})N$ (Ref. 14)

conductivity of $(Pu_{0.7}U_{0.3})C$ is lower than that of $(U_{0.8}Pu_{0.2})C$ at lower temperatures. This could be because of the defect structure of the former (similar to PuC_{1-x} [13]) compared to that of the latter. However, at higher temperatures both uranium and plutonium rich mixed carbides were found to have similar thermal conductivity values.

Mixed Nitride

Fig. 6 shows the thermal conductivity plots of uranium plutonium mixed nitrides of uranium and plutonium rich composition as a function of temperature. The values evaluated for $(U_{0.8}Pu_{0.2})N$ are in excellent agreement with that reported by Alexander et al. [14] The thermal conductivity of $(Pu_{0.7}U_{0.3})N$ was found to be lower than that of $(U_{0.8}Pu_{0.2})N$ at all temperatures.

CONCLUSIONS

1. The thermal conductivity of oxide fuels can be improved by modifying the microstructure in terms of large grain size and uniform distribution of nearly spherical pores in the diameter range of 2 – 5 μm. This can be achieved by using TiO_2 dopant and following the SGMP route of fuel fabrication.

2. The thermal conductivity of plutonium rich mixed monocarbide and mononitride $\underline{/}(U_{0.3}Pu_{0.7})C$ and $(U_{0.3}Pu_{0.7})N\underline{\smash{\big/}}$ are lower than their uranium rich counterparts $\underline{/}(U_{0.8}Pu_{0.2})C$ and $(U_{0.8}Pu_{0.2})N\underline{\smash{\big/}}$, particularly below 1200 K.

ACKNOWLEDGEMENT

Authors are grateful to Messrs. Arun Kumar, U. Basak and P. V. Hegde for fabrication of the titania doped oxide pellets, SGMP derived UO_2 pellets and the mixed carbide/nitride samples respectively and Mr. V. D. Pandey for metallography.

Authors are also thankful to Messrs. T. Jarvis and R. V. Kulkarni for their assistance in the experimental measurements of thermal diffusivity.

REFERENCES

1. H. S. Kamath, D. S. C. Purushotham, D. N. Sah and P. R. Roy, "Some aspects of plutonium recycle in thermal reactors", IAEA Specialists meeting on improved utilisation of water reactor fuel with special emphasis on extended burn-up and plutonium recycling, CEN/SCK, Mol, Belgium, 227 (1984).

2. C. Ganguly, E. Zimmer and E. Merz, "Sol-gel-microsphere-pelletisation process for fabrication of uranium and thorium based fuels", Trans. ANS, Vol. 52, 44 (1986).

3. C. Ganguly, P. V. Hegde, G. C. Jain, U. Basak, R. S. Mehrotra, S. Majumdar and P. R. Roy, "Development and fabrication of 70 % PuC - 30 % UC fuel for the fast breeder test reactor in India", Nuclear Technology, Vol. 72, 59 (1986).

4. W. J. Parker, R. J. Jenkins, C. P. Butler and G. L. Abbott, "Flash method of determining thermal diffusivity, heat capacity and thermal conductivity", J. Appl. Phys. 32(9), 1979-84 (1961).

5. J. A. Cape and G. W. Lehman, "Temperature and finite pulse time effects in the flash method for measuring thermal diffusivity", J. Appl. Phys. 34 (7), 1909-13 (1963).

6. J. K. Fink, M. G. Chasnov and L. Leibowitz, "Thermophysical properties of uranium dioxide", J. Nucl. Mater. 102, 17-25 (1981).

7. F. L. Oetting, "The chemical thermodynamic properties of plutonium compounds", Chemical Review, 67, 261-297 (1967).

8. C. E. Holley Jr., M. H. Rand and E. K. Storms, "The chemical thermodynamics of actinide elements and compounds, Part 6", The Actinite Carbides (IAEA, Vienna), 59 (1984).

9. Hj. Matzke, "Science of advanced LMFBR fuels", North Holland Physics Publishing, Amsterdam, 106 (1986).

10. Thermal conductivity of uranium dioxide, Report of the Panel on thermal conductivity of uranium dioxide, held in Vienna 1965, Technical Report Series No. 59 (IAEA, Vienna, 1966).

11. A. B. G. Washington, "Preferred values for the thermal conductivity of sintered ceramic fuel for fast reactor use", The Reactor Group, United Kingdom Atomic Energy Authority Report TRG-R-2236, (1973).

12. H. D. Lewis and J. F. Kerrisk, "Electrical and thermal properties of uranium and plutonium carbides", LA 6096 (1976).

13. A. K. Sengupta, K. B. Arora, S. Majumdar, C. Ganguly and P. R. Roy, "Thermal conductivity of sintered plutonium carbide", J. Nucl. Mater., 139, 282 (1986).

14. C. A. Alexander, J. S. Ogden and W. M. Pardue, "Thermophysical properties of (UPu)N", Nuclear Metallurgy, 17, Part I, 95 (1970).

THERMAL CONDUCTIVITY OF GALLIUM ARSENIDE AT HIGH TEMPERATURE

Ronald H. Bogaard and C. Y. Ho

Center for Information and Numerical Data Analysis
  and Synthesis (CINDAS)
Purdue University
2595 Yeager Road
West Lafayette, Indiana 47906

ABSTRACT

Experimental data on the thermal conductivity of gallium arsenide have been searched, compiled, and critically evaluated. The behavior at high temperature is examined for electronic and radiative contributions to the total thermal conductivity. The observations are discussed and a temperature dependence for the lattice conductivity is suggested. Remaining problems are identified.

INTRODUCTION

The thermal conductivity of gallium arsenide at high temperatures is of interest to the modeling of crystal growing processes. Our present interest is to examine the available experimental data in order to gain an understanding of the behavior of thermal conductivity at high temperatures. One of the gallium arsenide materials studied was of the semi-insulating type (room-temperature electrical resistivity $> 10^6$ ohm cm).

In this paper a brief review of the literature data is presented. Then, possible electronic and radiative contributions to the measured thermal conductivity are discussed. Finally, conclusions are drawn as to the temperature behavior and the magnitude of the lattice thermal conductivity at elevated temperatures.

LITERATURE DATA DISCUSSION

Experimental data on the thermal conductivity of gallium arsenide available in the literature are shown in Fig. 1. Since our present interest is in the behavior at higher temperatures, data are presented only for the down-slope region. It is observed that, with one or two exceptions, the data sets tend to lie parallel, indicating a similar temperature behavior [1-13]. Individual data sets are identified in Table 1. The materials were essentially state-of-the-art at the time of measurement; most exhibit n-type electrical conduction, with a carrier concentration characteristic of unintentionally doped starting-grade materials. One set of data for semi-insulating, chromium-compensated

Table 1. Identification of Gallium Arsenide Materials

Data Set	Ref.	Material Characteristics	Author(s)
1	1	----	Sirota and Berger
2	2	$n = 4.3 \times 10^{17} \text{cm}^{-3}$	Timberlake, Shilliday, and Davis
3	3	----	Stuckes
4	4	$\rho = 10^6 \ \Omega \ \text{cm}$, 700°C anneal reduced ρ by ~10%	Blanc, Bube, and Weisberg
5	5	$n = 7 \times 10^{16} \text{cm}^{-3}$	Holland
6	6	$p = 10^{15} \text{cm}^{-3}$ (Zn), $\rho = 22 \ \Omega \ \text{cm}$	Aliyev and Akhmedli
7	7	$n = 3.5 \times 10^{17} \text{cm}^{-3}$, $\rho = 0.0007 \ \Omega \ \text{cm}$	Amith, Kudman, and Steigmeier
8	7	$n = 7.7 \times 10^{17} \text{cm}^{-3}$ (Te)	Amith, Kudman, and Steigmeier
9	7	$p = 6.4 \times 10^{19} \text{cm}^{-3}$ (Zn)	Amith, Kudman, and Steigmeier
10	7	$n = 5 \times 10^{16} \text{cm}^{-3}$, $\rho = 0.031 \ \Omega \ \text{cm}$	Amith, Kudman, and Steigmeier
11	8	$n = 9 \times 10^{16} \text{cm}^{-3}$	Wagini
12	9	$n = 1.4 \times 10^{16} \text{cm}^{-3}$	Vook
13	10	high-purity	Carlson, Slack, and Silverman
14	10	$n = 1.4 \times 10^{16} \text{cm}^{-3}$	Carlson, Slack, and Silverman
15	11	$n = 2 \times 10^{16} \text{cm}^{-3}$	Aliev and Shalyt
16	12	$\rho = 3.65 \times 10^6 \ \Omega \ \text{cm}$ (10^{18}cm^{-3}, Cr)	Vuillermoz, Laugier, and Mai
17	13	$n = 1 \times 10^{17} \text{cm}^{-3}$ (Si), $\rho = 0.021 \ \Omega \ \text{cm}$	Hunt

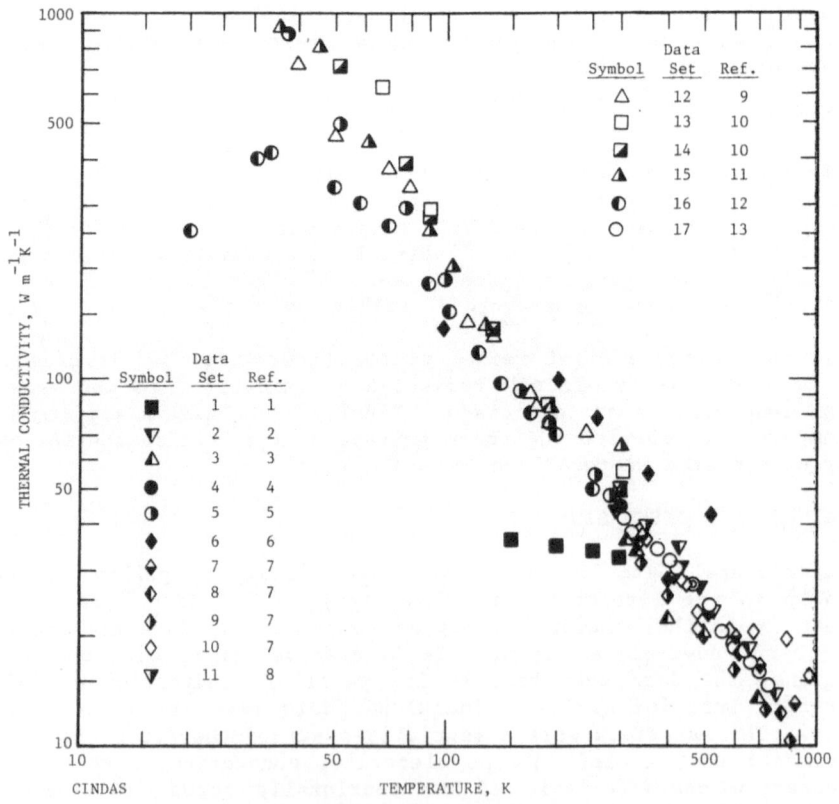

Fig. 1. Thermal Conductivity of Gallium Arsenide.

material [12] is shown, but it evidently contains sufficient chromium to reduce its thermal conductivity noticeably even for temperatures above 100 K.

The high-temperature region is shown in Fig. 2. It is observed that the data follow a temperature dependence of T^{-n}. A value of n = 1.28 for temperatures above about 500 K is consistent with the data reported by Amith et al. [7] (two specimens), by Stuckes [3], and by Hunt [13]. The two purer specimens reported by Amith et al. [7], which exhibit some extra conduction that will be discussed below, are exceptions. A discussion of the high temperature behavior should include the electronic thermal conductivity. This is reviewed below in the next section.

A theoretical analysis of the lattice thermal conductivity of gallium arsenide was recently reported by Dubey [14]. Following the approach of Holland [15] whereby contributions from longitudinal and transverse phonons are considered separately, an analysis was carried out for gallium arsenide and the results are included along with the literature data shown in Fig. 2. It is observed that the model fits the data fairly well up to room temperature or so, but results in a flatter slope toward higher temperatures. By comparison, the simple theory of three-phonon Umklapp processes due to Leibfried and Schloemann [16] predicts $T^{-1.0}$ behavior. Numerous measurements on other III-IV materials as well as the elemental semiconductors silicon and germanium indicate slopes steeper than $T^{-1.0}$. In particular, data reported by Hunt [13] on three gallium phosphide materials all show a slope change with the steeper slope beyond about 500 K. Mechanisms which may affect the high temperature thermal

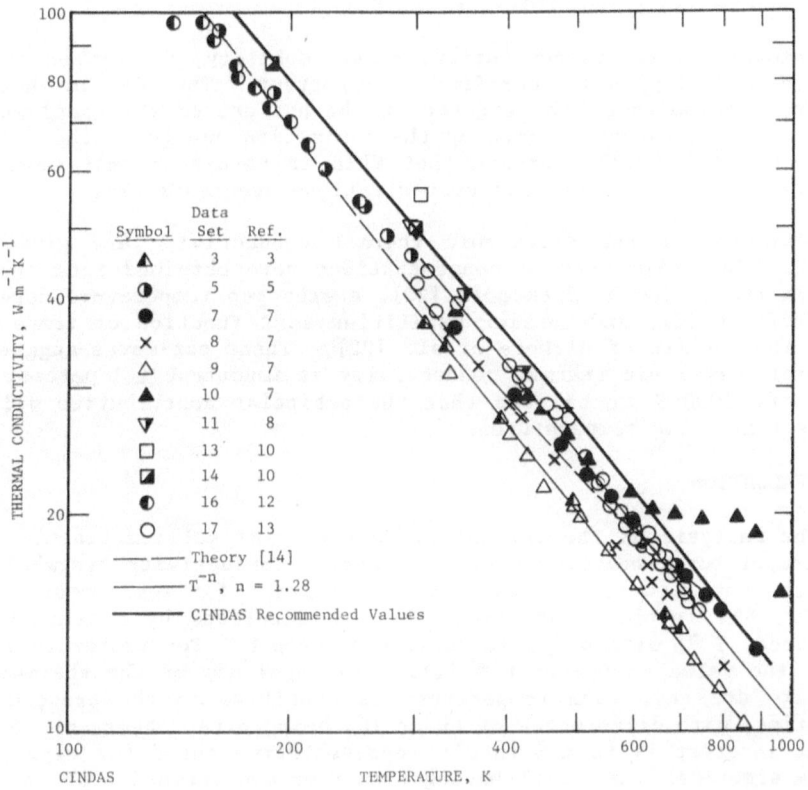

Fig. 2. High-Temperature Thermal Conductivity of Gallium Arsenide.

conductivity but which have not been considered at this point are electron-phonon effects [17] and thermal expansion effects [18].

ELECTRONIC THERMAL CONDUCTIVITY

The temperature region where the electronic thermal conductivity may be important for gallium arsenide can be found by examining the expression given by Drabble and Goldsmid [19]:

$$\lambda_e = \sigma T A (k_B/e)^2 \tag{1}$$

The electronic thermal conductivity, λ_e, is given in terms of the electrical conductivity, σ, and the absolute temperature, T. The constant A is about 2.5-3.0 for gallium arsenide at high temperatures [19]. Since the electrical conductivity may be due to both electrons and holes, the total electronic thermal conductivity may contain an ambipolar term [19]. For the intrinsic case,

$$\lambda_{e,total} = \lambda_e \left[1 + \frac{b}{(1+b)^2} \frac{1}{A} \left(4 + \frac{E_g}{k_B T} \right)^2 \right] \tag{2}$$

The ambipolar term, which depends on the energy gap, E_g, and ratio of electron-to-hole mobilities, b, may be significant.

For the extrinsic case,

$$\lambda_{e,total} = \lambda_e \left[1 + \frac{n_2 \mu_2}{n_1 \mu_1} \frac{1}{A} \left(4 + \frac{E_g}{k_B T} \right)^2 \right] \tag{3}$$

The ratio of carrier concentration, n, and mobility, μ, between minority, $n_2\mu_2$, and majority, $n_1\mu_1$, carriers is important. The idea is that electron and hole pairs may be created at the hot end of the specimen and, after traveling together, give up the ionization energy at the cold end. Equations (2) and (3) indicate that this is important only when the relative mobilities or carrier concentrations are favorable.

Estimates of the electronic thermal conductivity are given in Table 2. Intrinsic carrier concentrations were obtained from formulas given in the review by Blakemore [20], energy gap temperature dependence from Thurmond [21] and carrier mobilities as a function of temperature follow the results of Nichols et al. [22]. These estimates suggest that the total electronic thermal conductivity is important (>1 percent or so) only above 1000 K or so, and that the ambipolar contribution will be important at these temperatures.

DATA EVALUATION

The analysis of the thermal conductivity of gallium arsenide above room temperature indicates that the lattice conductivity overwhelmingly dominates any electronic contribution, at least for temperatures below 1000 K. The lattice conductivity may be described by a temperature dependence, T^{-n}, with n = 1.28 above about 450 K. For temperatures below 450 K, the value is nearer n = 1.1. The magnitude of the thermal conductivity at, say, room temperature is sensitive to the presence of impurities, with differences of about 10% being often observed. Since our present interest is to generate a representative curve for high purity gallium arsenide, a material having a carrier concentration of no more

Table 2. Electronic Thermal Conductivity of Gallium Arsenide*

[Temperature, T, K; Electronic Thermal Conductivity, λ_e, W m^{-1}K^{-1}; Total Electronic Thermal Conductivity, $\lambda_{e,total}$, W m^{-1}K^{-1}]

Temp. (K)	Intrinsic		Extrinsic ($n_{300K} = 5 \times 10^{16}$cm^{-3})	
	λ_e	$\lambda_{e,total}$	λ_e	$\lambda_{e,total}$
300	---	---	---	---
900	0.004	0.045	0.025	0.031
950	0.008	0.077	0.033	0.048
1000	0.014	0.12	0.046	0.075
1050	0.023	0.18	0.066	0.12
1100	0.036	0.25	0.090	0.17
1150	0.053	0.34	0.12	0.23
1200	0.075	0.43	0.16	0.31
1250	0.100	0.51	0.20	0.38

* Bipolar conduction formula from Drabble and Goldsmid [19].
  Intrinsic carrier concentration from Blakemore [20].
  Energy band-gap temperature dependence from Thurmond [21].
  Carrier mobility from Nichols et al. [22].

than 1.0×10^{16} cm^{-3} with minimal compensation was selected. The representative curve, drawn through a value of 50 W m^{-1}K^{-1} at 300 K, is shown in Fig. 2.

RADIATIVE THERMAL CONDUCTIVITY

The possibility of a radiative thermal conductivity is of interest to high-temperature applications. For gallium arsenide, the absorption coefficient determined by Spitzer and Whelan [23] is small at wavelengths of one micron or so, which is favorable. In fact, radiative thermal conductivity was suggested by Amith et al. [7] to explain an excess thermal conductivity that they had observed (data sets 7 and 10 in Fig. 2).

An estimate of the thermal radiative conductivity has been made based on a formula given by Klemens [24]:

$$\lambda_{rad} = 4\sigma n^2 T^3 \cdot \ell_{mfp} \tag{4}$$

in terms of the photon mean free path,

$$\ell_{mfp} = (\alpha + L^{-1})^{-1}_{ave.}, \tag{5}$$

the absolute temperature, T, and index of refraction, n. In Eq. (5), the absorption coefficient, α, and specimen thickness, L, are averaged over the spectral distribution as a function of temperature. The absorption coefficient was shown [23] to be sensitive to dopant content and wavelength. Further, since the temperature dependence has not been fully determined experimentally [23], theoretical estimates due to Jordan [25] were applied. Finally, index of refraction was obtained from the critical compilation of Seraphin and Bennett [26].

Estimates of thermal radiative conductivity were made for 5.0 mm thick samples having absorption coefficients appropriate to carrier concentrations of 5×10^{16} cm^{-3} and 3.5×10^{17} cm^{-3}, following Ref. [7]. The results are shown in Fig. 3. The good agreement between the excess thermal conductivity reported in Ref. [7] and the present estimate of thermal radiative conductivity is suprising given: (1) the sensitivity of the absorption coefficient to temperature and dopants, and (2) the dominating thickness dependence expressed by Eq. (5). In addition, the reduction in the energy gap with increasing temperature, which conceivably could lead to smaller radiative conductivity at sufficiently high temperatures, is apparently effective only above 1400 K, near the melting point.

CONCLUSIONS

In summary, a number of conclusions can be drawn:

1. The temperature behavior of the thermal conductivity of gallium arsenide is sufficiently well established to generate a representative curve for temperatures up to 1000 K. The material being characterized is high purity and uncompensated.

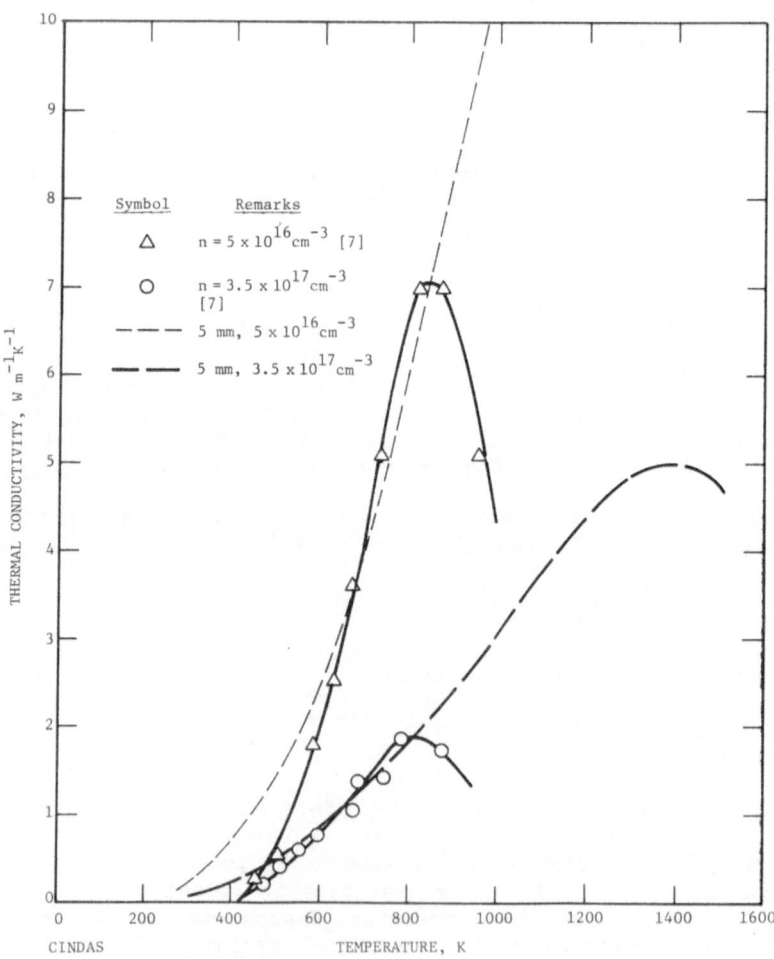

Fig. 3. Excess Thermal Conductivity and Radiative Thermal Conductivity of n-GaAs.

2. Electronic thermal conductivity is negligible or small below 1000 K, but becomes increasingly important (including an ambipolar contribution) at higher temperatures.

3. Radiative thermal conductivity depends strongly upon specimen thickness and upon the absorption coefficient sensitivity to temperature, wavelength, and dopants, but can be estimated.

4. Problem areas that are unresolved are: impurity (dopant) effects upon the magnitude of the thermal conductivity near room temperature, and the apparent change in temperature dependence (slope change) occurring above 500 K.

REFERENCES

1. N. N. Sirota and L. I. Berger, "Heat Conductivity of Indium and Gallium Arsenides, Indium Selenide, and Indium Telluride," Inzh. Fiz. Zh., Akad. Nauk Belorus. SSR, (11), 117-20, 1958.
2. A. B. Timberlake, T. S. Shilliday, and P. W. Davis, "Thermal Diffusivity Measurements," in Thermoelectric Power Generation and Related Phenomena, U.S. Navy Bureau of Ships Final Report, 19-33, 1961. [AD-265 147]
3. A. D. Stuckes, "Measurement of Thermal Conductivity of Semiconductors At High Temperatures," Brit. J. Appl. Phys., 12(12), 675-9, 1961.
4. J. Blanc, R. H. Bube, and L. R. Weisberg, "Evidence For The Existence of High Concentrations of Lattice Defects In GaAs," Phys. Rev. Letts., 9(6), 252-4, 1962.
5. M. G. Holland, "Phonon Scattering in Semiconductors From Thermal Conductivity Studies," Phys. Rev., 134(2A), A471-A480, 1964.
6. M. I. Aliyev and G. T. Akhmedli, "Investigating Thermal Conduction of Alloyed Gallium Arsenide," in News of the Academy of Sciences of the USSR Physics Series, U.S. Air Force Foreign Technology Division Translation FTD-TT-65-218, 13-17, 1965. [AD-621 056]
7. A. Amith, I. Kudman, E. F. Steigmeier, "Electron And Phonon Scattering In GaAs At High Temperatures," Phys. Rev., 138(4A), A1270-A1276, 1965.
8. H. Wagini, "Die Warmeleitfahigkeit von GaAs Oberhalb Zimmertemperatur," Z. Naturforsch., 20A(3), 494, 1965.
9. F. L. Vook, "Change In Thermal Conductivity Upon Low-Temperature Electron Irradiation: GaAs," Phys. Rev., 135(6A), A1742-A1749, 1964.
10. R. O. Carlson, G. A. Slack, and S. J. Silverman, "Thermal Conductivity of GaAs and GaAs$_{1-x}$P$_x$ Laser Semiconductors," J. Appl. Phys., 36(2), 505-7, 1965.
11. S. A. Aliev and S. S. Shalyt, "Thermal Conductivity and Thermoelectric Power of Gallium Arsenide At Low Temperatures," Sov. Phys. - Solid State, 7(12), 2986-7, 1966.
12. P. L. Vuillermoz, A. Laugier, and C. Mai, "Thermal Conductivity of Chromium Doped Gallium Arsenide At Low Temperature," J. Appl. Phys., 46(11), 4623-6, 1975.
13. D. R. Hunt, "The Dependence Upon Temperature of Thermal and Electrical Properties Of III-V Semiconducting Compounds," Ph.D. Thesis, University College of Swansea, UK, 122 pp. + Appendices, 1974.
14. K. S. Dubey, "Analysis of Lattice Thermal Conductivity of GaAs at High Temperatures," J. Thermal Analysis, 14, 213-9, 1978.
15. M. G. Holland, "Analysis of Lattice Thermal Conductivity," Phys. Rev., 132(6), 2461-71, 1963.
16. G. Leibfried and E. Schloemann, Nachr. Akad. Wiss. Gottingen Math-Physik Kl., 2A, 71, 1954.
17. E. F. Steigmeier and B. Abeles, "Scattering of Phonons by Electrons in Germanium-Silicon Alloys," Phys. Rev., 136(4A), 1149-55, 1964.

18. D. J. Ecsedy, "Four-Phonon Processes and the Thermal Expansion Effects in the Thermal Resistivity of Crystals at High Temperatures," in Thermal Conductivity 14 (P. G. Klemens and T. K. Chu, Editors), Plenum Press, New York, NY, 195-200, 1976.

19. J. R. Drabble and H. J. Goldsmid, Thermal Conduction in Semiconductors, Pergamon Press, New York, NY, Sec. 4.1-2, 1961.

20. J. S. Blakemore, "Semiconducting and Other Major Properties of Gallium Arsenide," J. Appl. Phys., 53, R123-R181, 1982.

21. C. D. Thurmond, "The Standard Thermodynamic Functions for the Formation of Electrons and Holes in Ge, Si, GaAs, and GaP," J. Electrochem. Soc., 122(8), 1133-41, 1975.

22. K. H. Nichols, C. M. L. Yee, and C. M. Wolfe, "High-Temperature Carrier Transport in n-Type Epitaxial GaAs," Solid-State Electronics, 23, 109-116, 1980.

23. W. G. Spitzer and J. M. Whelan, "Infrared Absorption and Electron Effective Mass in n-Type Gallium Arsenide," Phys. Rev., 114, 59-63, 1959.

24. P. G. Klemens, "Radiative Heat Transfer in Composites," High Temp.-High Pressures, 17, 381-5, 1985.

25. A. S. Jordan, "Determination of the Total Emittance of n-Type GaAs With Application to Czochralski Growth," J. Appl. Phys., 51(4), 2218-27, 1980.

26. B. O. Seraphin and H. E. Bennett, "Optical Constants," in Semiconductors and Semimetals, Volume 3, Optical Properties of III-V Compounds (R. K. Willardson and A. C. Beer, Editors), Academic Press, New York, NY, 512-23, 1967.

SESSION 4

FLUIDS

MEASUREMENT OF THERMAL CONDUCTIVITY AND THERMAL DIFFUSIVITY

OF FLUIDS OVER A WIDE RANGE OF DENSITIES*

Hans M. Roder

Thermophysics Division
National Engineering Laboratory
National Bureau of Standards
Boulder, Colorado 80303

Carlos A. Nieto de Castro

Centro de Quimica Estrutural
Complexo I, IST
1096 Lisboa Codex, Portugal

ABSTRACT

    A transient hot wire apparatus was revised to make simultaneous meas-
urements of thermal conductivity and thermal diffusivity of fluids for a
wide range of densities. In addition, the theory of the method had to be
revised to allow correct evaluation of the thermal diffusivity. We illustrate the new method and theory with data for argon along 6 isotherms
between 172 and 325 K with pressures up to 70 MPa and densities up to 30
mol/L. For thermal conductivity, the precision is about 0.3% and the
accuracy better than 1.0%. For thermal diffusivity the accuracy is estimated to be 5%. From the two variables measured and from the density
taken from the equation of state we obtain values of the specific heat,
Cp, of the fluid. The actual status of this application is described and
present limitations are analyzed.

INTRODUCTION

    The transient hot wire method has been used extensively during the
past ten years to measure thermal conductivity. The theory of the method
has been developed to a point where it is now considered a standard, absolute method. This technique is recognized today as the method of highest
accuracy for the measurement of the thermal conductivity of fluids for
conditions removed from the critical region. For example, data obtained
near the saturation lines for toluene, water and n-heptane with instruments of this type have been proposed recently as standard reference data
for the thermal conductivity of liquids near ambient temperatures [1].
At cryogenic temperatures, measurements on liquid methane [2,3] and on
liquid argon [4,5], made at two different laboratories with different

instruments, agree to within 0.5%; the argon values have also been recommended as standard reference data [5]. In addition, it has been known for some time that, with appropriate apparatus design, it should be possible to obtain the thermal diffusivity as well as the thermal conductivity [6,7,8]. Early attempts to measure the thermal diffusivity have been restricted to liquids, such as toluene and n-heptane, where the corrections that have to be applied are small.

The theory of the transient hot-wire technique was developed by Healy et al. [9]. These authors considered the corrections that affect the determination of the thermal conductivity. Thermal diffusivity, however, depends on the absolute value of the temperature rise in the wire and it became necessary to review all the corrections to the ideal model. A complete analysis of the application of the transient hot-wire technique to the simultaneous measurement of the thermal conductivity and thermal diffusivity has been performed [10]. This analysis shows that several time independent terms in the corrections have to be included to measure the thermal diffusivity accurately, making it possible to obtain values of the thermal diffusivity with accuracies of 5%.

We report here thermal conductivity and thermal diffusivity measurements on argon that cover a wide range of thermodynamic states including the dilute and dense gas, and conditions near the vapor-liquid critical point. The measured values of the thermal conductivity and of the thermal diffusivity were used, along with density data from an equation of state [11], to obtain heat capacity values, C_p. Heat capacity in a practical sense, has more practical value than thermal diffusivity. We use argon as the sample fluid, because then a direct check on the accuracy of the C_p measurements is possible, by extrapolating the C_p's measured at low density to zero density and comparing the result to the theoretical value, $C_p^0 = 5/2$ R. The apparatus used for measuring both thermal conductivity and thermal diffusivity of fluids was described earlier [12], but it incorporates a number of changes to improve the accuracy in both properties.

WORKING EQUATIONS

The transient hot-wire method has been defined as an absolute primary instrument for the measurement of thermal conductivity [13]. The working equation is based on a specific solution of Fourier's law, which can be found in standard texts [14] for an infinite line source of heat per unit length q, initiated at time t = 0 and immersed in a fluid of constant properties and infinite extent. The temperature increase of the wire of radius a, $\Delta T_{id}(a,t)$, is governed by the equation

$$\Delta T_{id} (a,t) = T (a,t) - T_O = \frac{q}{4\pi\lambda} E_1 \left(\frac{a^2}{4\kappa t}\right) \tag{1}$$

where T_O is the initial, uniform temperature of the fluid and the wire, λ the thermal conductivity of the fluid, κ the thermal diffusivity,

$$\kappa = \frac{\lambda}{\rho C_p} \tag{2}$$

and $E_1(x)$ is the exponential integral [14].
In eq. (2) ρ is the density and C_p the heat capacity of the fluid. For small values of $a^2/4\kappa t$, the exponential integral in (1) can be expanded to yield

$$\Delta T_{id} (a,t) = \frac{q}{4\pi\lambda} \ln \frac{4\kappa t}{a^2 C} + 0 \left(\frac{a^2}{4\kappa t}\right) \cong I + S \ln t \tag{3}$$

where C=1.781..., and 0(x) is the function order of x.

174

If the wire radius is chosen such that the o(x) term in the right hand side of eq. (3) is less than 0.01% of ΔT_{id}, the thermal conductivity of the fluid can be obtained from the slope S of the regression line of ΔT_{id} vs ln t,

$$\lambda = \frac{q}{4\pi S} \qquad (4)$$

The thermal diffusivity is obtained at a particular value of ΔT_{id} at a specific time t', by the relation

$$\kappa = \frac{a^2 C}{4t'} \exp\left[\frac{\Delta T_{id}(t')}{S}\right] \qquad (5)$$

If t' is chosen to be one second ΔT_{id} (1 s) is the intercept I from the least squares analysis of the linear fit in eq. (3).

In any practical instrument the source of heating is provided by a finite length of metallic wire which also serves as a resistance thermometer. As a consequence, the measured temperature rise of the wire, ΔT, departs from the ideal value of eq. (3), even at the surface of the wire. Most of the possible causes of departure of the measured temperature rise from the ideal value have been investigated by Healy et al. [9]. Their results were reanalysed by Nieto de Castro et al. [10] to account for all effects that influence both the value of S, and therefore the measured thermal conductivity, and the value of I, and consequently the measured thermal diffusivity. The new results are expressed as small corrections, δT_i, in the way proposed by Healy et al. [9], to be applied to the measured temperature rise, ΔT_w, so that

$$\Delta T_{id} = \Delta T_w + \Sigma_i \, \delta T_i \qquad (6)$$

The values of the δT_i's have been fully described in ref. [10]. The temperature T_r to which the thermal conductivity must be referred is given by

$$T_r = T_0 + \delta T_1^* \qquad (7)$$

where δT_1^* has been given elsewhere [9,15]. Nieto de Castro et al. [10] have shown that the thermal diffusivity evaluated from eq. (5) must be refered to the equilibrium temperature, T_0. In summary

$$\lambda = \lambda (T_r, \rho_r)$$

$$\rho_r = \rho (T_r, P_0)$$

$$\kappa = \lambda (T_0, \rho_0)/\rho_0 (C_p)_0 \qquad (8)$$

$$\rho_0 = \rho (T_0, P_0)$$

$$(C_p)_0 = C_p(T_0, P_0)$$

where P_0 is the equilibrium pressure, at time t = 0. A detailed analysis of the corrections [10] has shown that the assigned value of time, t', in equation (5), must be of the order of one second to minimize the errors in the thermal diffusivity evaluation.

APPARATUS

The apparatus used for the measurements is the general purpose system discussed earlier [12]. It includes the following elements: the hot wires, the high pressure cell with the wire supports, the Wheatstone bridge, the cryostat, the measuring and control circuitry, the sample handling system, and the microcomputer. The measurements are made with a 12.7 μm diameter platinum wire at times of up to one second. The hot wire and a short compensating hot wire are arranged in different arms of a Wheatstone bridge. The cell containing the core of the apparatus is designed to accommodate pressures from nearly zero to 70 MPa and temperatures from 78 to 330 K. The data aquisition system is controlled by a microcomputer and includes several programmable digital voltmeters. We measure the temperature rise in the hot wire at 250 fixed times 4 ms apart with a modified Wheatstone bridge, and use linear regression to arrive at the two coefficients of the straight line, I and S.

To obtain accurate results for the thermal diffusivity and ultimately for the heat capacity, some changes had to be made in the apparatus. These changes improve the measurement of resistance for both the bridge and the wires, improve the nulling of the bridge prior to applying the power, improve the timing of the experiment, provide some redundancy in the measurement capability, and, finally, reduce the noise level in the voltage measurements across the bridge. The Wheatstone bridge circuit was changed to improve the accuracy with which the arms of the bridge, the hot wire resistances, and the balance of the bridge can be measured. Each arm of the bridge is now about 110 Ω at ambient temperature and includes a series of new precision decade resistances. Each side of the bridge includes a calibrated standard resistor. We added a digital voltmeter to the system, capable of measuring voltages to 0.5 μV at the 50 mV level. As a result, we can now measure each resistance with an accuracy of 4 mΩ, nearly a factor 10 better than in the earlier instrument, and the bridge balance is accurate to 0.5 μV. With the new voltmeter, the voltage applied to the bridge can be measured directly, eliminating a standard resistor that was in series with the bridge and several corrections that had to be applied to the bridge power. During a powered run an instrumentation amplifier with a gain of 100 and a filter is used with the digital voltmeter, resulting in a final noise level of 3 μV. The corresponding precision in the temperature rise measurement has improved to 0.2% as inferred from the linear regression statistics. Since the use of a filter results in a certain time lag, the A/D acquisition capability of the computer is still used but without amplification or filter. The less precise set of voltages then forms the basis for correcting the time lag of the more precise set. Improvements in the timing are based on using the clock of the computer to provide simultaneous triggers for both computer A/D and digital voltmeter. Further, the power switch is activated by the very first and the very last trigger coming from the clockboard. Finally, the delay time for the closing of the power switch was measured very accurately with a digital oscilloscope after the entire system had been assembled.

RESULTS

We have tested the new arrangement by measuring six isotherms in the single phase, supercritical region of argon. Approximately 800 points were measured. The measurements are distributed along the six isotherms where the nominal isotherm temperatures are 170, 200, 220, 275, 300 and 325 K, the first three isotherms having been measured with the new R-C filter and amplifier. For these temperatures the densities range from the dilute gas to 2.2 times critical density while the heat capacity varies by a factor of 7 from the ideal gas value. There are about 100 points per isotherm taken at about 30 different pressure levels. At each pressure replicate measurements made at the same cell temperature, but with different applied power, verify the absence of convection. The transient hot wire technique is unique because the onset of convection can be observed experimentally with this method [12]. The pressure, temperature and applied power are measured directly, while the density is calulated from an equation of state [11] using the measured pressure and temperature. Since the temperature of each measurement varies with the the applied power, each experimental point must be adjusted sightly to obtain values on isotherms. These corrections were accomplished using a recent thermal conductivity correlation for argon [16].

We describe here only three of the tests performed to assess the adequacy of the changes introduced in the apparatus. First, the measurements of the thermal conductivity at 300 K were compared with data

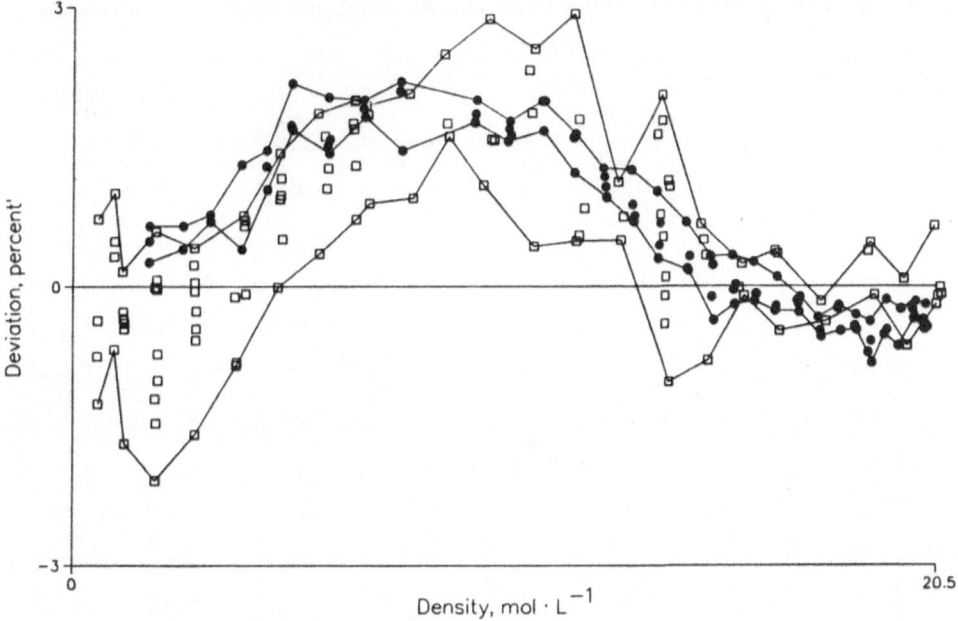

Fig. 1. Comparison of the scatter in the data obtained for T = 300.65 K previous version (1981) - □, present apparatus - ●.

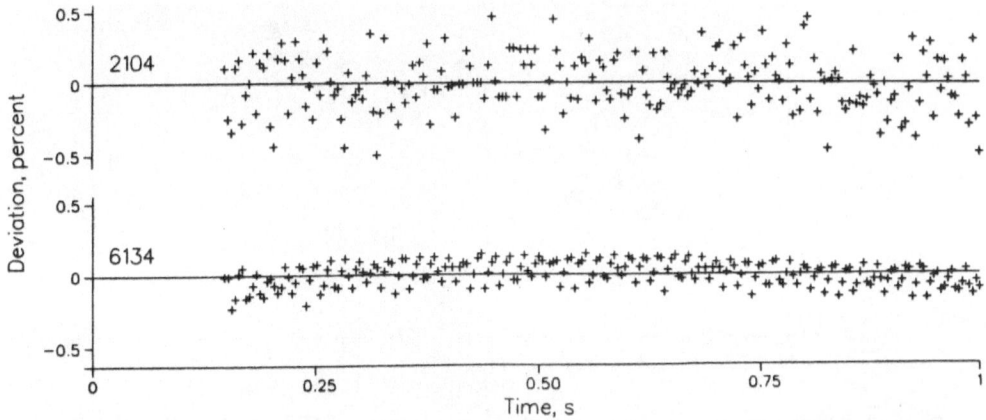

Fig. 2 Deviation from the straight-line of eq. (3) for two runs: 2104 - without the filter and amplifier; 6134 - with both.

obtained with the earlier version of the instrument [17], at 300.65 K.
Figure 1 displays the percentage difference between experimental values
and the background thermal conductivity for both sets of data, as a
function of density. The scatter in the results has decreased from a
band of 2% to a band of 0.6%, in the present version. It is also clear
that the existence of a critical enhancement previously reported at a
reduced temperature of 1.99 [17,18] is confirmed by the present runs.
Secondly, we display in figure 2 a typical deviation plot for two runs at
301.10 K and 222.55 K, at about 6 mol/L, with the same temperature rise,
before and after adding the R/C filter and the operational amplifier to
the bridge. The precision of the temperature rise measurement improved
from 0.4% to 0.2%. Finally, the data obtained for the isotherms at 170
K, 220 K and 300 K at low densities, are compared with data obtained in a
different laboratory [19], with reported accuracy of 0.5%, in figure 3.

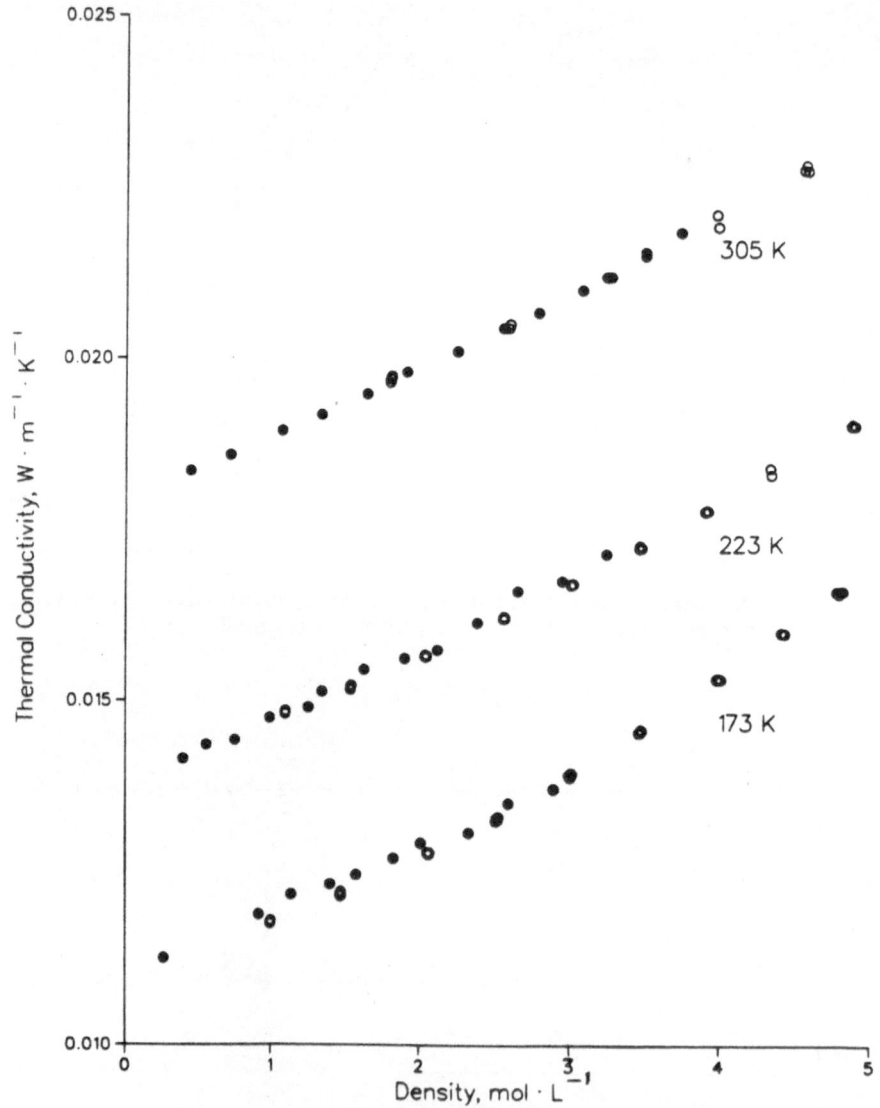

Fig. 3. Comparison of present data with ref. [19] at 173 K, 223 K and
305 K. ● - ref. [19], o - present work.

The maximum deviation between the sets of data occurs in the 173 K isotherm, at low densities and its is of the order of 1.3%. The average deviation for all three isotherms is of the order of 0.5%. This agreement seems to indicate that our present data on thermal conductivity are accurate to better than 1%.

The entire set of thermal conductivities is shown vs. density in figure 4. Shown as lines in figure 4 are the isotherms calculated from the correlation developed under the auspicies of IUPAC [16]. The other variables obtained from the data reduction program are thermal diffusivity and specific heat.

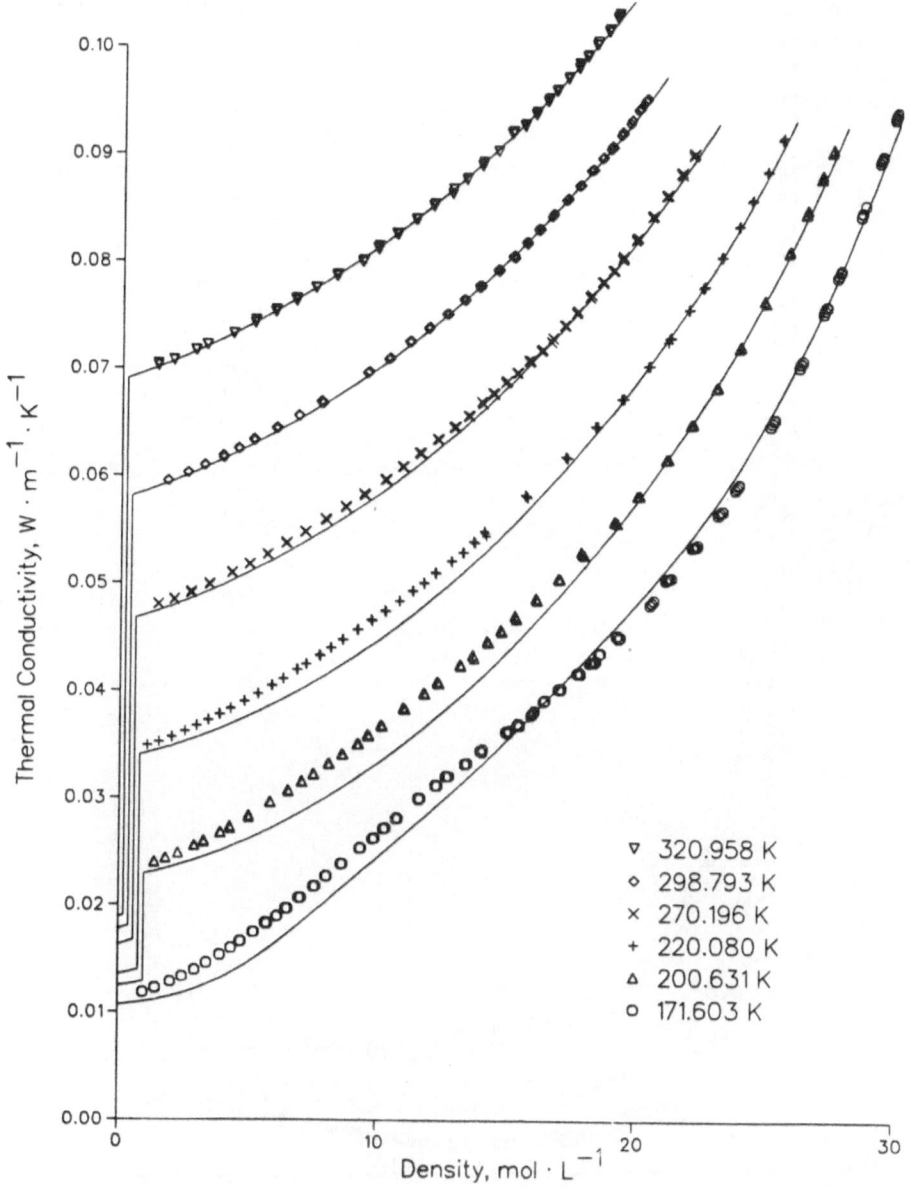

Fig. 4. Thermal conductivity of argon as a function of density, for the different isotherms. Each isotherm is displaced by 0.01 Wm⁻¹K⁻¹ from the adjacent isotherm.

The entire set of thermal diffusivities is shown vs. density in figure 5.
The lines in figure 5 are calculated from the correlated values of thermal conductivity [16] taking density and specific heat from the equation
of state [11]. In the case of the heat capacities, the values obtained
with different power levels were averaged for a given pressure level.
These points are shown as a function of density in figure 6. The bars
connect maximum and minimum values of C at each pressure level. The
lines in figure 6 are specific heats taken directly from the equation of
state.

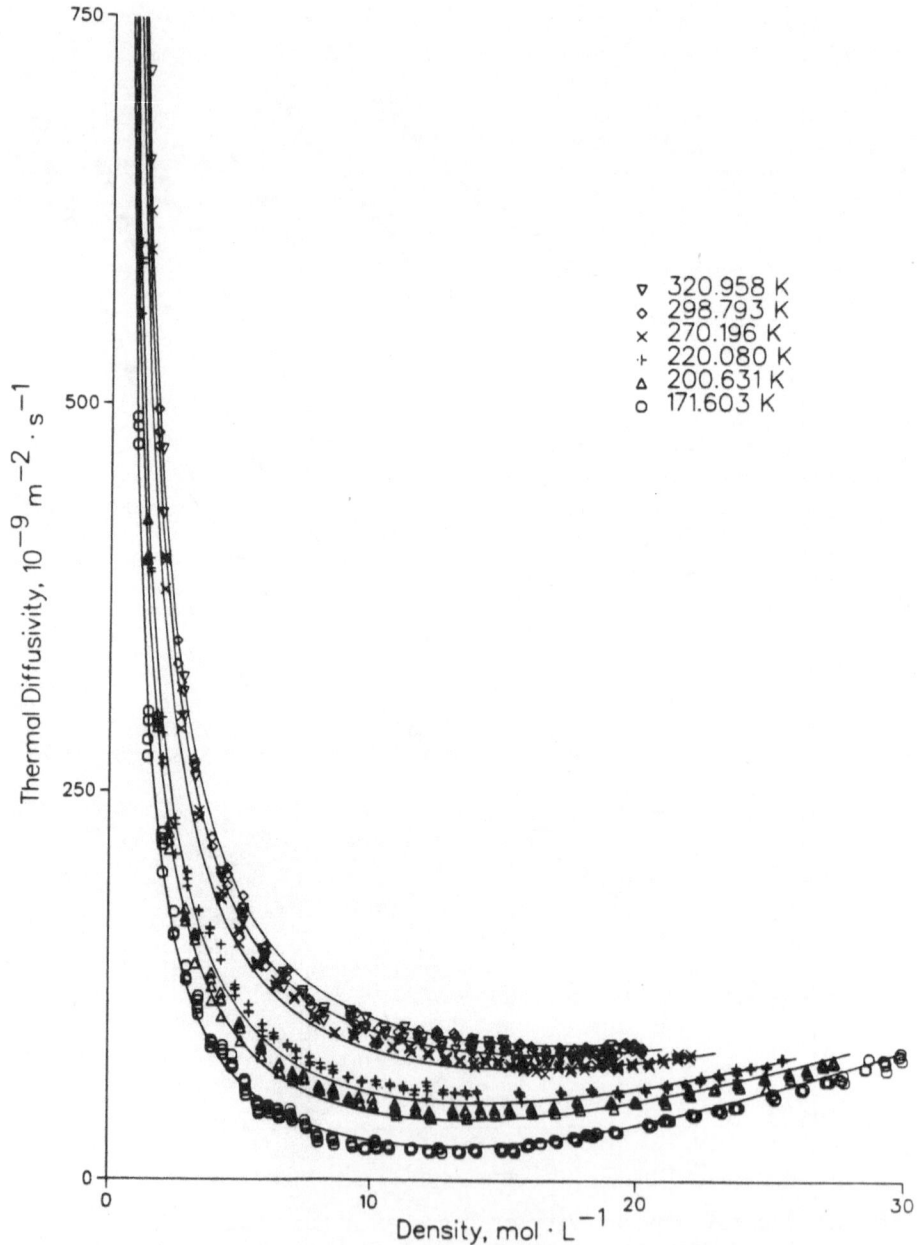

Fig. 5. Thermal diffusivity of argon as a function of density.

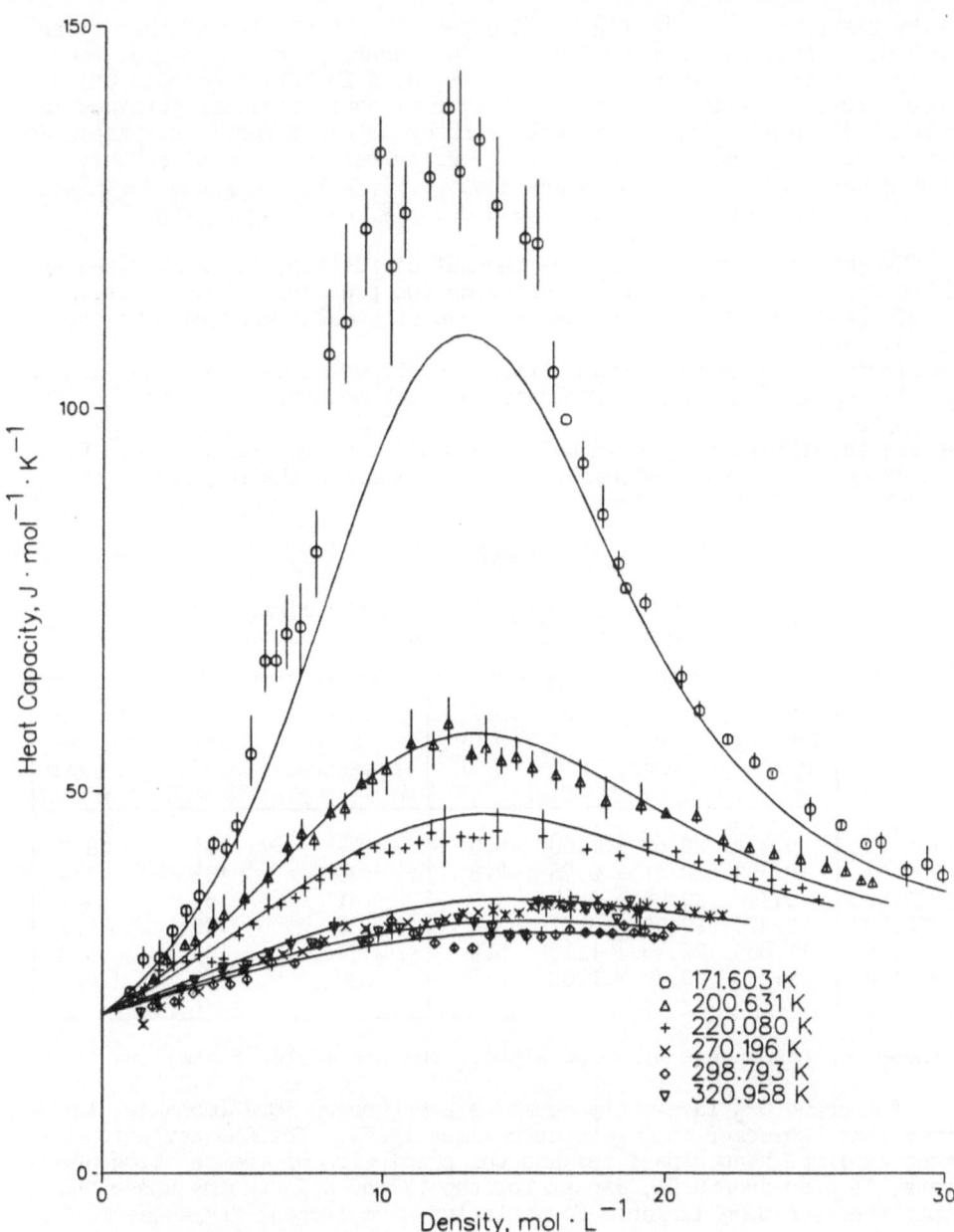

Fig. 6. Heat capacity of argon as a function of density. The data
taken with different power levels were corrected to the
nominal temperatures displayed and averaged for a given
pressure level. The bars connect maximum and minimum values
of C_p for each pressure level.

DISCUSSION

We report here new measurements of the thermal conductivity, thermal
diffusivity, and derived heat capacity of supercritical argon, at six temperatures and with pressures up to 70 MPa. In these measurements the precision of thermal conductivity improved to 0.2% as determined by reproduci-

bility tests, as shown by fig. 1, and from the statistics of the linear
regression fits, as shown by fig. 2. The accuracy of the thermal conductivity is estimated to be better than 1%, at a 2σ level, because the
present results agree very well with thermal conductivities obtained in
Lisbon [19], confirming the results already obtained for liquid argon and
liquid methane [3,5], and fully discussed in ref. 5. The differences
between these two sets of data are about 0.5%, with a maximum deviation
of 1.3% at the lowest temperature and density, as shown by fig. 3.

We estimate the error in the thermal diffusivity and heat capacity
in two different ways. First, following the procedure already outlined
in ref. [20], we extrapolate the heat capacity data, corrected to the
same averaged
equilibrium temperature, with a linear function and obtained the value of
$C_p^0 = \lim\limits_{\rho \to 0} C_p(\rho)$. Second, we evaluate the root mean square deviation (RMS)
between the adjusted experimental values and the equation of state [11]
prediction, along each isotherm. Table 1 displays the results of this
analysis.

TABLE 1

Statistics for the zero density extrapolation of
heat capacity and for the entire isotherms

T/K	Max density mol/L	$C_p^0 \pm 2\sigma$ J/mol^{-1}K^{-1}	% error in C_p^0	number of pressures	number of points	RMS for entire isotherm %
171.60	5.331	20.63 ± 1.00	5.0	47	181	18.7
200.63	12.304	20.21 ± 0.96	4.8	39	153	3.9
220.08	9.746	20.58 ± 0.65	3.2	41	152	4.1
270.20*	13.375	20.44 ± 0.71	3.5	36	120	4.7
298.79*	11.865	20.48 ± 1.10	5.4	30	103	4.3
320.96*	9.845	20.70 ± 1.00	4.8	30	82	6.7

* these isotherms were measured without the R-C filter installed.

For argon C_p^0 is exactly equal to 2.5 R or 20.78 J/(mol·K). Table 1
shows that the error in C_p^0 is approximately 5%. The RMS deviation between measured heat capacities and the prediction of the equation of
state, is also around 5%, except for the 171.60 K isotherm, where the
departures are much larger. For this isotherm large differences in C_p
between measured values and the equation of state are found in the middle
of the density range. The failure of a modified BWR equation of state to
represent heat capacity data in the region close to the critical point
has been noted previously, particularly for ethane [21]. From these
tests we infer that errors should be as follows: precision and accuracy
of thermal conductivity 0.3% and 1.0% respec tively; accuracy of thermal
diffusivity and specific heat around 5%.

ACKNOWLEDGEMENTS

C.A. Nieto de Castro wants to thank the Faculty of Sciences of the
University of Lisbon for a leave of absence and the Thermophysics
Division of NBS, USA, for their invitation as a guest scientist during
1986/87.

REFERENCES

1. C.A. Nieto de Castro, A. Nagashima, R.D. Trengove and W.A. Wakeham, Standard reference data for the thermal conductivity of liquids, J. Phys. Chem. Ref. Data, 15, 1073 (1986).
2. H.M. Roder, Thermal conductivity of methane for temperatures between 110 and 310 K with pressures to 70 MPa, Int. J. Thermophys., 6, 119 (1985).
3. U.V. Mardolcar and C.A. Nieto de Castro, The thermal conductivity of liquid methane, Ber. Bunsenges. Phys. Chem., 91, 152 (1987).
4. J.C. Calado, U.V. Mardolcar, C.A. Nieto de Castro, H.M. Roder and W.A. Wakeham, The thermal conductivity of liquid argon, Physica, 143A, 314 (1987).
5. H.M. Roder, C.A. Nieto de Castro and U.V. Mardolcar, The thermal conductivity of liquid argon for temperatures between 110 and 140 K with pressures to 70 MPa, Int. J. Thermophys., 8, 521 (1987).
6. A. Ciochina and R. Vilcu, Transient hot wire method for absolute and simultaneous measurement of thermal conductivity and thermal diffusivity of fluids, Revue Roumaine de Chimie, 28, 795 (1983).
7. Y. Nagasaka and A. Nagashima, Simultaneous measurement of the thermal conductivity and the thermal diffusivity of liquids by the transient hot-wire method, Rev. Sci. Instrum., 52, 229 (1981).
8. P.G. Knibbe and R.D. Raal, Simultaneous measurement of the thermal conductivity and thermal diffusivity of liquids, Int. J. Thermophys., 8, 181 (1987).
9. J.J. Healy, J.J. de Groot and J. Kestin, The theory of the transient hot-wire method for measuring thermal conductivity, Physica, 82C, 392 (1976).
10. C.A. Nieto de Castro, B. Taxis, H.M. Roder and W.A. Wakeham, Thermal diffusivity measurements by the transient hot-wire technique: a reappraisal, submitted to Int. J. Thermophys., 1987.
11. R.B. Stewart, R.T. Jacobsen and J.H. Becker, Center for Applied Thermo. Studies Report No. 81-3, University of Idaho (1981).
12. H.M. Roder, A transient hot wire thermal conductivity apparatus for fluids, J. Res. Natl. Bur. Stand. (U.S.), 86, 457 (1981).
13. C.A. Nieto de Castro, The measurement of transport properties of fluids – a critical appraisal, invited paper, Proc. ASME/JSME Thermal Engineering Conf., Honolulu, Hawaii, March 22-26, (1987), page 327.
14. H.S. Carslaw and J.C. Jaeger, Heat conduction in solids, 2nd Ed., Oxford University Press, (1959), page 261.
15. J.C.G. Calado, J.M.N.A. Fareleira, C.A. Nieto de Castro and W.A. Wakeham, Reference state in thermal conductivity measurements, Rev. Port. Quim., 26, 173 (1984).
16. B.A. Younglove and H.J.M. Hanley, The viscosity and thermal conductivity coefficients of gaseous and liquid argon, J. Phys. Chem. Ref. Data, 15, 1323 (1986).
17. C.A. Nieto de Castro and H.M. Roder, Absolute determination of the thermal conductivity of argon at room temperature and pressures up to 68 MPa, J. Res. Natl. Bur. Stand. (U.S.), 86, 293 (1981).
18. C.A. Nieto de Castro and H.M. Roder, Thermal conductivity of argon at 300.65 K. Evidence for a critical enhancement?, Proceedings of the 8th Symposium on Thermophysical Properties, J.V. Sengers, ed., ASME, New York, Vol. I, 1982, page 241.
19. U.V. Mardolcar, C.A. Nieto de Castro and W.A. Wakeham, Thermal conductivity of argon in the temperature range 107 to 423 K, Int. J. Thermophys., 7, 259 (1986).

20. H.M. Roder and C.A. Nieto de Castro, Heat capacity, Cp, of fluids from transient hot wire measurements, Cryogenics, 27, 312, (1987).

21. K. Bier, J. Kunze and G. Maurer, Thermodynamic properties of ethane from calorimetric measurements, J. Chem. Thermodynamics, 8, 857 (1976).

THERMAL CONDUCTIVITY OF LIQUID WATER AT HIGH PRESSURES

M. Dix, W.A. Wakeham and M. Zalaf

Department of Chemical Engineering
Imperial College
London SW7 2BY
U.K.

ABSTRACT

The paper describes a new instrument for the measurement of the thermal conductivity of electrically-conducting liquids for use at pressures up to 700 MPa. The instrument is based upon the transient hot-wire principle which has been adapted to the special needs of electrically-conducting liquids. The instrument has been designed to secure an accuracy of $\pm$ 0.3% in the thermal conductivity. The results of measurements of the thermal conductivity of water in the temperature range 300-324 K are presented in order to confirm the proper operation of the instrument. A comparison with the results of· earlier measurements reveals good agreement although the present data are more accurate.

INTRODUCTION

There have been rapid developments in the measurement of the thermal conductivity of electrically-insulating, non-polar liquids in the last ten years[1]. However, the transient hot-wire technique that has been at the forefront of these developments has not been so readily applicable to electrically-conducting or polar fluids. Because fluids of this type have a number of applications involving heat transfer[2,3] and because it is important to determine their thermal conductivity with high accuracy, the extension of the technique is evidently necessary. To this end a new version of the transient hot-wire instrument suitable for use with electrically-conducting fluids has been developed[4,5], capable of an accuracy of $\pm$ 0.3% over a wide range of thermodynamic states including pressures up to 700 MPa. In this paper we report the final stages of this development and present the first experimental data for liquid water at pressures up to 100 MPa.

EXPERIMENTAL

Theory of the Technique

In the transient hot-wire technique for the measurement of the thermal conductivity of fluids the time history of the temperature of a thin metallic wire surrounded by the fluid is determined following the initiation of heat flux within it. According to the theory of the method the temperature rise of the wire, ΔT_w, is related to the fluid thermal conductivity, λ, by the equation[4]

$$\Delta T = \Delta T_w(t) + \sum_{i=1}^{n} \delta T_i = \frac{q}{4\pi\lambda(T_r, \rho_r)} \ell n\left[\frac{4\kappa t}{a^2 C}\right] \tag{1}$$

Here, q denotes the heat flux per unit length in the wire, t the time from initiation of the heat flux, κ the thermal diffusivity of the fluid, a the wire radius and C a numerical constant. The terms δT_i represent a series of small, additive corrections to be applied to the measured temperature rise of the wire. The thermal conductivity determined from the slope of the line ΔT υs $\ell n t$ refers to the thermodynamic state (T_r, ρ_r) where the reference temperature T_r is

$$T_r = T_0 + \tfrac{1}{2}(\Delta T(t_1) + \Delta T(t_2)) + \delta T_c^* \tag{2}$$

and ρ_r (T_r, P) is the corresponding fluid density for a measurement performed at a pressure P for an equilibrium temperature T_0. The times t_1 and t_2 correspond to the first and last times of a measurement of the temperature rise[4]. Finally, δT_c^* is a further small correction to the reference temperature[5].

In most practical realizations of the technique it is easily arranged that many of the corrections δT_i are negligible, amounting to no more than $\pm$ 0.01% of the temperature rise. However, *four* corrections must always be considered since they may be as large as $\pm$ 0.1% of the temperature rise. Of these, the corrections arising from the finite heat capacity of the metallic wire and from the presence of an outer boundary to the fluid around the wire are present in all applications[4]. But in the case to be considered here there are additional corrections arising from the fact that the metallic wire is coated with an electrically-insulating layer of radius a_L in order to permit measurements on electrically-conducting fluids. The corrections may be derived from the work of Nagasaka and Nagashima[6]. For the temperature rise of the wire we have

$$
\begin{aligned}
\delta T_c = \frac{q}{4\pi\lambda} &\left[(1-\lambda/\lambda_L)\ell n\ r^2 - \frac{1}{t}\left\{ \frac{a^2}{8}\left[\frac{(\lambda-\lambda_L)}{\lambda_W}\left[\frac{1}{\kappa_W} - \frac{1}{\kappa_L}\right] + 4\left[\frac{1}{\kappa_L} - \frac{1}{\kappa}\right] \right. \right.\right.\\
&+ 4r^2\left[\frac{1}{\kappa} - \frac{1}{\kappa_L}\right] + \frac{8}{\lambda_L}\left[\frac{\lambda_L}{\kappa_L} - \frac{\lambda_W}{\kappa_W}\right] \ell n\ r \\
&+ \frac{4}{\lambda}\left[\left[\frac{\lambda_L}{\kappa_L} - \frac{\lambda}{\kappa}\right](1-r^2)\ell n\left[\frac{4\kappa t}{a^2 C}\right] - \ell n\ r^2\left[\left[\frac{\lambda_L}{\kappa_L} - \frac{\lambda_W}{\kappa_W}\right] + r^2\left[\frac{\lambda}{\kappa} - \frac{\lambda_L}{\kappa_L}\right]\right]\right]\right]\right\}\right]
\end{aligned}
\tag{3}
$$

and for the reference temperature

$$\delta T_c^* = -\frac{q}{8\pi\lambda}\left\{2\ell n\ r^2 - \frac{a^2}{4}\left\{\left[\frac{1}{t_1}+\frac{1}{t_2}\right]\left[\frac{1}{\kappa_L}+\frac{1}{\kappa_W}\right] + r^2\left[\frac{1}{\kappa}-\frac{1}{\kappa_L}\right]\right.\right.$$

$$+\frac{2}{\lambda_L}\left[\frac{\lambda_L}{\kappa_L}-\frac{\lambda_W}{\kappa_W}\right]\ell n\ r - \frac{2}{\lambda}\left[\left[\frac{\lambda_L}{\kappa_L}-\frac{\lambda_W}{\kappa_W}\right]+r^2\left[\frac{\lambda}{\kappa}-\frac{\lambda_L}{\kappa_L}\right]\right]\ell n\ r$$

$$\left.\left.+\frac{2}{\lambda}\left[\left[\frac{\lambda_L}{\kappa_L}-\frac{\lambda_W}{\kappa_W}\right]+r^2\left[\frac{\lambda}{\kappa}-\frac{\lambda_L}{\kappa_L}\right]\right]\left[\frac{1}{t_1}\ell n\ \frac{4\kappa t_1}{a^2 C}+\frac{1}{t_2}\ell n\ \frac{4\kappa t_2}{a^2 C}\right]\right\}\right\} \tag{4}$$

In these expressions the subscript W refers to properties of the metallic wire while the subscript L refers to properties of the insulating coating.

It can be seen from equations (3) and (4) that the corrections arising from the insulating layer increase rapidly as the coating radius increases. Consequently, if these corrections are to be kept small, so that their estimation does not unduly influence the results, the insulating layer must be very thin.

The Thermal Conductivity Cells

In order to satisfy the electrical requirements that the metallic wires be electrically-insulated from the surrounding fluid by a layer for which $|a_L - a| \ll a$, with a $\sim$ 10 μm, we have chosen to make the hot-wires of tantalum with a $\simeq$ 12.7 μm. This choice permits the deposition of an anodic oxide film (Ta_2O_5) on the surface of the wire with a thickness of order 7 nm, by a technique described elsewhere[7]. Under these circumstances the correction δT_c is never more than 0.05% of ΔT within the measurement period while δT_c^* amounts to no more than 0.001 K. Because both corrections may be estimated with an uncertainty of no more than $\pm$ 10%, neither contributes significantly to the error in the thermal conductivity.

Figure 1 shows the design of the thermal conductivity cells for application at pressures up to 700 MPa. The cells are machined as two cylindrical cavities within a single cylinder of Inconel 625 alloy chosen to mitigate corrosion. The inconel cylinder is divided into two halves along a diametral plane so that one half carries the tantalum wires 1 whereas the other forms a cover. Within each cell the tantalum wire (25 μm nominal diameter) is supported along the axis. One wire is 15 cm long while the other is 6 cm long so that the use of two wires enables the effects of the ends of the wires to be eliminated[4]. At the upper end each wire is spot-welded directly to a tantalum hook 2 set in a ceramic insulator 3, while at the bottom the connection to a similar hook 4 is made through a weak tantalum loop 5 for electrical continuity. At its bottom end the tantalum wire carries a 100 mg tantalum weight 6 which serves to maintain the wire taut and vertical during transient heating. This arrangement allows the anodic oxide film to be deposited in situ on the entire wire assembly.

For operation at high pressures the cell is mounted on the top cap of a 700 MPa pressure vessel in the manner described elsewhere[8]. The pressure vessel is thermostatted in an oil bath with a stability better than $\pm$ 0.01 K over several hours. The pressurisation is performed by an hydraulic pump, and the hydraulic oil and the test fluid are separated by a flexible diaphragm as described earlier[5].

The Measurement System

The initiation of the heat flux within the tantalum wires and the simultaneous measurement of wire temperature rise are performed with a computer-controlled automatic bridge described earlier[5]. The bridge permits the measurement of the time, after initiation of the transient heating, at which the resistance difference of the long and short tantalum wires attains a number (~ 100) of certain preset values. From these measurements the temperature rise of a notional, finite segment of

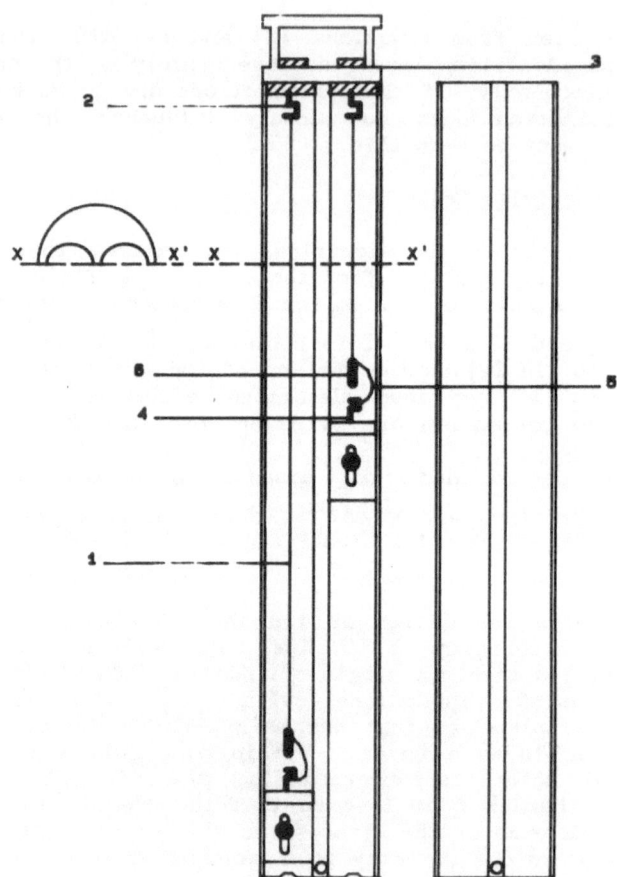

Fig. 1. The thermal conductivity cells

an infinite tantalum wire may be deduced from an appropriate calibration of the resistance temperature characteristic of the tantalum wire[5]. The bridge employed for the present measurements has been designed to yield a precision of $\pm$ 0.05% in a temperature rise of the wire of approximately 4 K. The entire measurement is completed in a time of no more than 1 second so that there is no observable or significant effect of convective heat transfer on the results.

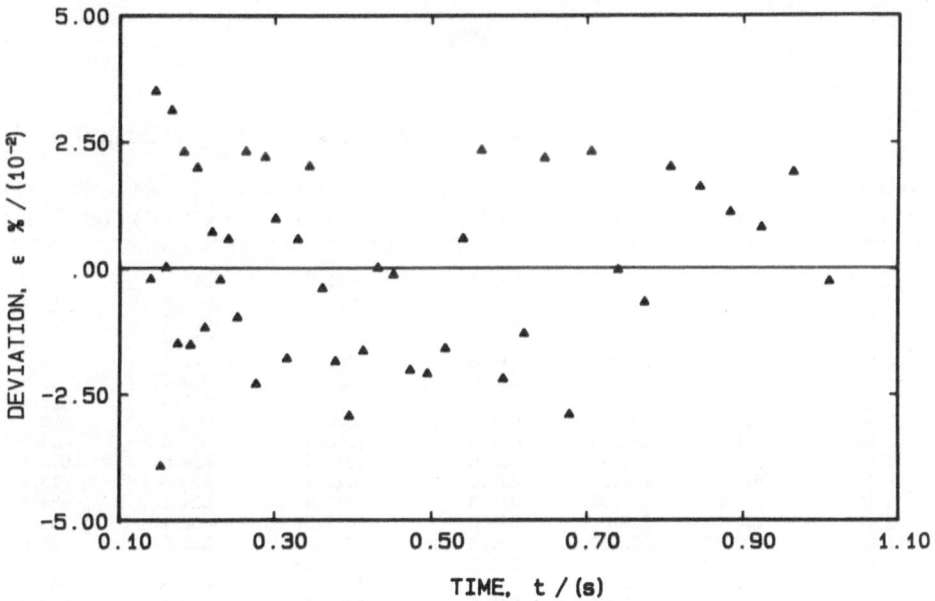

Fig. 2. Deviations of the corrected, experimental temperature rise as a function of time from a linear fit in ℓn t. A measurement in water at T = 302.65 K and P = 102.7 MPa.

RESULTS

One important characteristic of the transient hot-wire technique is that it is straightforward to establish that the instrument operates in accord with the mathematical model of it. This is because only if this condition is met is the corrected temperature rise of the wire a linear function of ℓn t as predicted by equation (1). Effects such as convective[4] or radiative[9] heat transfer cause departures from linearity if they are significant. In order to demonstrate that the present instrument operates satisfactorily, we have carried out a series of measurements in liquid water at pressures up to 100 MPa along two isotherms at 302.65 K and 324.15 K. The water employed for the measurements was triply distilled, deionized and degassed immediately before use. The thermodynamic properties of water, required for the application of corrections in the data reduction have been taken from the work of Haar et al.[10].

As an example of the performance of the instrument, Figure 2 contains the results obtained for a measurement performed in water at a temperature of 302.65 K and a pressure of 102.7 MPa. The figure shows the deviations of the measured values of ΔT (ℓn t) from a linear representation of the data. It can be seen that no point deviates by more than ± 0.04% from the linear fit, while the standard deviation is one of ± 0.02% which is consistent with the estimated precision of the measurements. Furthermore, there is no evidence of any systematic curvature in the data. Plots of this kind are taken as evidence that the instrument operates in accordance with the theoretical model of it.

TABLE 1. The thermal conductivity of water at T_{nom} = 302.65 K

Pressure, P (MPa)	T_o (K)	T_r (K)	$\lambda (T_r, \rho_r)$ (mW m^{-1} K^{-1})	$\rho_r(P, T_r)$ (kg m^{-3})	$\lambda(T_{nom}, \rho_r)$ (mW m^{-1} K^{-1})	$\lambda(T_{nom}, P)$ (mW m^{-1} K^{-1})
0.1013	297.57	302.71	611.7	995.7	611.7	611.7
0.1013	297.52	302.68	613.5	995.7	613.4	613.5
3.74	297.52	302.67	614.7	997.4	614.7	614.7
9.72	297.60	302.73	616.1	1000.0	615.9	617.0
16.01	297.57	302.69	618.0	1002.7	617.9	617.9
20.75	297.62	302.73	620.9	1004.7	620.7	620.8
25.49	297.62	302.76	623.6	1006.6	623.4	623.4
25.49	297.62	302.80	623.7	1006.6	623.5	623.5
28.99	297.62	302.83	624.6	1008.1	624.2	624.3
41.96	297.57	302.67	628.5	1013.4	628.5	628.5
41.96	297.57	302.67	630.6	1013.4	630.6	630.6
52.15	297.57	302.59	634.0	1017.5	634.1	634.1
52.15	297.57	302.61	635.6	1017.5	635.7	635.7
62.63	297.62	302.69	638.9	1021.6	638.8	638.8
62.63	297.62	302.70	638.2	1021.6	638.1	638.1
70.44	297.60	302.67	644.9	1024.6	644.9	644.9
82.14	297.57	302.54	648.5	1029.1	648.7	648.7
92.82	297.60	302.56	651.8	1033.1	652.0	652.0
92.82	297.60	302.58	652.6	1033.1	652.7	652.7
102.66	297.62	302.57	655.4	1036.7	655.6	655.5
0.1013	297.60	302.79	613.5	995.7	613.3	613.3

Table 2. The thermal conductivity of water at T_{nom} = 324.15 K

Pressure, P (MPa)	T_o (K)	T_r (K)	$\lambda (T_r, \rho_r)$ (mW m^{-1} K^{-1})	$\rho_r(P, T_r)$ (kg m^{-3})	$\lambda(T_{nom}, \rho_r)$ (mW m^{-1} K^{-1})	$\lambda(T_{nom}, P)$ (mW m^{-1} K^{-1})
0.1052	319.58	324.29	645.4	987.5	645.2	645.2
6.22	319.55	324.30	648.5	990.1	648.3	648.3
6.22	319.55	324.32	649.6	990.1	649.4	649.4
9.00	319.55	324.29	649.6	991.3	649.4	649.4
19.82	319.50	324.20	655.6	995.9	655.5	655.5
19.82	319.50	324.16	655.3	995.9	655.3	655.3
19.82	319.50	324.23	654.0	995.8	654.0	654.0
33.52	319.53	324.15	658.3	1001.4	658.3	658.3
33.52	319.53	324.15	659.5	1001.4	659.5	659.5
41.55	319.55	324.22	663.0	1004.6	662.9	662.9
50.91	319.55	324.11	668.2	1008.3	668.2	668.2
61.29	319.55	324.14	672.2	1012.3	672.2	672.2
61.29	319.55	324.14	673.1	1012.3	673.1	673.1
72.70	319.55	324.20	676.9	1016.5	676.8	676.8
72.70	319.55	324.15	677.3	1016.5	677.3	677.3
82.76	319.50	324.07	681.6	1020.3	681.7	681.6
92.82	319.55	324.29	687.8	1023.8	687.8	687.8
92.82	319.55	324.18	686.4	1023.9	686.4	686.4
92.82	319.55	324.23	686.5	1023.9	686.4	686.5
102.56	319.53	324.18	689.5	1027.4	689.5	689.5
102.56	319.53	324.08	688.8	1027.4	688.9	688.8
38.77	319.50	324.12	663.6	1003.6	663.6	663.6
0.1052	319.53	324.32	644.0	987.5	644.0	644.0

Tables 1 and 2 list the present experimental data for the thermal conductivity of water along two isotherms. As well as the thermal conductivity at the reference density we list the thermal conductivity at the experimental pressure. In both cases the thermal conductivity has been converted to a nominal temperature by the application of a small, linear correction based upon the IAPS formulation for the thermal conductivity of water substance. The quoted density is, in each case, based upon the equation of state of Haar *et al.*[10]. The tables include check measurements carried out at the end of each isotherm.

The precision of the measurement of thermal conductivity, based upon the statistical uncertainty in the slope of the line ΔT vs ℓn t is ± 0.1%. However, the accuracy is rather worse because it involves the uncertainty in the temperature coefficient of resistance of tantalum. It is estimated that the overall accuracy of the data is one of ± 0.3%.

DISCUSSION

As a result of a critical evaluation of the experimental data for the thermal conductivity of water, the International Association for the Properties of Steam have issued a recommended correlation[11] for scientific use. Figure 3 compares the present experimental data with the recommendations along the two isotherms of interest as a function of pressure. In this region of states the IAPS correlation has an estimated uncertainty of ± 1.5–2.5% and it can be seen that the present data lie within this band. For the lower isotherm at 302.65 K the present data are systematically approximately 0.4% below the correlation. At a temperature of 324.15 K the data are distributed about the correlation. In neither case does the deviation exceed ± 0.7% while the standard deviation over the entire set of data is 0.3%.

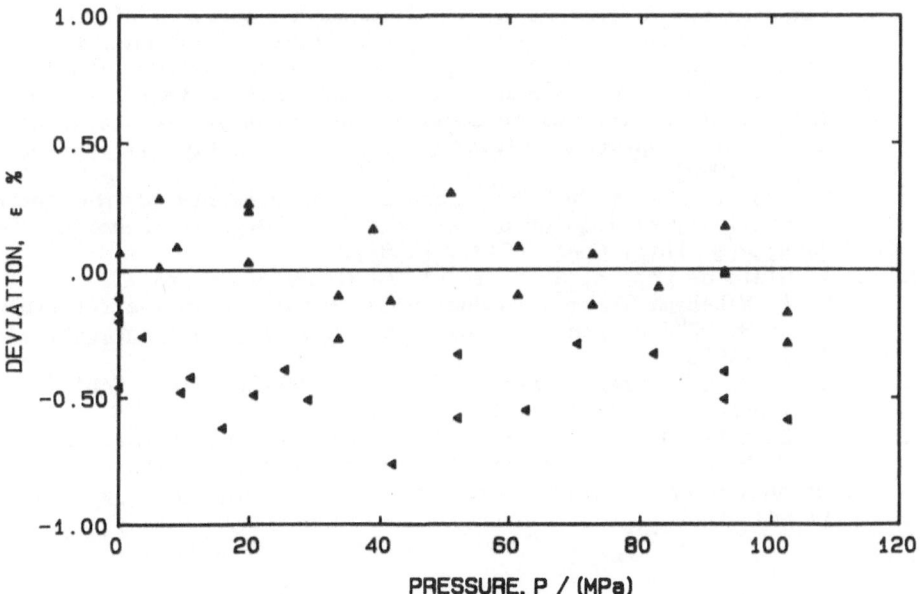

Fig. 3. Deviations of the present experimental data for the thermal conductivity of water from the IAPS correlation[11].
◄: 302.65 K; ▲: 324.15 K.

The IAPS correlation[11] therefore provides a reasonable representation of the present data although the latter are more accurate. Further measurements are in progress to extend the ranges of temperature and pressure of the present measurements, which should eventually allow the improvement of the international correlation.

CONCLUSIONS

The results presented in the paper have demonstrated that the new transient hot-wire instrument operates in accord with the theory of it and yields satisfactory thermal conductivity data for the highly-polar fluid water over a wide range of thermodynamic states. The experimental data produced have an accuracy superior to that of any earlier measurements. The application of the instrument to aqueous, and alcoholic mixtures should provide much-needed accurate experimental data.

REFERENCES

1. C. A. Nieto de Castro, S. F. Y. Li, A. Nagashima, R. D. Trengove and W. A. Wakeham, Standard reference data for the thermal conductivity of liquids, J. Phys. Chem. Ref. Data. 15:1073 (1986).
2. W. B. Gosney, "Principles of Refrigeration", Cambridge University Press, Cambridge (1982).
3. M. Hoshi, T. Omotani and A. Nagashima, Transient method to measure the thermal conductivity of high temperature melt using a liquid metal probe, Rev. Sci. Instrum. 52:755 (1981).
4. J. J. Healy, J. J. de Groot and J. Kestin, The theory of the transient hot-wire method for measuring thermal conductivity, Physica 82C:392 (1976).
5. W. A. Wakeham and M. Zalaf, The thermal conductivity of some electrically-conducting liquids, Fluid Phase Equilib. (in press) (1987).
6. Y. Nagasaka and A. Nagashima, Simultaneous measurement of the thermal conductivity and thermal diffusivity of liquids by the transient hot-wire method, Rev. Sci. Instrum. 52:788 (1981).
7. A. Alloush, W. B. Gosney and W. A. Wakeham, A transient hot-wire instrument for thermal conductivity measurements in electrically-conducting liquids at elevated temperatures, Int. J. Thermophys. 3:225 (1982).
8. J. Menashe and W. A. Wakeham, Absolute measurements of the thermal conductivity of liquids by the transient hot-wire technique, Ber. Bunsenges. Phys. Chem. 85:340 (1981).
9. C. A. Nieto de Castro, S. F. Y. Li, G. C. Maitland and W. A. Wakeham, Thermal conductivity of toluene in the temperature range 35-90°C at pressures up to 600 MPa, Int. J. Thermophys. 4:311 (1983).
10. L. Haar, J. S. Gallagher and G. S. Kell, "NBS/NRC Steam Tables", Hemisphere, Washington D.C. (1984).
11. J. V. Sengers, J. T. R. Watson, R. S. Basu, B. Kamgar-Parsi and R. C. Hendricks, Representative equations for the thermal conductivity of water substance, J. Phys. Chem. Ref. Data 13:893 (1984).

RADIATION CORRECTION AND ONSET OF FREE CONVECTION IN THE TRANSIENT LINE-SOURCE TECHNIQUE (EXPERIMENTAL OBSERVATIONS IN FLUIDS)

G. Wang[1], R. C. Prasad[2] and J. E. S. Venart[1]

1. University of New Brunswick
   Fredericton, N.B., Canada

2. University of New Brunswick
   Saint John, N.B., Canada

ABSTRACT

The paper presents experimental observations on the radiation effects of lightly absorbent fluids in the transient line-source thermal conductivity measurement technique. The method of correction recommended by de Castro et al [13] was attempted. However, the present investigation shows that these corrections, which utilize a constant absorption coefficient (κ), do not correct all the measurements over the total measurement period. For a given value of κ, corrected measurements at large times as compared to those at short times do not agree. It is believed that a radiation correction model utilizing κ which is a function of the conduction boundary layer thickness is more appropriate. An analysis of measurements in toluene is presented which shows the effect of radiation correction on temperature-time [ΔT-$\ln(t)$] data for various values of absorption coefficient. The results are shown to be convection free by extension of the measurement period to very large times and powers and the criterion for the onset of free convection is obtained for liquid toluene.

INTRODUCTION

The transient line-source technique provides a rapid and accurate method for the measurement of fluid thermal conductivities [1-27]. In this method, several corrections are applied to account for the deviation of the actual hot-wire from the ideal line-source method. Although the relative magnitude of the radiation effect in this method is significantly less than that in other methods of measurement, suitable radiation correction techniques have not been fully developed. Some attempts [11,28] have been made but these do not appear to have resolved the issue. The present work illustrates the precise nature of the radiation effect in the transient line-source measurement. Also, an attempt has been made to develop a suitable procedure to correct for radiation effect with measurements in toluene.

The transient line-source technique utilizes a thin wire, usually a 5-50 micron diameter high purity platinum wire, which is immersed in the fluid. The thermal conductivity of the fluid is obtained from the transient temperature rise of the wire following a step change in power. The transient temperature change of the wire is described by:

$$\Delta T = [q/(4\pi\lambda)].\ln[4\alpha t/(a^2 C)] \tag{1}$$

However, due to non-ideal conditions, the actual temperature rise differs from that of the ideal. This difference is given by:

$$\Sigma\delta T = \Delta T_i - \Delta T_{actual} = \delta T_{cp} + \delta T_{end} + \delta T_r + \delta T_{conv} + \delta T_{st} + \delta T_{others} \tag{2}$$

The terms in the right hand side of equation-2 represent deviations due to finite specific heat of the wire, end effects, radiation, convection, stress in wire and the others such as the finite thermal conductivity of the wire.

Corrections arising due to the finite heat capacity of the wire are available and have been successfully applied [3-6] in all measurements. The end effects are minimized by a suitably designed cell utilizing either a compensative wire [14-19] or potential leads [3,21-24]. Adequate radiation corrections, which are significant for semi-transparent absorbing- emitting fluids, must be applied in order to obtain accurate thermal conductivities (Figure-1). The straight line fit of T-ln(t) data, necessary to calculate the thermal conductivity of the test fluid, results in a deviation plot shown in Figure-2 which is in agreement with observations of de Castro et al [13].

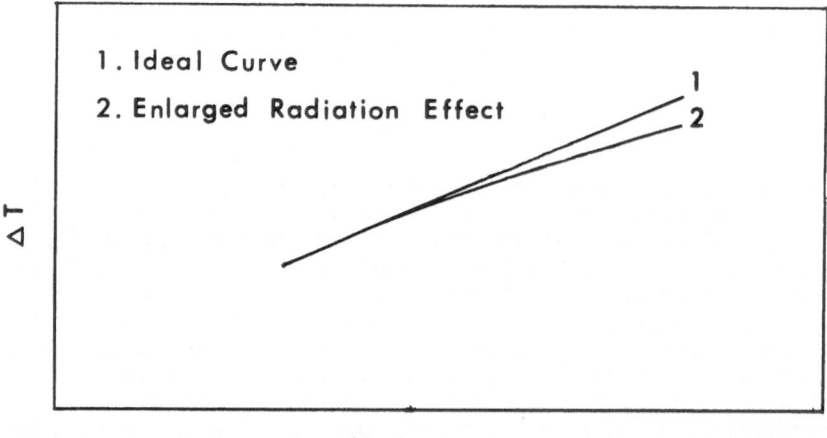

Figure-1   Plot of temperature rise ~ time

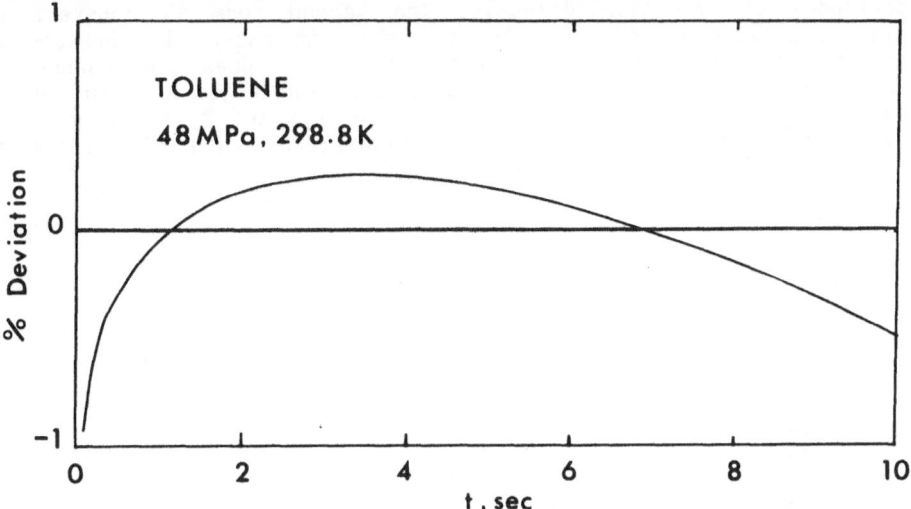

Figure-2 Deviation from linear fit due to radiation effect

De Castro et al [13] as well as Menashe and Wakeham [29] present numerical solutions for the conduction-radiation equation for the transient line-source technique in order to obtain radiation-free thermal conductivity measurements. The reliability of these corrections, however, is considerably reduced since many optical properties of the fluid and its bounding surfaces, necessary for the evaluation of the correction, cannot be measured under the experimental conditions. In addition, these numerical methods are very time consuming and expensive. Therefore, it is not practical to apply these methods on a routine basis. Thus, a simplified correction was suggested.

EXPERIMENTAL OBSERVATION

The radiation correction δT_r, represented by de Castro et al [13], is expressed as:

$$\delta T_r = -[q/(4\pi\lambda)].B.\{[a^2/(4\alpha)].\ln[4\alpha t/(a^2 C)] + a^2/(4\alpha)-t\}$$ (3)

where

$$B = [16/(\rho C_p)]\kappa n^2 \sigma T_o^3$$ (4)

κ and n are the volumetric absorption coefficient and the refractive index respectively. Both of these properties are considered constant during the entire transient heating period. The first term in equation-3 is proportional to $\ln(t)$, or the temperature rise while the third term is proportional to the time directly. The second term is constant and relatively small compared with the other two. At short time periods, the first term is the most significant one, which does not change the temperature-time curve behaviour. Thus, equation-3 is able to correct (Figure-3) the radiation effect shown by figures-1 and 2. At large times, however, the third term is much more significant than the others and changes the T-ln(t) curve behaviour (Figure-4).

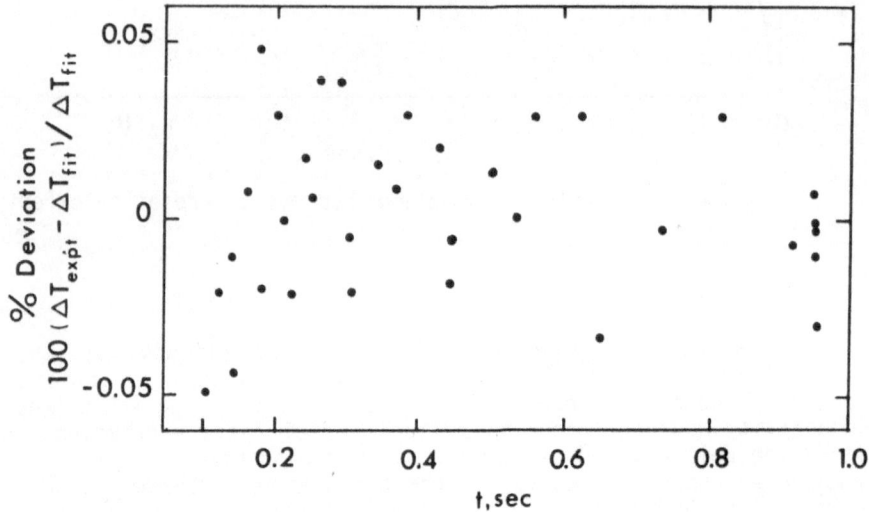

Figure-3   Deviation with radiation correction [13]

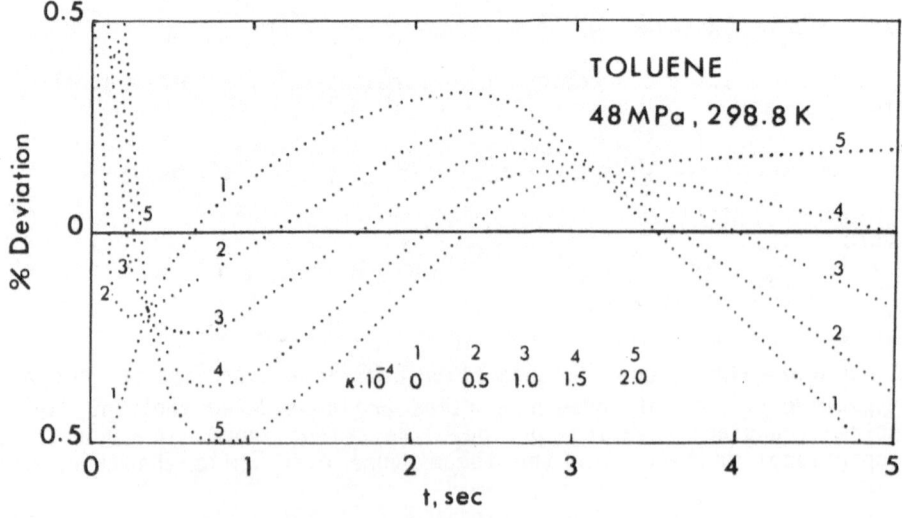

Figure-4   Deviation with various κ values

This correction works satisfactorily at short times but unsatisfactorily at large times where radiation effect becomes significantly larger (Figure-5).

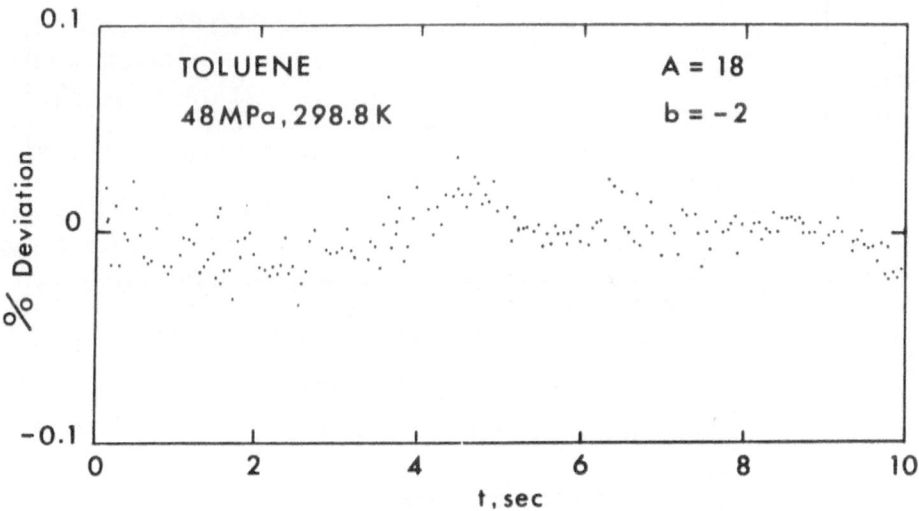

Figure-5 Radiation corrected deviation plot

PROPOSED RADIATION CORRECTION

The proposed radiation correction is based on the concept that the volumetric absorption coefficient κ is strongly dependent upon the thermal boundary layer thickness, δ [30]. Around the line source dissipating heat into the surrounding fluid medium, its conductive boundary layer develops with a complex interaction between the conductive and radiative heat fluxes. A dependence of κ on the thermal boundary layer thickness δ is suggested in a power law form as below:

$$\kappa = F(\delta) = A/\delta^b \tag{5}$$

where A is referred to as an "apparent extinction coefficient" and b is an exponent. In the present investigation, both A and b were determined from the experimental observations.

Tests were conducted with liquid toluene at high pressure (up to 48 MPa) as well as at the atmospheric pressure. Radiation correction over the entire test duration was successfully applied (Figure-6) to these measurements with equations-3 and 5. In the present investigation, a constant value of b=-2 was determined togehter with a suitable value of A (Figure-7). The coefficient A was found to be a function of the input heating power and/or the fluid temperature rise generated by this input heat flux (Figure-8). Selected pair of A and b provides a suitable radiation correction model which results in a nearly perfect straight line fit with the deviation less than 0.04% (Figure-6). This correction was successfully applied for the measurement up to 30 seconds in liquid toluene.

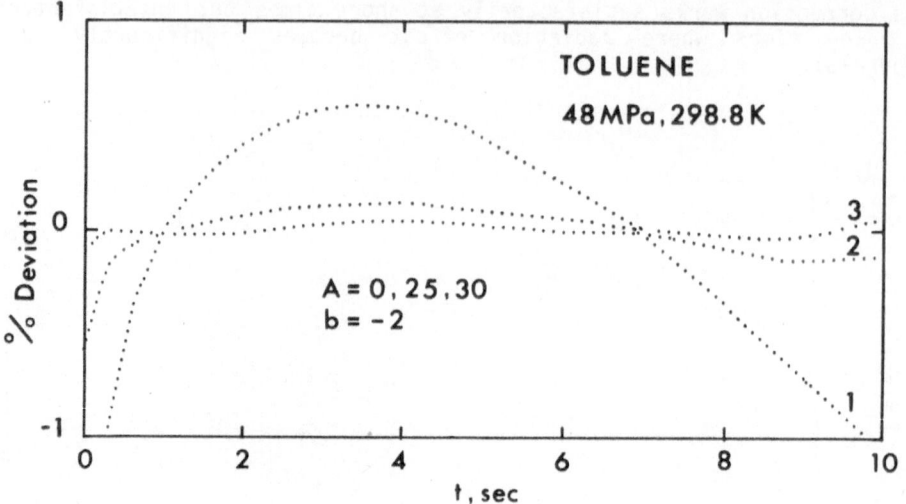

Figure-6   Effect of varying A

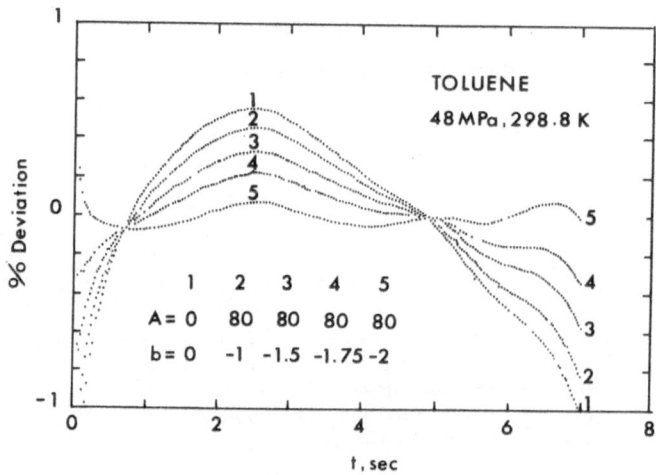

Figure-7   Effect of varying b

NATURAL CONVECTION

With the transient line-source instrument, convection-free measurements are obtained for some initial time period. At large times, the convection effect, however, appears and results in a reduction on the temperature rise of the wire. The effect of convection of the ΔT-ln(t) plot appears to be similar to that produced by radiation effect and hence it may, at times, be difficult to distinguish between the two effects. By adopting a suitable procedure to accurately correct the radiation effect, it is possible to determine the onset of natural convection in measurements with this instrument. Figure-9 shows the deviation due to convection of ΔT-ln(t) data from straight line fit after the application of radiation correction. The time for onset of convection can be seen   in

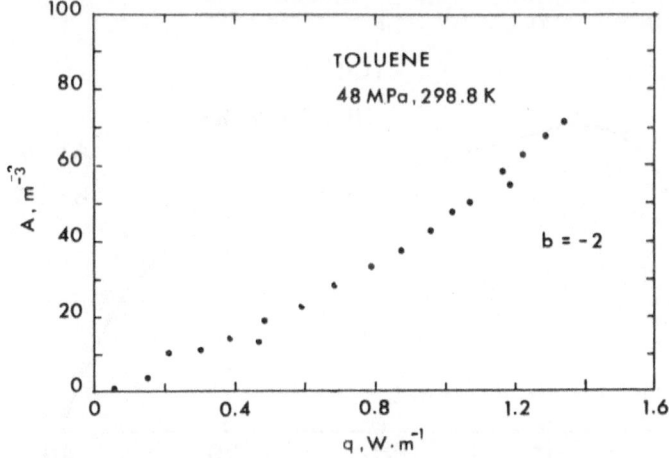

(a) Apparent absorption coefficient ~ power input

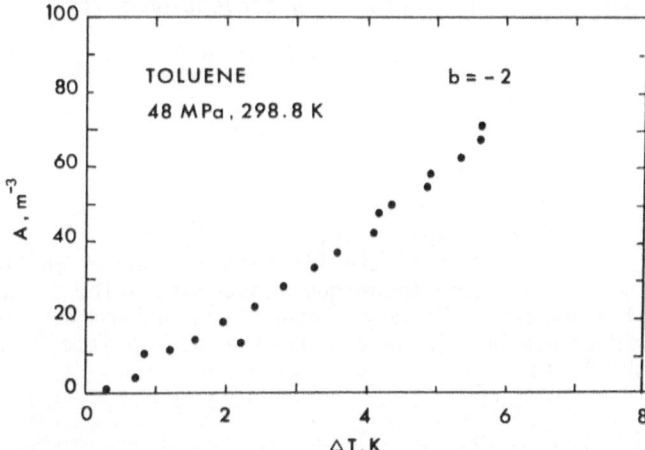

(b) Apparent absorption coefficient ~ temperature rise

Figure-8   Variation of apparent absorption coefficient (A)

Figure-9 as 33 seconds. Figure-10 shows the onset time of free convection as the function of input power and/or the temperature rise for toluene at 48 MPa, 298.15K with a 12.5 micron diameter vertical wire.

The previous works [3,31,32] suggested that for the onset of natural convection, the criterion is as:

$$Ra = Gr \cdot Pr = g\beta\Delta T \; \delta^3/(\nu\alpha) < 1000 \qquad (6)$$

where δ is the thermal boundary layer thickness given by [3]:

$$\delta = (28\alpha t)^{\frac{1}{2}} \qquad (7)$$

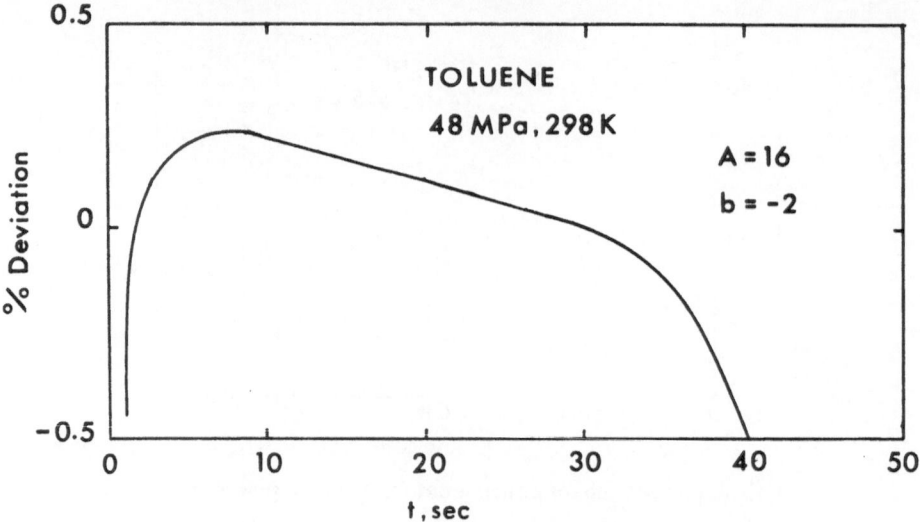

Figure-9   Deviation with convection effect

Experimental data show that the limiting Ra number in this criterion is too small for the hot-wire technique measurement. The actual onset time of the natural convection is much later than determined by equation-6 according to data obtained (Figure-10). The appropriate criterion found from this figure is as:

$$Ra = g\beta\Delta T\delta^3/(\nu\alpha) \leq 10^5 \qquad (8)$$

The normalized onset time of natural convection (τ) versus the normalized power input (q^*)   is plotted in Figure-11. It shows an excellent agreement with the suggested criterion (equation-8). Quantities τ and q^* are defined as following:

$$\tau = \nu\, t_v/a^2 \qquad (9)$$

$$q^* = g\beta d^3 q/(4\pi\lambda\nu\alpha) \qquad (10)$$

RESULTS

The radiation correction proposed in previous sections was applied to the measurement of thermal conductivity of toluene at 48 MPa and 298K-305K. These measurements represent the radiation-free, convection-free thermal conductivity. Figure-12 shows the comparison of the present results with the data reported by de Castro et al [13] and shows deviations of 1.8%-2.4% with the data of Shulga's [33].

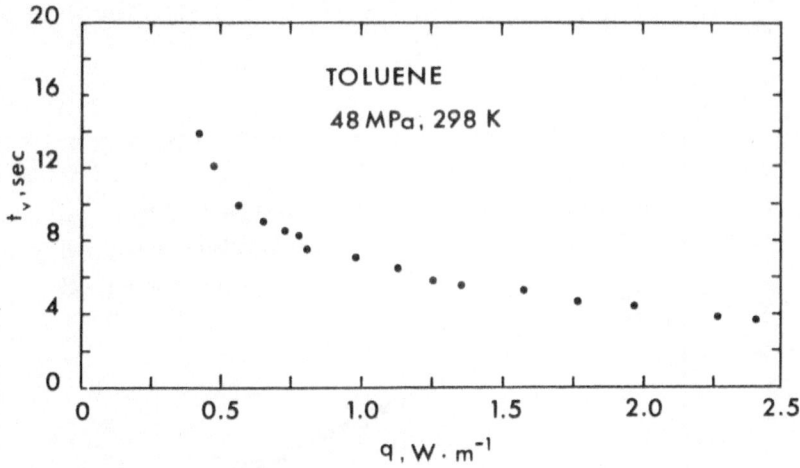

(a) Time for onset of natural convection ~ power input

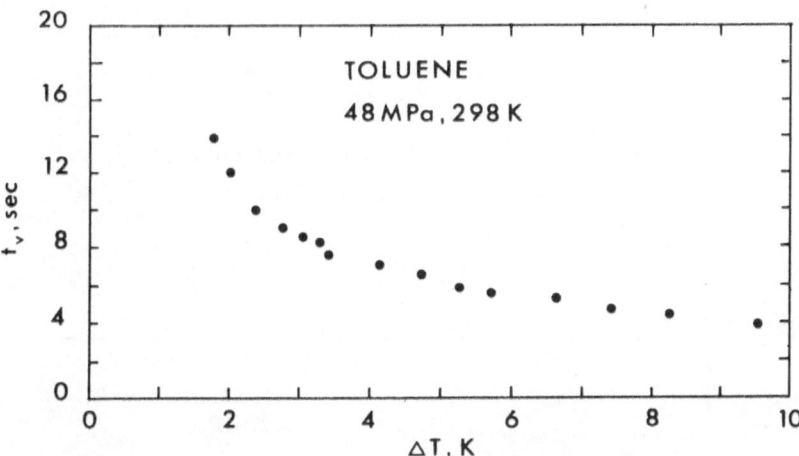

(b) Time for onset of natural convection ~ temperature rise

Figure-10 Onset of natural convection-experimental observation

DISCUSSION AND CONCLUSION

Radiation correction in transient line-source technique has been presented in a semi-empirical form. Preliminary investigations indicate that the proposed correction model adequately corrects for radiative effect at short times as well as at large times. This procedure includes a variation in the volumetric absorption coefficient with the thermal boundary layer thickness. This variation of δ is found to be proportional to the square of the boundary layer thickness.

Further investigation of this procedure to correct for the radiation effect at higher temperatures as well as for other fluids is necessary to further validate this complex problem of radiation-convection effect in the transient line-source technique.

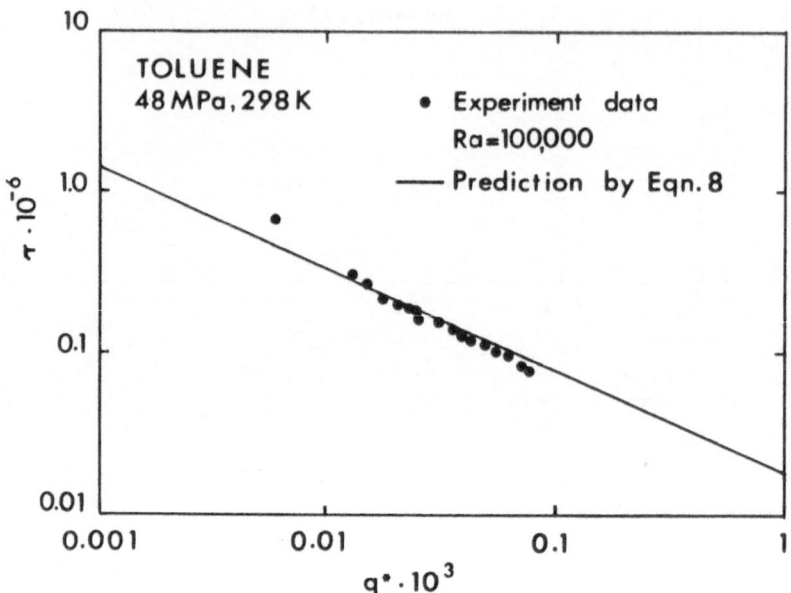

Figure-11 Rayleigh number and convection

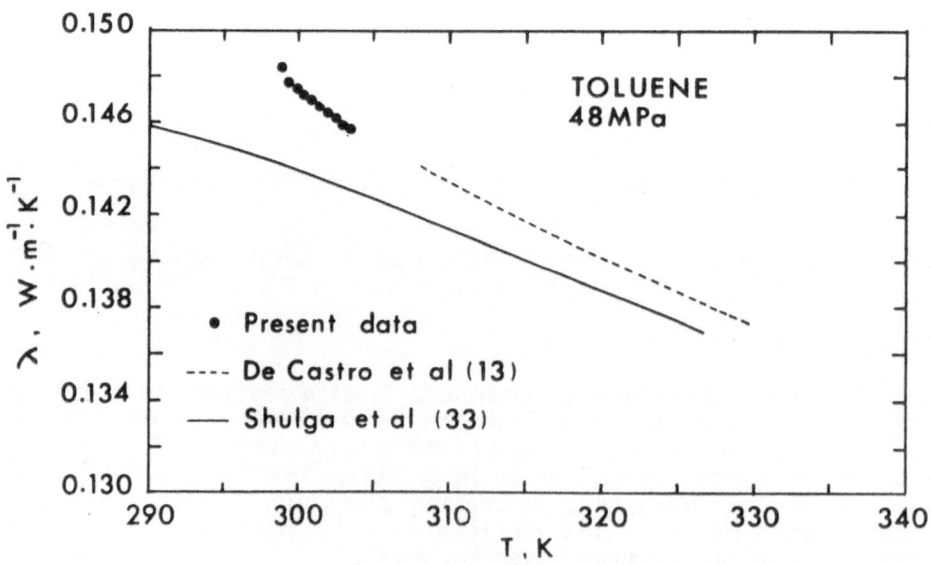

Figure-12 Thermal conductivity of toluene

NOMENCLATURE

a   Radius of the hot wire, m

A   Apparent absorption coefficient, m^{-3}

b   Exponent in the $\kappa \sim \delta$ relation

B   A term in radiation correction, equation-4

C   Euler constant, 1.781

Cp   Specific heat, $J.Kg^{-1}.K^{-1}$

d   Hot-wire diameter, m

g   Acceleration due to gravity, $9.8 \ m.s^{-2}$

n   Refractive index

q   Input heating power, $W.m^{-1}$

q*   Normalized input power

Ra   Rayleigh Number

t   Experimental time, s

Ti   Ideal temperature rise, K

To   Environment temperature, K

t_v Convection onset time, s

α Thermal diffusivity, $m^2.s^{-1}$

β Thermal expansion coefficient, K^{-1}

δ Boundary layer thickness, m

δT Temperature corrections, K

ΔT Temperature rise, K

ν Kinetic viscosity, $m^2.s^{-1}$

κ Volumetric absorption coefficient, m^{-1}

λ Thermal conductivity, $W.m^{-1}.K^{-1}$

ρ Density, $Kg.m^{-3}$

σ Steffan-Boltzmann constant, $5.669 \times 10^{-8} \ W.m^{-2}.K^{-4}$

τ Dimensionless time

ACKNOWLEDGEMENTS

This work was performed under a program of studies funded by the Natural Sources and Engineering Research Council, Canada.

REFERENCES

1. J.F.T. Pittman, Thesis (University of London, London, 1968)
2. J.W. Haarman, Thesis (Technische Hogeschool, Delft, 1969)
3. N. Mani, Thesis (University of Calgary, Canada 1971)
4. J.J. de Groot, J. Kestin, H. Sookiasian, Physica 75:454 (1974)
5. J.J. Healy, J.J. de Groot, J. Kestin, Physica 82:392 (1976)
6. J.J. de Groot, J. Kestin, H. Sookiasian, Physica 92A:117 (1978)

7. J. Kestin, R. Paul, A.A. Clifford, W.A. Wakeham, Physica 100:349 (1980)
8. J. Kestin, W.A. Wakeham, Physica 92A:102 (1978)
9. J. Kestin, A.A. Clifford, W.A. Wakeham, Physica 100a:370 (1980)
10. M.J. Assael, M. Dix, A. Lucas, W.A. Wakeham, J. Chem. Soc. Faraday Trans. 77:439 (1981)
11. C.A.N. de Castro, J.C.G. Calado, W.A. Wakeham, M. Dix, J. Phys. E. Sci. Instr. 9:1073 (1976)
12. C.A.N. de Castro, J.C.G. Calado, W.A. Wakeham, M. Dix, Proc. 7th Symp. Thermophysical Properties, A. Cezairliyan, ed (ASME, New York), 1977, pp. 730-738
13. C.A.N. de Castro, S.F.Y. Li, G.C. Maitland, W.A. Wakeham, Int. J. Thermophys. 4:311, (1983)
14. H.M. Roder, J. Res.NBS 86:457 (1981)
15. H.M. Roder, J. Res.NBS 87:279 (1982)
16. H.M. Roder, C.A.N. de Castro, J. Chem. Data 27:12 (1982)
17. C.A.N. de Castro, W.A. Wakeham, Thermal Conductivity-15, V.V. Mirkovich, ed., p. 236, Plenum Press, New York, (1978)
18. C.A.N. de Castro, H.M. Roder, J. Res. NBS 86:293 (1981)
19. J. Kestin, A.A. Clifford, W.A. Wakeham, Physica 97A:187, (1979)
20. M.J. Assael, W.A. Wakeham, J. Chem. Soc. Faraday Trans. 77:697 (1981)
21. R.C. Prasad, N. Mani, J.E.S. Venart, Int. J. Thermophys., 5:265 (1984)
22. R.C. Prasad, J.E.S. Venart, Int. J. Thermophys., 5:367 (1984)
23. R.C. Prasad, J.E.S. Venart, N. Mani,   Thermal Conductivity-18, T. Ashworth and D.R. Smith, eds, p. 81, Plenum Press, New York (1985)
24. R.C. Prasad, G. Wang, J.E.S. Venart, 1st Asian Thermophysical Properties Conference, Beijing, China (1986)
25. E.F. Buyukbicer, M.Sc.E. Thesis (University of New Brunswick, 1984)
26. G. Wang, M.Sc.E. Thesis (University of New Brunswick, Canada, 1985)
27. Y. Nagasaka, A Nagashima, Rev. Sci. Instrum. 52:229 (1981)
28. A. Saito, J.E.S. Venart, Proc. 6th Int. Heat Transfer Conf., 3:79 (1978)
29. J. Menashe, W.A. Wakeham, Int. J. Heat. Mass. Transfer., 25:661 (1982)
30. G. Schodel, U. Grigull, 4th Int. Heat Transfer Conf., 3:R22 (1970)
31. E.R. Peterson, Thesis, (Oklahoma State University, USA, 1985)
32. R.J. Goldstein, D.G. Briggs, Trans. ASME J. Heat Transfer, 86:490 (1964)
33. V.M. Shulga, F.G. Eldarov, Yu. A. Atanov, A.A. Kuyumchv, Inter. J. Thermophys.,7:1147 (1986)

THERMAL CONDUCTIVITY OF LIQUIDS IN THE REDUCED TEMPERATURE RANGE 0.3 TO 0.8 FROM SATURATION LINE TO 350 MPa

G. Latini,[*] F. Marcotullio,[*] P. Pierpaoli[**] and A. Ponticiello[*]

[*] Dipartimento di Energetica, Università de L'Aquila, Italia
[**] Dipartimento di Energetica, Università di Ancona, Italia

SUMMARY

An empirical correlation is proposed to predict the liquid thermal conductivity as a function of temperature and pressure. The correlation involves explicitly the reduced temperature, the reduced pressure (or more simply the pressure) and three factors A, a, b (or b') which are characteristic for each substance and, with a very close approximation, also pressure and temperature independent.

The equation was checked through experimental thermal conductivity values, generally available with an accuracy within ±1.5%, of 18 alkanes, 9 aromatics and 2 elements. Its effectiveness in the whole temperature range from normal melting point to $T_r = 0.8$ and in the reduced pressure range from the vapour pressure to nearly 350 MPa was also clearly proved. The deviations between predicted and experimental conductivity data are ordinarily less than 2.0% whilst the maximum deviations are likewise better than 3%. Relations are finally provided to properly predict A and b' factors.

INTRODUCTION

Experimental thermal conductivity data of liquid substances at different temperatures and pressures are scarcely available and in many instances there is quite a lack of them. Under these unfavourable circumstances it is often necessary to resort to prediction methods which relate the thermal conductivity λ to the density ϱ and the temperature T or, alternatively, to the pressure P and the temperature T.

Correlations of the form $\lambda = \lambda(\varrho, T)$ are usually to be preferred from a theoretical standpoint being based on simple physical assumptions or because they involve terms admitting own interpretation; such correlations were proposed, for example, by Roder, Nieto de Castro, Wakeham and their coworkers [1,2,3,4,5]: they cover large pressure and temperature ranges and generally reproduce the measured data within the experimental uncertainties.

Although these results are undoubtedly important, it would be preferable however to provide an equation such as $\lambda = \lambda(P, T)$ which is more suitable for engineering purposes. Actually, a correlation making

---

Work supported by M.P.I. - Ministero Pubblica Istruzione (ITALY)

use of the density ϱ or the molar volume V requires experimental knowledge for both these quantities and, as it is practically impossible to achieve such information at any temperature and pressure, appropriate equations of state should be provided to transform the measured pressures into equivalent densities. This approach is rather awkward and as yet it appears practicable with an acceptable accuracy only in few cases.

On the other hand, a correlation involving P and T, which are quantities easily measured with excellent accuracy, allows the evaluation of the thermal conductivity in a very simple way. In this case, the related errors are presumably due only to the limitations of the prediction method itself so they should be clearly identified and their magnitude explicitly stated.

This approach will be followed in the present work and an empirical correlation relating the thermal conductivity of liquids to the reduced temperature $T_r = T/T_c$ and the reduced pressure $P_r = P/P_c$ (or more simply P so that the critical pressure P_c is not required) will be provided. The formula is performed through an extensive analysis of selected experimental data and should be regarded as an improvement and generalization of the existing correlation proposed by Latini et al. [7], which holds at atmospheric pressure or along the saturation line:

$$\lambda = A \frac{(1-T_r)^{0.38}}{T_r^{1/6}} \tag{1}$$

where the temperature independent factor A is characteristic for each substance and can be evaluated by means of appropriate relations involving the molecular weight M, the critical temperature T_c and the normal boiling point T_b.

Equation (1) was usefully proved in the reduced temperature range from 0.3 to 0.8 and the mean deviations between predicted and experimental thermal conductivity values were found to be less than 3% for liquids of the most important organic families such as alkanes, olefins, cycloparaffins, aromatics, alcohols, ethers, ketones, aldehydes, organic acids, esters as well as for several inorganic compounds and elements.

It should be noted that while the correlation is based on experimental data retrieved from literature reviews up to 1982 showing an accuracy within ±5% or worse and only seldom better than ±2%, recent measurements of thermal conductivity for a wide variety of substances cited in scientific reports, normally exhibit accuracies within ±1.5% for large temperature and pressure ranges. This circumstance allowed going deep into the check of the correlation and attention was focused particularly to the following questions:

(a) the value of 0.38 assigned to the exponent of the term $(1-T_r)$ was kept constant for all the investigated substances even though a specific physical reason to support this result did not strictly exist: it arose rather from mathematical computations on the experimental data available at that time. So far, a more extensive analysis of equation (1) could be developed on the basis of further and more accurate information thus providing the appropriate exponent values for each organic family, inorganic compounds and elements as well;

(b) as the correlation does not explicitly involve the pressure P it may be used without appreciable errors only up to 3 to 4 MPa (within this range, the influence of pressure upon the thermal conductivity of liquids is usually negligible provided that one stands sufficiently far from the critical point). However, the availability of accurate measurements up to 600 to 700 MPa for enough substances, suggests to investigate for a generalization of the equation by involving explicitly P and T in a unique and compact formula.

The present work is mainly concerned with the above mentioned problems which are analyzed and discussed separately. Equation (1) is now rewritten as follows:

$$\lambda = A \; \frac{(1-T_r)^n}{T_r^{1/6}} \qquad (2)$$

THE THERMAL CONDUCTIVITIES OF LIQUIDS ALONG THE SATURATION LINE: THE EXPONENT n AND THE FACTOR A IN EQUATION (2).

In order to investigate on both the exponent n and the factor A in equation (2), two organic families (alkanes and aromatics) and two elements (argon and oxygen) were considered. This choice was justified, as earlier emphasized, by the availability over wide temperature and pressure ranges of very accurate thermal conductivity information for these substances; in addition, the data used in the analysis were selected possibly from the same source or from groups of researchers which used similar experimental techniques: this criterion was followed to avoid any mixing of data sets which could differ from one another by more than the mutual claimed measurement uncertainties.

Table I lists the substances under investigation with their molecular weights, the normal boiling points, the critical temperatures and the sources of the experimental thermal conductivity data at atmospheric pressure or along the saturation line.

The analysis of equation (2) performed on the basis of the above data series, led to the following results:

1) the best fitting is obtained when n is characteristic of the particular investigated family and A of the particular investigated compound;

2) n and A are temperature independent with a very close approximation;

3) n can be set at 0.50 for alkanes and aromatics while the value of 0.45 seems to be more suitable for argon and oxygen;

4) the values of A vary from 0.1301 W/m.K (2,2,4-trimethylpentane) to 0.2618 W/m.K (methane);

5) the mean deviations between the predicted series obtained from equation (2) and the experimental data are generally less than 1.0% with maximum deviations within 2%, i.e., the deviations of the predicted values are comparable with the measurement uncertainties stated in the sources of Table I.

The above statements are summarized in the same Table I where the mean $\overline{\Delta}$ % and the maximum Δ_M % deviations refer to the reduced temperature ranges ΔT_r: substantially, equation (2) may be efficiently employed when the reduced temperature varies from normal melting point to ≈0.8, which is quite above the normal boiling point (for the investigated compounds this value is usually smaller than 0.7).

If the reduced temperature is greater than 0.8 but lower than 0.9, the deviations $\Delta\%$ between predicted and experimental values are generally smaller than 10% while in the vicinity of the critical temperature they rapidly grow to about 20% and hence equation (2) ceases to be accurate.

The last question to be examined was concerned with setting the appropriate values of factor A for alkanes and aromatics which should likewise be determined for completeness of the prediction method.

In the old equation (1) the recommended values for A were referred to molecular weights ranging from 50 to 250. In the present case, instead, the above restriction is eliminated and the following equation seems to fit correctly for both the alkanes and aromatics families;

$$\lambda = 0.5199 \; \frac{T_b^{8.40}}{M^{2.00} T_c^{6.61}} \; \frac{(1-T_r)^{0.50}}{T_r^{1/6}} \qquad \text{(Alkanes)} \qquad (3)$$

$$\lambda = 1.8464 \; \frac{T_b^{0.05}}{M^{0.41} T_c^{0.32}} \; \frac{(1-T_r)^{0.50}}{T_r^{1/6}} \qquad \text{(Aromatics)} \qquad (4)$$

The results obtained at this stage of the analysis can be summarized as follows:

1) when few selected experimental thermal conductivity data (theoretically even only one) are available at atmospheric pressure or along the saturation line, the factor A is immediately evaluated and equation (2) properly used for alkanes and aromatics by setting n=0.5. The expected deviations between predicted and experimental set of values are ordinarily less than 1.0% in the reduced temperature range 0.3 to 0.8;

2) when experimental thermal conductivity data are not available at the above mentioned conditions, the factor A takes the values set in equations (3) and (4) for alkanes and aromatics and the same equations may be used in the reduced temperature range 0.3 to 0.8 but, in this case, the agreement between predicted and experimental data is somewhat reduced (deviations are within ±3%).

THE THERMAL CONDUCTIVITY OF LIQUIDS AS A FUNCTION OF PRESSURE

The second step of the analysis examined the possibility of introducing explicitly the pressure P in equation (2). In order to attain this goal, all the substances listed in Table I were considered for which conductivity measurements at different pressures ensured accuracy within at least ±2%. Table II exhibits now the sources of the experimental values at different P and T for 17 investigated substances.

Moreover equation (2), with n=0.5 for alkanes and aromatics and 0.45 for argon and oxygen, performed so satisfactory results along the saturation line that it was fairly reasonable to retain it also when the pressure dependence is considered. At this stage, two opposite assumptions were examined:

a) n is pressure dependent whilst A is independent;
b) n is pressure independent whilst A is dependent.

In the former case the output sample is clearly represented by the function $n=n(P)$ and in the latter by $A=A(P)$; although these two assumptions are theoretically equivalent, the best results were achieved by choosing the first alternative and, following this logical path, some outstanding conclusions were outlined:

1) at any given temperature, from the vapour pressure to nearly 60 MPa, a linear dependency between n and P showed very good agreement

$$n = 0.50 - b' \cdot P \qquad \text{(Alkanes and Aromatics)} \qquad (5)$$
$$n = 0.45 - b' \cdot P \qquad \text{(Argon and Oxygen)} \qquad (6)$$

where b' is a characteristic parameter for each substance as shown in Table II which also contains the mean $\overline{\Delta}$ % and the maximum Δ_M% deviations between the thermal conductivity values arising from equations (2) and (5), or from equations (2) and (6), and the experimental data; Δ % is usually less than 1.5% and Δ_M% within 3.5%, omitting only the n-pentane data;

2) when pressure increases beyond 60 MPa, the relations (5) and (6) are no longer accurate but up to 350 Mpa the following formulae can be used satisfactorily:

$$n = 0.50 - b \cdot P + a \cdot P^2 \qquad \text{(Alkanes and Aromatics)} \qquad (7)$$
$$n = 0.45 - b \cdot P + a \cdot P^2 \qquad \text{(Argon and Oxygen)} \qquad (8)$$

the best values for a and b parameters are also shown in Table II together with the mean $\overline{\Delta}$ % and maximum Δ_M% deviations of predicted values determined according to relations (2) and (7) or (2) and (8). As it can be observed, $\overline{\Delta}$ % is usually better than 2% and Δ_M% within 4%, once again with the only exception represented by n-pentane since the marked deviations of n-hexane, propane and 2,2,4-trimethylpentane arise from single scattered data;

3) finally, for pressures higher than 350 MPa relations (7) were likewise no longer reliable; however, the range of validity of equation (2) should be considered sufficiently well covered since a further term in formulae (7) and (8) yielded mean deviations well beyond 6% thus exceeding the limits of the working range dealt within the present paper.

At this stage, it seemed necessary to seek an appropriate correlation for the estimation of b in order to achieve a complete prediction method, formally represented by the function $\lambda = \lambda(P,T)$, at least for the alkanes family. The lack of accurate information in the range 0.1 to 60 MPa for aromatics, deprived of benzene and toluene, discouraged from pursuing the same aim. The following relation

$$n = 0.50 - 0.1537 \frac{M^{0.98}}{T_b^{1.78}} P \qquad \text{(Alkanes)} \qquad (9)$$

was found to usefully substitute equation (5) for alkanes; thereby, when the relations suggested for predicting n and A are employed together with equation (2), the mean deviations from experimental data will naturally increase but, with the exception of ethane, the mean deviation Δ % will usually never exceed 4%.

Certainly, it would have been of the utmost interest to provide a procedure for estimating the parameters a and b of equations (7) and (8) in terms of physical properties of the different substances. Unfortunately this objective was practically unreachable, since the available experimental information on alkanes over the whole range of vapour pressure up to 350 MPa is somewhat incomplete and inhomogeneous for correlation purposes.

CONCLUSIONS

The correlation (1), already proposed in the past to assess the thermal conductivity at atmospheric pressure or along the saturation line and thoroughly checked through more than 150 substances in the liquid phase, has been extensively investigated on the basis of further and more accurate thermal conductivity data retrieved from relatively recent literature.

The analysis was carried out in two successive steps, the first aimed to improve the prediction accuracy of the correlation itself along the saturation line, the other one to extend its use over a wide range of pressure variation.

The results obtained yielded the following summarized conclusions:

- along the saturation line the exponent value of 0.38 in equation (1) should be successfully replaced by 0.50 for alkanes and 0.45 for argon and oxygen; thus, equation (1) is substituted by equation (2) but holding the preceding values of the temperature independent exponent which is to be considered as characteristic for each organic family, or inorganic compound, or element as well;

- the factor A of equation (2) which is obviously different from the corresponding one set in equation (1), was recalculated for alkanes, aromatics, argon and oxygen as shown in table I; the conclusion was that it had to be considered as strongly temperature independent and hence appropriate equations were derived for the alkanes and aromatics families. These relations coupled to equation (2) (into which n is set equal to 0.5) provide, by means of formulae (3) and (4), a complete prediction procedure along the saturation line and in the reduced temperature range 0.3 to 0.8;

- when pressure increases beyond 0.4 to 0.5 MPa, the factor A standing in equation (2) retains the value referred to the saturation line whereas the exponent n is now significantly pressure dependent in the reduced temperature range 0.3 to 0.8, as shown by relations (5) and (6) and by relations (7) and (8) that hold up to 60 MPa and 350 MPa respectively;

- finally, in the pressure range from 0.1 to 60 MPa, where n depends almost linearly on pressure, a suitable relation for alkanes such as equation (9) was also derived.

Future work for other organic families as well as for inorganic compounds and elements is foreseen.

210

Table I. General table of the investigated substances with molecular weight M, normal boiling point Tb, critical temperatures Tc, sources of experimental λ data at 0.1 MPa or along the saturation line, values of exponent n and factor A of eq(2) and mean $\bar{\Delta}$ and maximum ΔM deviations between thermal conductivity values predicted according to eq(2) and the experimental ones in the reduced temperature ranges ΔTr.

SUBSTANCE	M	Tb(K)	Tc(K)	SOURCE of exp.λ	n eq.(2)	A eq.(2) (W/mK)	$\bar{\Delta}$ (%)	ΔM (%)	ΔTr
METHANE	16.043	111.7	190.6	[5]	0.50	0.2618	0.9	2.6	0.58 to 0.79
ETHANE	30.070	184.5	305.4	[8][9]	0.50	0.2480	1.5	2.3	0.37 to 0.77
PROPANE	44.097	231.1	369.8	[10]	0.50	0.2030	0.8	1.2	0.41 to 0.60
n-BUTANE	58.124	272.7	425.2	[11]	0.50	0.1848	0.3	0.7	0.35 to 0.58
n-PENTANE	72.151	309.2	469.6	[11]	0.50	0.1730	0.8	1.2	0.34 to 0.60
n-HEXANE	86.178	341.9	507.4	[10]	0.50	0.1704	0.3	0.7	0.36 to 0.60
2,3-DIMETHYLBUTANE	86.178	331.2	499.9	[14]	0.50	0.1469	0.8	1.2	0.62
n-EPTANE	100.205	371.6	540.2	[13]	0.50	0.1668	1.2	2.9	0.35 to 0.68
n-OCTANE	114.232	398.8	568.8	[12]	0.50	0.1664	0.6	1.4	0.50 to 0.60
2,2,4-TRIMETHYLPENTANE	114.232	372.4	543.9	[12]	0.50	0.1301	0.4	0.8	0.54 to 0.66
n-NONANE	128.259	424.0	594.6	[12]	0.50	0.1658	0.2	0.5	0.47 to 0.61
n-DECANE	142.286	447.3	617.6	[16]	0.50	0.1628	0.6	1.1	0.47 to 0.60
n-UNDECANE	156.313	469.1	638.8	[15][12]	0.50	0.1606	0.2	0.4	0.45 to 0.59
n-DODECANE	170.340	489.5	658.3	[17]	0.50	0.1575	0.7	1.5	0.45 to 0.57
n-TRIDECANE	184.367	508.6	675.8	[18]	0.50	0.1580	0.3	0.6	0.45 to 0.69
n-TETRADECANE	198.394	526.7	694.0	[12][15]	0.50	0.1621	0.6	0.9	0.41 to 0.53
n-PENTADECANE	212.421	543.8	707.0	[15]	0.50	0.1632	0.7	1.3	0.40 to 0.51
n-HEXADECANE	226.448	560.0	717.0	[15]	0.50	0.1648	1.0	1.7	0.41 to 0.51
BENZENE	78.114	353.3	562.1	[19]	0.50	0.1870	0.3	0.5	0.52 to 0.61
TOLUENE	92.141	383.8	591.7	[13]	0.50	0.1658	0.1	0.2	0.39 to 0.61
ETHYLBENZENE	106.168	409.3	617.1	[17]	0.50	0.1586	0.4	1.0	0.49 to 0.58
o-XYLENE	106.168	417.6	630.2	[17]	0.50	0.1594	0.5	1.1	0.47 to 0.57
m-XYLENE	106.168	412.3	617.0	[17]	0.50	0.1621	0.3	0.7	0.48 to 0.58
p-XYLENE	106.168	411.5	616.2	[17]	0.50	0.1596	0.3	0.4	0.48 to 0.58
ISOPROPYLBENZENE	120.195	425.6	631.0	[17]	0.50	0.1486	0.9	1.6	0.47 to 0.57
CHLOROBENZENE	112.559	404.9	632.4	[17]	0.50	0.1521	0.5	1.2	0.47 to 0.55
BROMOBENZENE	157.010	429.2	670.0	[17]	0.50	0.1330	0.8	1.8	0.45 to 0.56
ARGON	39.948	87.3	150.9	[20][21]	0.45	0.1654	0.3	0.5	0.71 to 0.83
OXYGEN	31.999	90.2	154.6	[22]	0.45	0.2042	1.1	3.2	0.35 to 0.81

Table II. Table showing the substances investigated at various pressures, the sources of the experimental λ data, the factor b' of eq.(5) or (6) together with the mean $\bar{\Delta}$ and the maximum ΔM deviations between predicted and experimental data, the factors a and b of eq.(7) or (8) together with the mean and the maximum deviations between predicted and experimental λ data and the reduced temperature ranges ΔTr.

SUBSTANCE	SOURCE of exp. λ	b'·10³ (bar⁻¹) eq.(5) or (6) P≤60 MPa	$\bar{\Delta}$% Eq.(2) with eq.(5) or (6)	ΔM% with Eq.(2) or (6)	b·10³(bar⁻¹) Eq.(7) or (8) P≤350 MPa	a·10⁸(bar⁻²)	$\bar{\Delta}$% Eq.(2) with eq.(7) or (8)	ΔM% with Eq.(2) or (8)	ΔTr
METHANE	[2][5][9]	0.526	0.7	3.3	0.662	30.029	0.3	1.4	0.58 to 0.71
ETHANE	[8][9]	0.446	1.7	4.8	0.522	17.980	1.7	3.9	0.36 to 0.77
PROPANE	[9][23]	0.367	1.6	6.0	0.389	5.102	1.6	5.6	0.30 to 0.81
n-BUTANE (*)	[24]	0.397	0.8	1.2	----	-----	---	---	0.70
n-PENTANE	[25]	0.436	2.8	8.6	0.360	5.963	3.1	9.0	0.65 to 0.77
n-HEXANE	[26]	0.415	0.9	3.2	0.350	5.056	2.2	5.4	0.61 to 0.71
2,3-DIMETHYLBUTANE	[14]	0.445	1.7	3.4	0.381	5.917	2.9	8.9	0.62 to 0.72
n-EPTANE	[27]	0.357	0.5	0.9	0.351	4.875	1.3	3.4	0.57 to 0.64
n-OCTANE	[26]	0.359	0.5	1.1	0.333	4.422	1.2	3.0	0.54 to 0.64
2,2,4-TRIMETHYLPENTANE	[14]	0.465	1.9	3.4	0.387	5.509	2.0	5.5	0.58 to 0.65
n-NONANE	[28]	0.346	0.3	0.7	0.343	4.674	1.0	2.5	0.52 to 0.61
n-UNDECANE	[28]	0.399	0.5	1.2	0.371	5.010	1.2	2.6	0.48 to 0.55
n-TRIDECANE	[29]	0.423	0.8	1.3	0.404	6.282	1.1	3.8	0.46 to 0.51
BENZENE	[30]	0.292	0.4	0.9	0.286	4.109	0.7	2.6	0.55 to 0.64
TOLUENE	[31][32][33]	0.331	0.4	1.0	0.333	3.925	1.1	2.8	0.46 to 0.63
ARGON	[20][21][34]	0.525	1.3	2.3	0.702	40.070	0.7	2.3	0.71 to 0.78
OXYGEN	[11][35]	0.453	1.4	2.8	0.568	24.640	0.8	2.9	0.50 to 0.78

(*) Experimental λ data up to 50 MPa in the liquid phase in the reduced temperature range 0.3 to 0.8

REFERENCES

1. H. R. Roder, The Thermal Conductivity of Oxygen, J. Res. Nat. Bur. Stand. (U.S.) 86 (5):279 (1982).
2. H. R. Roder, Thermal Conductivity of Methane for Temperatures Between 110 and 310 K with Pressures to 70 MPa, Int. J. Thermophys. 6 (2):119 (1985).
3. S. F. Y. Li, R. D. Trengove, W. A. Wakeham and M. Zalaf, The Transport Coefficients of Polyatomic Liquids, Int. J. Thermophys. 7 (2):273 (1986).
4. S. F. Y. Li, G. C. Maitland and W. A. Wakeham, The Thermal Conductivity of Liquid Hydrocarbons, High Temp.-High Pressures 17:241 (1985).
5. U. V. Mardolcar and C. A. Nieto de Castro, The Thermal Conductivity of Liquid Methane, Chem. Phys. (Holland) in press.
6. R. C. Reid, J. M. Prausnitz and B. E. Poling, Chapter 10 in "The Properties of Gases and Liquids" 4th ed. McGraw-Hill, New York (1987).
7. G. Latini, C. Baroncini, P. Pierpaoli, Liquids under Pressure: an Analysis of Methods for Thermal Conductivity Prediction and a General Correlation, High Temp.-High Pressures 19 (1987) in press.
8. H. M. Roder and C. A. Nieto de Castro, Thermal Conductivity of Ethane at Temperatures Between 110 and 325 K and Pressures to 70 MPa, High Temp.-High Pressures 17:453 (1985).
9. H. M. Roder, Experimental Thermal Conductivity Values For Hydrogen, Methane, Ethane and Propane, Nat. Bur. Stand. (U.S.) NBSIR 84-3006, May 1984.
10. V. P. BryKov, G. Kh. Mukhamedzyanov and A. G. Usmanov, Experimental Investigation of the Thermal Conductivity of Organic Liquids at Low Temperatures (in Russian), Inzh-fiz. Zh. 18 (1):82 (1970).
11. J. F. T. Pittman, Fluid Thermal Conductivity Determination by the Transient Line Source Method, Ph. D. Thesis, Imperial College of Science and Technology, Department of Chemical Engineering, London (1968).
12. J. C. G. Calado, J. M. N. A. Fareleira, C. A. Nieto de Castro and W. A. Wakeham, Thermal Conductivity of Five Hydrocarbons Along the Saturation Line, Int. J. Thermophys. 4 (3):193 (1983).
13. C. A. Nieto de Castro, S. F. Y. Li, A. Nagashima, R. D. Trengave and W. A. Wakeham, Standard Reference Data for the Thermal Conductivity of Liquids, J. Phys. Chem. Ref. Data 15 (3):1073 (1986).
14. J. M. N. A. Fareleira, S. F. Y. Li, G. C. Maitland and W. A. Wakeham, Thermal Conductivity of Two Branched Alkanes in the Temperature Range 36-88 °C up to 0.6 GPa, High Temp.-High Pressure 16:427 (1984).
15. Y. Wada, Y. Nagasaka and A. Nagashima, Measurements and Correlation of the Thermal Conductivity of Liquid n-Paraffin Hydrocarbons and Their Binary and Ternary Mixtures, Int. J. Thermophys. 6 (3):251 (1985).
16. C. A. Nieto de Castro, J. C. G. Calado and W. A. Wakeham, Thermal Conductivity of Organic Liquids Measured by a Transient Hot-Wire Technique, High Temp.-High Pressure 11:551 (1979).
17. H. Kashiwagi, M. Oishi, Y. Tanaka, H. Kubota and T. Makita, Thermal Conductivity of Fourteen Liquids in the Temperature Range 298-373 K, Int. J. Thermophys. 3 (2):101 (1982).
18. Yu. L. Rastorguev, G. F. Bogatov and B. A. Grigor'ev, Thermal Conductivity of Hydrocarbons at High Pressures and Temperatures (in Russian), Teplo. Svoistva. Zhidk; Mater. Vses. Teplofiz. Konf. Svoistva. Vesch. Vys. Temp. 3rd, pp. 88-91 (1970).

19. J. K. Horrocks, E. McLaughlin and A. R. Ubbenlohde, Liquid-phase Thermal Conductivities Isotopically Substituted Molecules, Tras. Faraday Soc. 59 (5):1110 (1963).

20. J. C. G. Calado, U. V. Mardolcar, C. A. Nieto de Castro, H. M. Roder and W. A. Wakeham, The Thermal Conductivity of Liquid Argon, Physica, in press.

21. H. M. Roder and C. A. Nieto de Castro, The Thermal Conductivity of Liquid Argon for Temperatures Between 110 and 140 K with Pressures to 70 MPa, Int. J. Thermophys., in press.

22. Y. S. Touloukian and C. Y. Ho, "Properties of Nonmetallic Elements" McGraw-Hill, New York (1981).

23. H. M. Roder and C. A. Nieto de Castro, The Thermal Conductivity of Liquid Propane, J. Chem. Eng. Data 27 (1):12 (1982).

24. C. A. Nieto de Castro, R. Tufeu and B. Le Neindre, Thermal Conductivity Measurement of n-Butane Over Wide Temperature and Pressure Range, Int. J. Thermophys. 4 (1):11 (1983).

25. A. M. F. Palavra, W. A. Wakeham and M. Zalaf, Thermal Conductivity of Normal Pentane in the Temperature Range 306-360 K at Pressures up to 0.5 GPa, Int. J. Thermophys. 8 (3):305 (1987).

26. S. F. Y. Li, G. C. Maitland and W. A. Wakeham, The Thermal Conductivity of n-Exane and n-Octane at Pressures up to 0.64 GPa in the Temperature Range 34-90°C, Ber. Bunsenges. Phys. Chem. 88:32 (1984).

27. J. Menashe and W. A. Wakeham, Absolute Measurements of the Thermal Conductivity of Liquids at Pressures up to 500 MPa, Ber. Bunsenges. Phys. Chem. 85:340 (1981).

28. J. Menashe and W. A. Wakeham, The Thermal Conductivity of n-Nonane and n-Undecane at Pressures up to 500 MPa in the Temperature Range 35-90°C, Ber. Bunsenges. Phys. Chem. 86:541 (1982).

29. M. Mustafa,,M. Sage and W. A. Wakeham, The Thermal Conductivity of n-Tridecane at Pressures up to 500 MPa in the Temperature Range 35-75°C, Int. J. Thermophys. 3 (3):217 (1982).

30. S. F. Y. Li, G. C. Maitland and W. A. Wakeham, Thermal Conductivity of Benzene and Cyclohexane in the Temperature Range 36-90°C at Pressures up to 0.33 GPa, Int. J. Thermophys. 5 (4):351 (1984).

31. C. A. Nieto de Castro, S. F. Y. Li, G. C. Maitland and W. A. Wakeham, Thermal Conductivity of Toluene in the Temperature Range 35-90°C at Pressures up to 600 MPa, Int. J. Thermophys. 4 (4):311 (1983).

32. C. A. Nieto de Castro, R. D. Trengave and W. A. Wakeham, The Density Dependence of the Thermal Conductivity of Toluene, Rev. Port. Quim. (Abril 1985)

33. H. Kashiwagi, H. Hashimoto, Y. Tanaka, H. Kubota and T. Makita, Thermal Conductivity and Density of Toluene in the Temperature Range 273-373 K at Pressures up to 250 MPa, Int. J. Thermophys. 3 (3):201 (1982).

34. U. V. Mardolcar, C. A. Nieto de Castro and W. A. Wakeham, Thermal Conductivity of Argon in the Temperature Range 107 to 423 K, Int. J. Thermophys. 7 (2):259 (1986).

35. H. M. Roder, Transport Properties of Oxygen, NASA Reference Publication 1102, NBSIR 82-1672, (April 1983).

A MODIFIED GENERAL CORRESPONDING STATES

EQUATION FOR POLAR LIQUID MIXTURES

Chi Zheng

Peimin Tang

Shanghai Pharmaceutical
Ind. Design Institute
Shanghai, P.R.China

Shanghai Second
Education Institute
Shanghai, P.R.China

INTRODUCTION

Great attention has been paid to the use of corresponding states principle for predicting the properties of fluids. The equation proposed by Pitzer et al.[1], introducing ascentric factor (ω) as a third parameter, is mostly well known. As ω accounts mainly for the shape and size of a molecule, so Pitzer's equation gives relatively large deviations for polar fluids with dipole moments. In view of this problem, several corresponding states equation with a fourth parameter have been developed. Among these, the one with reduced dipole moment (μ_R) proposed by O'Connell and Prausnitz [2] has some theoretical ground. But, because μ_R enters as a logarithm term in their equation for calculating the second virial coefficient, they include a factor of 10^5 in μ_R, thus their equation has limitations. Recently, Wang Rengyoan[3] has made rather systematic review of the development of corresponding states principle.

Based on the fundamental principle of corresponding states and utilizing the handy fluid parameters, the authors of this paper proposed a reasonably simplified corresponding states model. The modified general corresponding states equation developed has been applied to predict the viscosities of polar liquid mixtures satisfactorily[4]. The purpose of this paper is to use this modified general corresponding states equation to predict the thermal conductivities of polar liquid mixtures.

EQUATIONS

1. Pure Liquids

The reduced thermal conductivity of liquid is expressed as $\lambda\phi$, where $\phi = Vc^{2/3} Tc^{-1/2} M^{1/2}$.

Suppose $\lambda\phi$ is composed of two contributions, one is non-polar part and the other is polar part. This simplified model can be expressed by the equation

$$\lambda\phi = (\lambda\phi)^0_{NP} + \omega(\lambda\phi)^i_P \tag{1}$$

Let the non-polar contribution part

$$(\lambda\phi)_{NP} = (\lambda\phi)^{\circ}_{NP} + \omega(\lambda\phi)^{1}_{NP} \tag{2}$$

where, $(\lambda\phi)^{\circ}_{NP}$ = non-polar thermal conductivity function of liquid with spherical molecules;

$(\lambda\phi)^{1}_{NP}$ = non-polar thermal conductivity deviation function of objective liquid.

Let the polar contribution part

$$(\lambda\phi)_{P} = (\lambda\phi)^{\circ}_{P} + \mu_{R}(\lambda\phi)^{1}_{P} \tag{3}$$

where, $(\lambda\phi)^{\circ}_{P}$ = polar thermal conductivity function of liquid with spherical molecules;

$(\lambda\phi)^{1}_{P}$ = polar thermal conductivity deviation function of objective liquid;

μ_{R} = reduced dipole moment.

Substitute eq.(2) and eq.(3) into eq.(1), we get

$$(\lambda\phi) = [(\lambda\phi)^{\circ}_{NP} + (\lambda\phi)^{\circ}_{P}] + [\omega(\lambda\phi)^{1}_{NP} + \mu_{R}(\lambda\phi)^{1}_{P}] \tag{4}$$

Suppose $(\lambda\phi)^{1}_{P} = \alpha(\lambda\phi)^{1}_{NP}$, and let $(\lambda\phi)^{\circ}_{NP} + (\lambda\phi)^{\circ}_{P} = (\lambda\phi)^{\circ}$. Drop the subscript NP, then eq.(4) is simplified to

$$(\lambda\phi) = (\lambda\phi)^{\circ} + (\omega + \alpha\mu_{R})(\lambda\phi)^{1} \tag{5}$$

Eq.(5) is the modified general corresponding states equation for the thermal conductivities of pure liquids. Where, α is defined as the polar contribution factor of pure liquid, $(\lambda\phi)^{\circ}$ is the effective thermal conductivity function of liquid with spherical molecules, $(\lambda\phi)^{1}$ is the effective conductivity deviation function of objective liquid.

2. Liquid Mixtures

When the modified general corresponding states equation is applied to binary liquid mixtures, eq.(5) can be separated into non-polar part and polar part, i.e.

$$(\lambda\phi)_{NP} = (\lambda\phi)^{\circ}_{NP} + \omega(\lambda\phi)^{1} \tag{6}$$

and

$$(\lambda\phi)_{P} = (\lambda\phi)^{\circ}_{P} + \alpha\mu_{R}(\lambda\phi)^{1} \tag{7}$$

Select two reference liquids, as proposed by Teja et al.[5,6,7], the corresponding non-polar part and polar part of the reduced thermal conductivity of a liquid mixture can be obtained from eq.(6) and eq.(7) respectively:

$$(\lambda\phi)_{m,NP} = (\lambda\phi)_{1,NP} + \frac{\omega_{m} - \omega_{1}}{\omega_{2} - \omega_{1}} \cdot [(\lambda\phi)_{2,NP} - (\lambda\phi)_{1,NP}] \tag{8}$$

$$(\lambda\phi)_{m,P} = (\lambda\phi)_{1,P} + \frac{\alpha_{m}\mu_{Rm} - \alpha_{1}\mu_{R1}}{\alpha_{2}\mu_{R2} - \alpha_{1}\mu_{R1}} \cdot [(\lambda\phi)_{2,P} - (\lambda\phi)_{1,P}] \tag{9}$$

Suppose

$$(\lambda\phi)_{2,P} - (\lambda\phi)_{1,P} = \beta[(\lambda\phi)_{2,NP} - (\lambda\phi)_{1,NP}] \tag{10}$$

then

$$(\lambda\phi)_{2,NP} - (\lambda\phi)_{1,NP} = \frac{1}{1+\beta}[(\lambda\phi)_{2} - (\lambda\phi)_{1}] \tag{11}$$

After substituting eq.(10) and eq.(11) into eq.(8) and eq.(9), add to obtain the following modified general corresponding states equation for the thermal conductivities of liquid mixtures:

$$(\lambda\phi)m = (\lambda\phi)_{1} + \frac{1}{1+\beta}[\frac{\omega_{m} - \omega_{1}}{\omega_{2} - \omega_{1}} + \beta(\frac{\alpha_{m}\mu_{Rm} - \alpha_{1}\mu_{R1}}{\alpha_{2}\mu_{R2} - \alpha_{1}\mu_{R1}})] \cdot [(\lambda\phi)_{2} - (\lambda\phi)_{1}] \tag{12}$$

Where, β is defined as the polar contribution factor of liquid mixture, and is a function of composition.

3. Mixing Rules

The mixing rules used in this paper are:

$$Tcm = (X_1^2 \cdot \frac{Tc_1^2}{Pc_1} + X_2^2 \cdot \frac{Tc_2^2}{Pc_2} + 2X_1X_2 \cdot \frac{Tc_{12}^2}{Pc_{12}})/$$
$$(X_1^2 \cdot \frac{Tc_1}{Pc_1} + X_2^2 \cdot \frac{Tc_2}{Pc_2} + 2X_1X_2 \cdot \frac{Tc_{12}}{Pc_{12}}) \tag{13}$$

$$Pcm = Tcm/(X_1^2 \cdot \frac{Tc_1}{Pc_1} + X_2^2 \cdot \frac{Tc_2}{Pc_2} + 2X_1X_2 \cdot \frac{Tc_{12}}{Pc_{12}}) \tag{14}$$

$$\omega m = [X_1^2 \omega_1 (\frac{Tc_1}{Pc_1})^{2/3} + X_2^2 \omega_2 (\frac{Tc_2}{Pc_2}) + 2X_1X_2 \omega_{12}(\frac{Tc_{12}}{Pc_{12}})^{2/3}]/$$
$$(\frac{Tcm}{Pcm})^{2/3} \tag{15}$$

$$Tc_{12} = \sqrt{Tc_1 \cdot Tc_2} \tag{16}$$

$$Pc_{12} = 8 Tc_{12}/[(\frac{Tc_1}{Pc_1})^{1/3} + (\frac{Tc_2}{Pc_2})^{1/3}]^3 \tag{17}$$

$$\omega_{12} = \frac{\omega_1 + \omega_2}{2} \tag{18}$$

$$\mu_{Rm} = X_1^2 \mu_{R1} + X_2^2 \mu_{R2} + 2X_1X_2 \sqrt{\mu_{R1} \cdot \mu_{R2}} \tag{19}$$

$$\alpha_m = X_1^2 \alpha_1 + X_2^2 \alpha_2 + 2X_1X_2 \cdot (\frac{\alpha_1 + \alpha_2}{2}) \tag{20}$$

$$Vcm = X_1^2 Vc_1 + X_2^2 Vc_2 + 2X_1X_2 Vc_{12} \tag{21}$$

$$Vc_{12} = \psi_{12} (\frac{Vc_1^{1/3} + Vc_2^{1/3}}{2})^3 \tag{22}$$

where, ψ_{12} is the interactive coefficient.

CACULATION

The procedure of calculating the thermal conductivity of a binary liquid mixture is as follows:

1. Calculation of $(\lambda\phi)^0$ and $(\lambda\phi)^1$ of reference liquids

Choose two non-polar liquids ($\mu_R=0$) with ω values close to those of objective liquids (i.e. components) as reference liquids. Calculate $(\lambda\phi)^0$ and $(\lambda\phi)^1$ at different reduced temperatures by means of eq.(5).

2. Calculation of μ_R of polar components

In view of the fact that the Stockmayer parameter $\delta=\mu/2\epsilon\sigma^3$ appears as δ^2 in the Stockmayer collision integral[8], we take the reduced dipole moment proportional to δ^2, as $\mu_R= \mu^4 Pc^2/Tc^4$.

Calculate μ_{R1} and μ_{R2} of polar components.

3. Calculation of α of polar components.

Substitute ω , μ_R and thermal conductivity data at different temperatures of polar components, as well as the corresponding $(\lambda\phi)^0$ and $(\lambda\phi)^1$ values calculated in step 1, into eq.(5), calculate the corresponding α values.

4. Calculation of β of the liquid mixture and ψ_{12} in the mixing rule

Let $\beta = X_1X_2 \cdot [A + B (X_1 - X_2)]$ (23)

Use equations (12) to (22) by substituting in relevant data, and adopt the Powell method to optimize the calculated values of ψ_{12}, A and B at different temperatures with the objective function

$$F = \sum_n [(\lambda_{ex} - \lambda_{ca})/ \lambda_{ex}]^2 \qquad (24)$$

5. Calculation of the root-mean-square deviation of the calculated thermal conductivities of the liquid mixture

$$RMS\lambda = \sqrt{\sum_n (\frac{\lambda_{ex} - \lambda_{ca}}{\lambda_{ex}})^2 / n} \qquad (25)$$

where, n =   number of data points at certain temperature.

RESULT

22 pairs of binary liquid mixtures have been calculated according to the procedure mentioned above. They are: ethanol-benzene, acetone-isobutanol, acetone-carbon tetrachloride, n-butanol-carbon tetrachloride, isobutanol-carbon tetrachloride, cyclohexane-carbon tetrachloride, dichloromethane-carbon tetrachloride, n-hexanol-carbon tetrachloride, n-hepthanol-carbon tetrachloride, methanol-carbon tetrachloride, ethyl ether-chloroform, ethanol-cyclohexane, dichloromethane-ethyl ether, aniline-n-hepthanol, ethylbenzene-methyl ethyl ketone, toluene-methyl ethyl ketone, cyclohexane-isopropanol, toluene-isopropanol, methanol-toluene, phenol-toluene, acetone-water, ethanol-water. The calculation result of 60 sets of data (350 data points in all) has a root-mean-square deviation range of 0.0055% to 8.0591%. Table 1 is the list of polar components with the corresponding reference liquids, Table 2 shows the calculation result of binary liquid mixtures.

Table 1. Polar Components with Corresponding Reference Liquids

Component	μ_R	ω	Ref.liq.	ω	Data source
$(CH_3)_2 CO$	2.285	0.309	$n-C_6H_{14}$	0.296	[9]
$C_6H_5 NH_2$	7.538×10^{-2}	0.382	$n-C_8H_{18}$	0.394	
$n-C_4H_9 OH$	0.199	0.590	$n-C_{12}H_{26}$	0.562	
$i-C_4H_9 OH$	0.165	0.588	$n-C_{12}H_{26}$	0.562	
$CHCl_3$	5.157×10^{-2}	0.216	$n-C_5H_{12}$	0.251	
$c-C_6H_{12}$	1.396×10^{-4}	0.213	C_6H_6	0.212	
CH_2Cl_2	0.559	0.193	C_6H_6	0.212	
C_2H_5OH	0.467	0.635	$n-C_{13}H_{28}$	0.632	
$C_6H_5C_2H_5$	2.237×10^{-4}	0.301	$n-C_6H_{14}$	0.296	
$(C_2H_5)_2 O$	7.759×10^{-2}	0.281	$n-C_6H_{14}$	0.296	
$n-C_6H_{13} OH$	0.121	0.560	$n-C_{12}H_{26}$	0.562	
$n-C_7H_{15} OH$	4.682×10^{-2}	0.560	$n-C_{12}H_{26}$	0.562	
CH_3OH	0.772	0.599	$n-C_{12}H_{26}$	0.562	
$CH_3 COC_2H_5$	2.422	0.329	$n-C_7H_{16}$	0.351	
C_6H_5OH	0.103	0.440	$n-C_9H_{20}$	0.444	
$i-C_3H_7OH$	0.276	0.624	$n-C_{13}H_{28}$	0.623	
$C_6H_5CH_3$	3.443×10^{-4}	0.257	$n-C_5H_{12}$	0.251	
H_2O	2.831	0.344	$n-C_7H_{16}$	0.351	

Table 2. Calculation Result of Binary Liquid Mixtures

System	T/K	α_1	α_2	X	RMSλ/%	Data source
$C_2H_5OH(1)$- $C_6H_6(2)$	279.15	-1.1165	-		0.8704	[10]
	293.15	-1.1070	-		0.7614	
	313.15	-1.0777	-	0.1585-	1.8333	
	333.15	-1.1053	-	0.9057	1.5058	
	348.15	-1.1026	-		1.2767	

$\psi_{12} = 1.3929$
$\beta = X_1X_2[1.6925-0.1264(X_1-X_2)]$

System	T/K	α_1	α_2	X	RMSλ/%
$(CH_3)_2CO(1)$- $i-C_4H_9OH(2)$	273.15	-0.0337	-2.1799	0.2984- 0.8362	0.5995

$\psi_{12} = 1.0597$
$\beta = X_1X_2[0.0896-2.7538(X_1-X_2)]$

System	T/K	α_1	α_2	X	RMSλ/%
$(CH_3)_2CO(1)$- $CCl_4(2)$	273.15	-0.0697	-	0.3984- 0.9137	1.9771

$\psi_{12} = 1.2254$
$\beta = -0.0498X_1X_2$

System	T/K	α_1	α_2	X	RMSλ/%
$n-C_4H_9OH(1)$- $CCl_4(2)$	273.15	-1.9850	-	0.3416- 0.8925	0.3423
	323.15	-1.9535	-		1.2707

$\psi_{12} = 1.3046$
$\beta = X_1X_2[1.0447+13.103(X_1-X_2)]$

System	T/K	α_1	α_2	X	RMSλ/%
$i-C_4H_9OH(1)$- $CCl_4(2)$	273.15	-2.5355	-	0.1874- 0.8616	0.7088
	288.15	-2.5192	-	0.1941- 0.8889	1.1431
	338.15	-2.4355	-	0.3416- 0.8925	2.7256

$\psi_{12} = 1.1357$
$\beta = X_1X_2[-1.6169+1.0242(X_1-X_2)]$

System	T/K	α_1	α_2	X	RMSλ/%
$c-C_6H_{12}(1)$- $CCl_4(2)$	293.15	-257.51	-	0.1483- 0.8527	0.6511
	313.15	-296.80	-		0.7511
	333.15	-271.56	-		1.1886

$\psi_{12} = 1.2318$
$\beta = X_1X_2[2.7201-6.0799(X_1-X_2)]$

System	T/K	α_1	α_2	X	RMSλ/%
$CH_2Cl_2(1)$- $CCl_4(2)$	273.15	-0.0738	-	0.3116- 0.8446	0.6063

$\psi_{12} = 0.9440$
$\beta = X_1X_2[-0.0025+0.3245(X_1-X_2)]$

System	T/K	α_1	α_2	X	RMSλ/%
$n-C_6H_{13}OH(1)$- $CCl_4(2)$	273.15	-2.4871	-	0.3341- 0.8187	0.5762
	323.15	-2.3875	-		1.6241

$\psi_{12} = 1.0618$
$\beta = X_1X_2[1.8756-4.2235(X_1-X_2)]$

System	T/K	α_1	α_2	X	RMSλ/%
$n-C_7H_{15}OH(1)$- $CCl_4(2)$	273.15	-5.8645	-	0.2486- 0.7988	1.1259
	323.15	-5.0080	-	0.3062- 0.7988	7.3765

$\psi_{12} = 1.1399$
$\beta = X_1X_2[0.0741-2.0488(X_1-X_2)]$

Table 2. Calculation Result of Binary Liquid Mixtures
(continued)

System	T/K	α_1	α_2	X	RMS$\sqrt{}$/%	Data source
$CH_3OH(1)$-$CCl_4(2)$	273.15	-0.6620	-	0.6253-0.9639	0.8918	
	323.15	-0.6683	-		2.6943	

ψ_{12} = 2.1965

$\beta = X_1X_2[2.0003-2.3064(X_1-X_2)]$

System	T/K	α_1	α_2	X	RMS$\sqrt{}$/%	Data source
$(C_2H_5)_2O(1)$-$CHCl_3(2)$	223.15	-0.1355	0.1186	0.2871-0.8656	0.1784	
	273.15	-0.1569	0.2176	0.3493-0.8285	8.0591	

ψ_{12} = 1.5115

$\beta = X_1X_2[-3.0672-1.7473(X_1-X_2)]$

System	T/K	α_1	α_2	X	RMS$\sqrt{}$/%	Data source
$C_2H_5OH(1)$-c-$C_6H_{12}(2)$	298.15	-1.1046	-181.74	0.2733-0.8679	0.4703	
	303.15	-1.0806	-174.61		2.9457	
	313.15	-1.0777	-160.17		2.4230	

ψ_{12} = 1.3040

$\beta = X_1X_2[-0.3417\times10^{-4}+0.0133(X_1-X_2)]$

System	T/K	α_1	α_2	X	RMS$\sqrt{}$/%	Data source
$CH_2Cl_2(1)$-$(C_2H_5)_2O(2)$	223.15	-0.1017	-0.4349	0.1791-0.7773	0.4535	
	273.15	-0.1127	-0.5669		6.8201	

ψ_{12} = 1.4258

$\beta = X_1X_2[-0.6603+1.5966(X_1-X_2)]$

System	T/K	α_1	α_2	X	RMS$\sqrt{}$/%	Data source
$C_6H_5NH_2(1)$-n-$C_7H_{15}OH(2)$	273.15	-1.4494	-4.5185	0.2937-0.7892	5.7221	
	348.15	-1.2563	-3.6649		1.7229	
	423.15	-0.9793	-3.5169		4.9673	

ψ_{12} = 1.1596

$\beta = X_1X_2[0.6163-4.2297(X_1-X_2)]$

System	T/K	α_1	α_2	X	RMS$\sqrt{}$/%	Data source
$C_6H_5C_2H_5(1)$-$CH_3COC_2H_5(2)$	293.15	-61.563	-0.0293	0.1034-0.8772	0.8956	
	303.15	-58.536	-0.0344		0.8543	
	313.15	-55.484	-0.0311		0.8472	
	323.15	-61.556	-0.0287		0.9730	

ψ_{12} = 1.1453

$\beta = X_1X_2[-0.9290\times10^{-2}-0.5485\times10^{-2}(X_1-X_2)]$

System	T/K	α_1	α_2	X	RMS$\sqrt{}$/%	Data source
$C_6H_5CH_3(1)$-$CH_3COC_2H_5(2)$	298.15	-38.538	-0.0353	0.0970-0.8824	0.3052	
	303.15	-37.990	-0.0354		0.3120	
	313.15	-36.880	-0.0326		0.8783	
	323.15	-35.754	-0.0316		0.9527	

ψ_{12} = 0.9133

$\beta = X_1X_2[0.0454+0.0678(X_1-X_2)]$

System	T/K	α_1	α_2	X	RMS$\sqrt{}$/%	Data source
c-$C_6H_{12}(1)$-i-$C_3H_7OH(2)$	298.15	-181.74	-1.8161	0.0951-0.9005	2.1811	
	303.15	-174.61	-1.8132		2.1533	
	313.15	-160.17	-1.8072		2.2034	
	323.15	-151.53	-1.8624		2.2148	

ψ_{12} = 0.9312

$\beta = X_1X_2[-0.9990\times10^{-2}+0.4553(X_1-X_2)]$

Table 2. Calculation Result of Binary Liquid Mixtures
(continued)

System	T/K	α_1	α_2	X	RMSλ%	Data source
$C_6H_5CH_3(1)-$	298.15	-42.472	-1.6178	0.2039-	0.0055	
$i-C_3H_7OH(2)$	303.15	-41.813	-1.6110	0.7680	0.4333	
	313.15	-40.482	-1.5972		0.5603	
	323.15	-39.134	-1.6278		0.1438	

$\psi_{12} = 0.8445$

$\beta = X_1 X_2 [-0.1419+0.7179(X_1-X_2)]$

System	T/K	α_1	α_2	X	RMSλ%	Data source
$CH_3OH(1)-$	273.15	-0.5798	-42.101	0.4182-	1.0643	
$C_6H_5CH_3(2)$				0.9200		
	323.15	-0.5690	-37.051	0.4894-	0.9876	
				0.8961		

$\psi_{12} = 1.3693$

$\beta = X_1 X_2 [-0.0610+0.2837(X_1-X_2)]$

System	T/K	α_1	α_2	X	RMSλ%	Data source
$C_6H_5OH(1)-$	333.15	-2.2902	-41.386	0.1530-	1.2757	
$C_6H_5CH_3(2)$				0.6824		

$\psi_{12} = 0.9511$

$\beta = X_1 X_2 [32776+4.4385(X_1-X_2)]$

System	T/K	α_1	α_2	X	RMSλ%	Data source
$H_2O(1)-$	273.15	-0.0637	-0.0288	0.0000-	0.5972	
$(CH_3)_2CO(2)$	293.15	-0.0588	-0.0369	1.0000	1.7345	
	313.15	-0.0541	-0.0340		3.5303	
	333.15	-0.0498	-0.0338		5.6071	

$\psi_{12} = 2.0967$

$\beta = X_1 X_2 [60.614-72.231(X_1-X_2)]$

System	T/K	α_1	α_2	X	RMSλ%	Data source
$H_2O(1)-$	273.15	-0.0939	-1.0808	0.0000-	0.2609	
$C_2H_5OH(2)$	293.15	-0.0871	-0.9805	0.9109	1.4769	
	313.15	-0.0807	-0.9498		5.6748	
	333.15	-0.0747	-1.0341		6.1907	
	253.15	-0.0691	-1.0313		6.4418	

$\psi_{12} = 1.5167$

$\beta = X_1 X_2 [-0.5960+0.5437(X_1-X_2)]$

DISCUSSION

The 22 pairs of binary liquid mixtures calculated deal with components of different categories, with a temperature range of -50°C to 150°C. The calculation result of 60 sets of data (350 data points in all) is satisfactory, with root-mean-square deviation range of 0.0055% to 8.0591%.

The result proves the feasibility of the proposed simplified model, i.e. the reduced thermal conductivity is assumed to be composed of non-polar contribution and polar contribution. In the simplified model, a concept of polar contribution factors (α, β) is introduced.

The polar contribution factors of pure polar liquids α are negative. This means that dipole moment restricts the vibration collision of molecules. The α value of a liquid varies not only with the reference liquid chosen for itself, but also with the reference liquid chosen for the other component liquid. That is why we choose non-polar liquids $(\mu_k=0)$ with ω values close to those of components as reference liquids, and define $(\lambda\phi)^\circ$ the effective thermal conductivity function of liquid with spherical molecules and $(\lambda\phi)^f$ the effective thermal conductivity deviation function of component liquid. This treatment seems arbitrary, but effective.

The modified general corresponding states equation proposed in this paper can be used very easily. Given one set of isothermal thermal conductivities data of a binary liquid mixture, any other set of thermal conductivities of this mixture can be predicted.

NOMENCLATURE

A,B = coefficients in eq. (23)
F   = objective function, eq. (24)
M   = molecular weight
n   = number of data points
P   = pressure
RMS = root-mean-square deviation
T   = temperature
V   = volume
X   = mole fraction

Greek

α = polar contribution factor of pure liquid
β = polar contribution factor of liquid mixture
λ = thermal conductivity
μ = dipole moment
ϕ = $Vc^{2/3} Tc^{-1/2} M^{1/2}$
ψ = interaction coefficient
ω = ascentric factor

Superscript

0   = function of spherical molecules
1   = deviation function

Subscript

1,2 = component number
c   = critical
ca = calculated
ex = experimental
m   = mixture
NP = non-polar
P   = polar
R   = reduced

REFERENCES

[1] Pitzer, K.S., D.Z.Lipmann, R.F.Curl, C.M.Huggins, P.E.Peterson, Ind. Eng. Chem., 50, 265 (1958).
[2] O'Connell, J.P., J.M.Prausnitz, I & EC Process Des. Dev., 6, 245 (1967).
[3] Wang Rengyuan: Master thesis, East China Institute of Chemical Technology (1986).
[4] Zheng, C., P.M.Tang: "A New Corresponding States Equation For Liquid Viscosities", presented at Chemical Engineering Thermodynamics Symposium, 1987, Hangzhou, P,R.China
[5] Teja, A.S., AIChE J. 26, 337 (1980).
[6] Teja, A.S., S.I.Sandler, AIChE J., 26, 341 (1980).
[7] Teja, A.S., N.C.Patel, S.I.Sandler, Chem.Eng.J., 21, (1981).

[8] Brokaw, R.S., I & EC Process Des. Dev., _8_, 240 (1969).

[9] Reid, R.C., J.M.Prausnitz, T.K.Sherwood, "The Properties of Gases and Liquids", 3rd ed., McGraw-Hill, New York (1977).

[10] Jamieson D.T., J.B.Irving, J.S.Tudhope, "Liquid Thermal Conductivity: A Data Survey to 1973", H.M. Stationery Office, Edinburgh (1975).

SESSION 5

OTHER MATERIALS AND EFFECTS

MEASUREMENT OF ANISOTROPIC BEHAVIOR OF

THERMAL DIFFUSIVITY OF POLYMERS

Masakazu Okuda   and   Akira Nagashima

Department of Mechanical Engineering
Keio University
Hiyoshi, Yokohama, 223, Japan

ABSTRACT

    When a polymer material is stretched, the chain molecules tend to
align along the stretch direction. As a result of this orientation of
polymer molecules, the polymer material is expected to have the anisotropic
thermal diffusivity which is extremely difficult to be measured by
conventional experimental methods. The forced Rayleigh scattering method
which has been developed by the present authors' group as one of the
reliable methods has a number of unique features such as high-speed, non-
contact technique and its ability to measure anisotropic behavior. In the
present study, the anisotropic behavior of the thermal diffusivity of
poly(vinyl chloride) films under uniaxially stretched condition was
precisely measured to demonstrate the application of the forced Rayleigh
scattering method to unusual substances. The stretch-ratio dependence and
the angular dependence of the thermal diffusivity were measured, and it was
confirmed that the molecular orientation causes a large anisotropy on the
thermal diffusivity of polymer materials. In addition, the relation between
the stress relaxation and anisotropy of the thermal diffusivity was
investigated, and the independence of them was confirmed. The accuracy of
the present measurement was estimated as ±5%.

INTRODUCTION

    Polymer materials are recently used in many diversified and new
fields. In such situations, more precise properties of polymer materials
have to be known. The oriented polymer shows anisotropy of many macroscopic
physical properties, especially of mechanical properties. It can be easily
expected that this orientation of polymer molecules also causes an
anisotropy of thermophysical properties like the thermal conductivity or
the thermal diffusivity. In regard to these properties, along the chain
axis, the atoms are covalently bonded, so the thermal conductivity or the
thermal diffusivity is expected to be large. However, perpendicular to the
chain axis, a smaller value is expected from the weaker van der Waals
interaction. Even though this anisotropy is of considerable scientific and
technological interest, it is difficult to measure quantitatively by
conventional experimental methods. Choy et al.[1] have measured the
anisotropy of the thermal conductivity of oriented polymers using a steady
heat-flow method, and Eiermann et al.[2] used a quasisteady state method.

In these previous studies, uniform heat flow in the sample was assumed by adjusting the sample shape and the way of heating, and the anisotropy of many polymers was measured over a range of temperature. However, the error factors of the experiment were not fully discussed. Because of the difficulty to satisfy the assumption exactly, it is considered very difficult to perform the measurement with high accuracy. More recently, Blum et al.[3] have measured the anisotropy ratio of the thermal conductivity of oriented polymers using the De Sénarmont method. Although this latter method is very useful and simple for measuring the ratio of anisotropy, it is difficult to evaluate the absolute values.

The forced Rayleigh scattering method which has been recently developed and improved to a reliable method by the present authors' group[4]-[9], is an optical measuring technique for the thermal diffusivity and has the unique features as follows : (1) capable of non-contact measurement, (2) very short measuring time (typically within 1ms), (3) requiring a very small sample volume, and (4) capability to measure the thermal diffusivity in any particular direction in anisotropic materials. Because of these characteristics, this method can be applied to measurements which have never been accomplished by other conventional methods. In the present study, the anisotropic behavior of the thermal diffusivity of Poly(vinyl chloride)(PVC) films under uniaxially stretched condition was measured to show one of the applications of the forced Rayleigh scattering method.

EXPERIMENTAL

Only a brief description is given here since the full details of the principle and the historical background were explained elsewhere [4][5][8].

The present experimental apparatus is shown in Figure 1. The thermal diffusivity measurements were made at room temperature under atmospheric pressure. The heating laser is an argon ion laser (1.8W). The heating laser beam is chopped by a rotating mechanical chopper into a short pulse, whose duration time is about 200μs, and is divided into two beams of equal intensity by a beam splitter. Divided beams cross in the thin layer of the sample to make an interference pattern. In the present study, the fringe spacing of the interference pattern was adjusted to be about 50μm.

The sample was dyed to absorb the wavelength of the heating laser beams. The resulting interference pattern gives a periodic temperature distribution in the sample. In the heated area, then, a spatially periodic distribution of the refractive index is created corresponding to the temperature distribution. This acts as an optical grating for another probing laser beam which has a different wavelength. The latter is not to be absorbed in the sample and we can get the diffracted light. After the very short heating time terminates, the temperature distribution in the sample is relaxed by heat conduction in the direction perpendicular to the fringes and consequently the intensity of the diffracted light decays. The intensity of the first order diffracted light I_1 is expressed as a function of time t;

$$I_1 \propto \exp \left(-\frac{2t}{\tau} \right) \qquad (1)$$

where τ is the time constant of heat conduction. It is defined as

$$\tau = \frac{1}{a} \left(\frac{d}{2\pi} \right)^2 \qquad (2)$$

where, a is the thermal diffusivity of the sample and d is the grating period, namely the fringe spacing of the interference pattern. Therefore we can calculate the thermal diffusivity of the sample by Eq.(2), if we observe the decay rate of the diffracted light intensity and measure the grating period.

As the probing laser, a He–Ne laser (5mW) was employed, and the first order diffracted beam was detected by a photomultiplier through a 500μm diameter pinhole and through an interference filter to cut the scattered light. The output signal of the photomultiplier was recorded by a digital memory, and the data were fed into a microcomputer to be analyzed. Figure 2 shows a typical example of the detected light signal.

In the process of relaxation of the temperature distribution after heating, the heat conduction is considered as one–dimensional heat conduction perpendicular to the interference pattern. So we measure the thermal diffusivity along this direction.

The sample used in this study was 0.4mm thick PVC films which contain 39wt% DOP as a plasticizer to be stretched easily at room temperature and 0.2wt% azo dye as coloring agent. For the measurements of the stress relaxation and the time dependence of the thermal diffusivity, we used a sample having about 0.1mm thickness and containing about 45wt% DOP for easier stretching. The width of the sample was 30mm and initial length (ℓ_0) was 20mm before stretching. The samples were stretched and held by the sample stretcher which was controlled by a microcomputer.

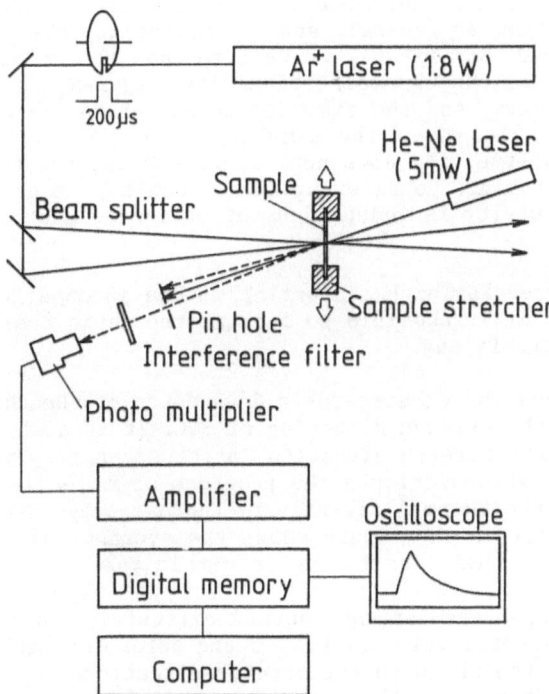

Figure 1. Schematic diagram of experimental apparatus.

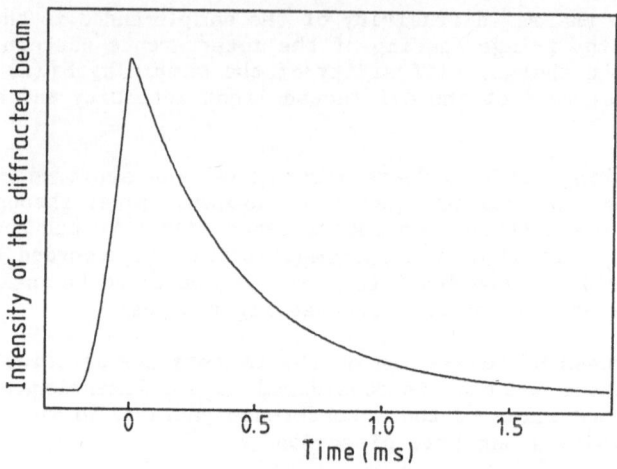

Figure 2. A typical diffracted light signal.

RESULTS AND DISCUSSIONS

In the present study, the molecular orientation was produced by stretching the sample. In general, stress relaxation take place when a polymer is stretched. So, firstly we want to know the relation between the stress relaxation and the thermal diffusivity. Figures 3 and 4 show the stress relaxation curve and the time dependence of the thermal diffusivity of PVC at $l/l_0=3$ (l:length of the sample), respectively. It was observed that the thermal diffusivity does not change whereas the stress decreases sharply after stretching. Considering these results, it can be concluded that thermal diffusivity is independent of stress relaxation in the case of PVC.

Since the forced Rayleigh scattering method is capable of very high-speed measurement, it is possible to measure the rapid change of the thermal diffusivity, if any.

Figure 5 shows the stretch–ratio dependence of the thermal diffusivity along the stretch direction of PVC. It is seen that the thermal diffusivity along the stretch direction increases sharply with increasing stretch–ratio l/l_0 which reflects the progress of molecular orientation. It was observed that the thermal diffusivity increases by more than 40% at $l/l_0=4$. Every plot in the figure shows the average value of about fifteen measured values, and their scattering is within ±4%.

The angular dependence of the thermal diffusivity of PVC is shown in Figure 6 which shows the value at $l/l_0=3$ and before stretching, respectively. θ is the angle to the stretch direction. The thermal diffusivity perpendicular to the interference pattern was measured by rotating the sample to the optical axis as shown in Figure 7. The stretched PVC shows a large anisotropy of the thermal diffusivity whereas the anisotropy of an unstretched sample is negligible. Moreover, an interesting feature is that the increase in the thermal diffusivity along the stretch

230

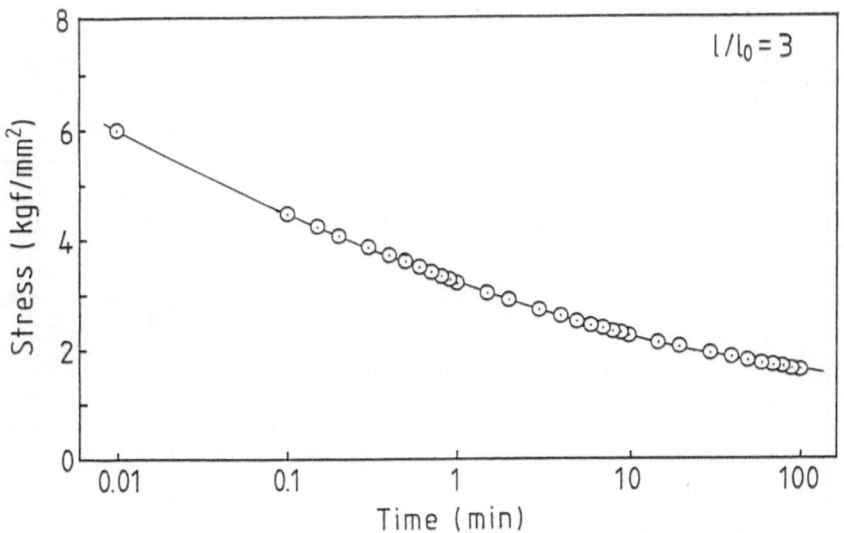

Figure 3.   The stress relaxation curve of PVC at
$\ell/\ell_0=3$, 22°C.

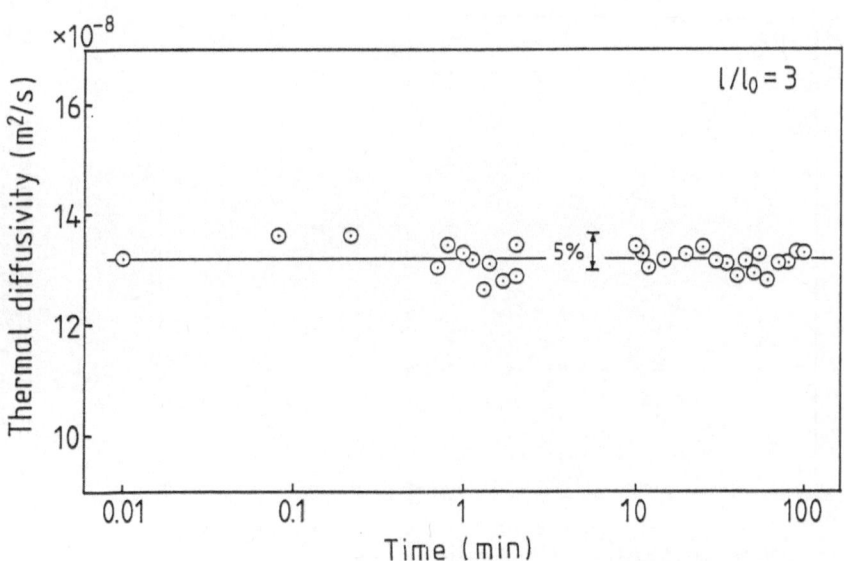

Figure 4.   Time-dependence of the thermal diffusivity
of PVC at $\ell/\ell_0=3$, 25°C.

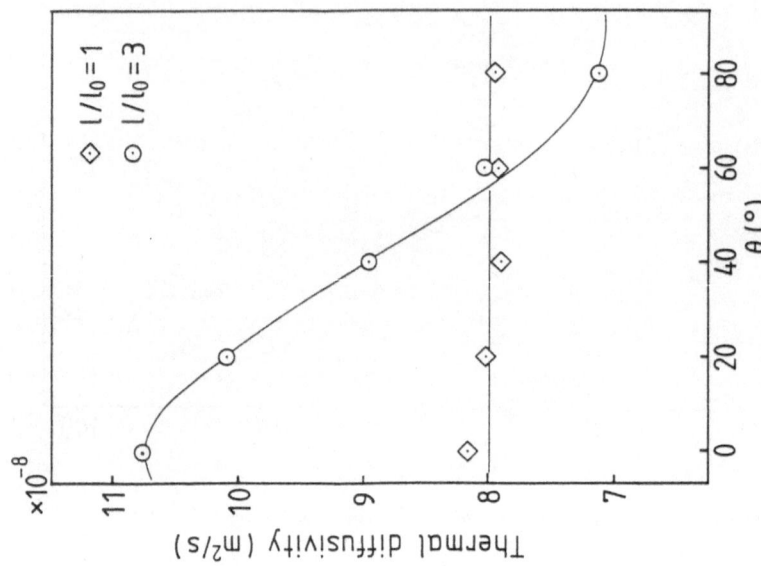
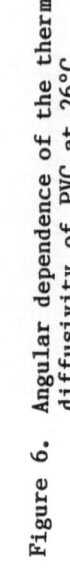

Figure 6. Angular dependence of the thermal diffusivity of PVC at 26°C.

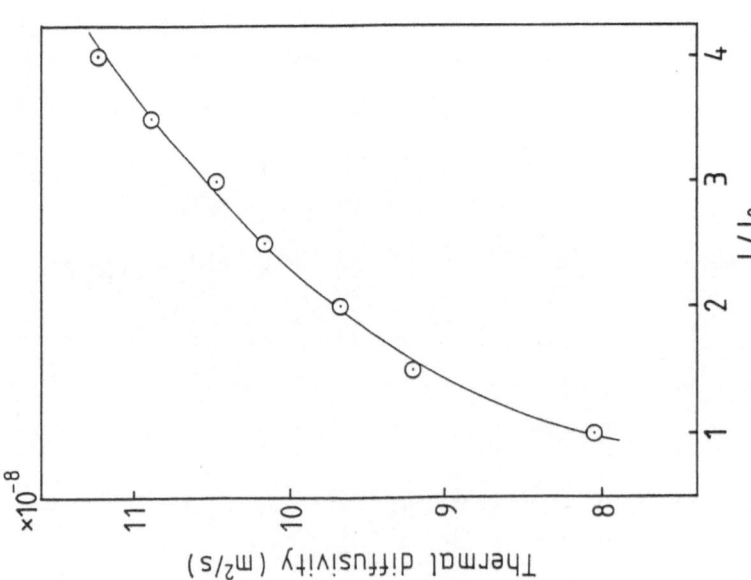

Figure 5. Stretch–ratio dependence of the thermal diffusivity of PVC at 26°C.

direction is larger than the decrease perpendicular to the stretch direction. Such behavior of oriented polymer agrees qualitatively with the results of Choy et al.[1] and Eiermann et al.[2].

Considering the accuracy of applying the same method to liquids, the accuracy of the present measurement is estimated to be better than ±5%.

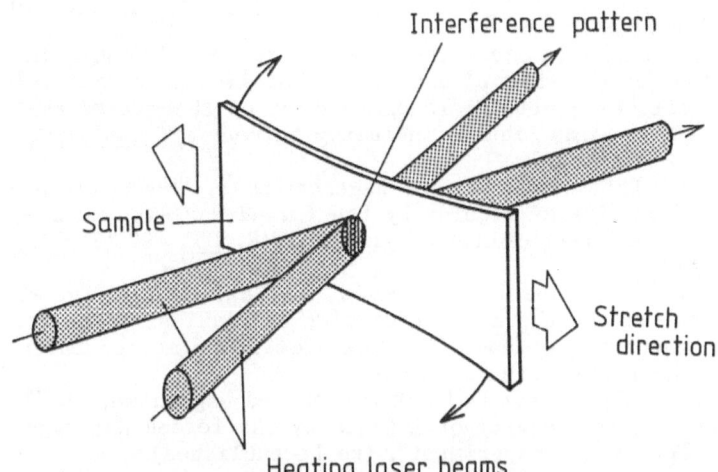

Figure 7.   The manner to measure the angular dependence of the thermal diffusivity. The thermal diffusivity along the stretch direction can be measured by this arrangement.

CONCLUSION

Using the forced Rayleigh scattering method, the thermal diffusivity of stretched PVC films has been measured. It was shown that the molecular orientation causes a large anisotropy in the thermal diffusivity of polymeric materials. The magnitude of this anisotropy was determined. Moreover, the relation between the thermal diffusivity and stress relaxation was investigated, and the independence of these two properties was confirmed. In the present study, the applicability of the method to measurement of oriented polymers has been shown, and the potential of the method was demonstrated.

ACKNOWLEDGMENTS

The authors sincerely wish to thank Mr. A. Watanabe and Mr. F. Ando of Okamoto industries, inc. for their cooperation to prepare the samples.

REFERENCES

1. Choy, C.L. and Greig, D., "The low temperature thermal conductivity of isotropic and oriented polymers", J. Phys. C: Solid State Phys., 10(1977), 169.
2. Eiermann, K. and Hellwege, K.H., "Thermal conductivity of high polymers from -180 C to 90 C", J. Polym. Sci., 57(1962), 99.
3. Blum, K., Kilian, H.G. and Pietralla, M., "A method for measuring the anisotropy ratio of the thermal conductivity of anisotropic solids", J. Phys. E: Sci. Instr., 16(1983), 807.
4. Hatakeyama, T., Nagasaka, Y. and Nagashima, A., "Measurement of the thermal diffusivity of liquids by the forced Rayleigh scattering method", Proc. ASME-JSME Thermal Engineering Joint Conf., Honolulu, (1987), Vol.4, 311.
5. Hatakeyama, T., Kadoya, K., Okuda, M., Nagasaka, Y. and Nagashima, A., "Measurement of the thermal diffusivity of liquids by the forced Rayleigh scattering method: I. Examination of the method and measurements on some liquids including toluene", Trans. JSME, B53-489 (1987),1590 (in Japanese).
6. Nagasaka, Y., Hatakeyama, T. and Nagashima, A., "Measurement of the thermal diffusivity of liquids by the forced Rayleigh scattering method: II. Analysis of error factors", Trans. JSME, B53-492 (1987), 2545 (in Japanese).
7. Hatakeyama, T., Miyahashi, Y., Okuda, M., Nagasaka, Y. and Nagashima, A., "Measurement of the thermal diffusivity of liquids by the forced Rayleigh scattering method: III. Measurement of molten salts", Trans. JSME, (to be published)(in Japanese).
8. Nagasaka, Y., Hatakeyama, T., Okuda, M. and Nagashima, A., "Measurement of the thermal diffusivity of liquids by the forced Rayleigh scattering method : Theory and experiment", (to be published).

L. Wieczorek, H.J. Goldsmid and G.L. Paul

School of Physics, University of New South Wales
P.O. Box 1,
Kensington, NSW 2033, Australia

ABSTRACT

A new technique has been introduced for the measurement of the thermal conductivity of thin amorphous films, following difficulties in extending the use of a previous laser-based method to low temperatures. Thin strips of gold are deposited on the film and on the bare substrate and, in each case, the resistance is determined as a function of the electric current. From the known temperature coefficient of the resistance, one can find the dependence of the temperature of each strip on the power input. The thermal conductivity of the amorphous material can be found from the difference between the temperature rises for the two strips. The method has been used to determine the thermal conductivity of amorphous silicon to below 10K.

INTRODUCTION

The measurement of the thermal conductivity of thin, amorphous, non-metallic films presents certain problems. Such films are invariably mounted on substrates that have a comparable or greater conductivity, so that accurate measurements can hardly be made with the flow of heat parallel to the surface. On the other hand, the thermal resistance per unit area is very small in a direction perpendicular to the surface because of the small thickness, usually about 1 μm.

Previously[1] we reported a technique for measuring the thermal conductivity, in the perpendicular direction, using a CW laser to provide the necessary heat flux. However, we encountered severe problems in extending the experiments to low temperatures, and we have now adopted a strip heating method.

Identical thin metallic strips are deposited on the amorphous film and the bare substrate, the latter being a good conductor of heat and being mounted on a heat sink at a controlled temperature. The temperature dependence of the electrical resistance of these strips is first determined, from liquid helium temperature up to about 300 K, using a very low current so as to avoid any significant heating effect. Then, a much larger current is passed and the potential difference across each of the strips is observed, as a function of time, at each temperature. The measured current and potential difference allow us to determine the power input to a strip and its temperature.

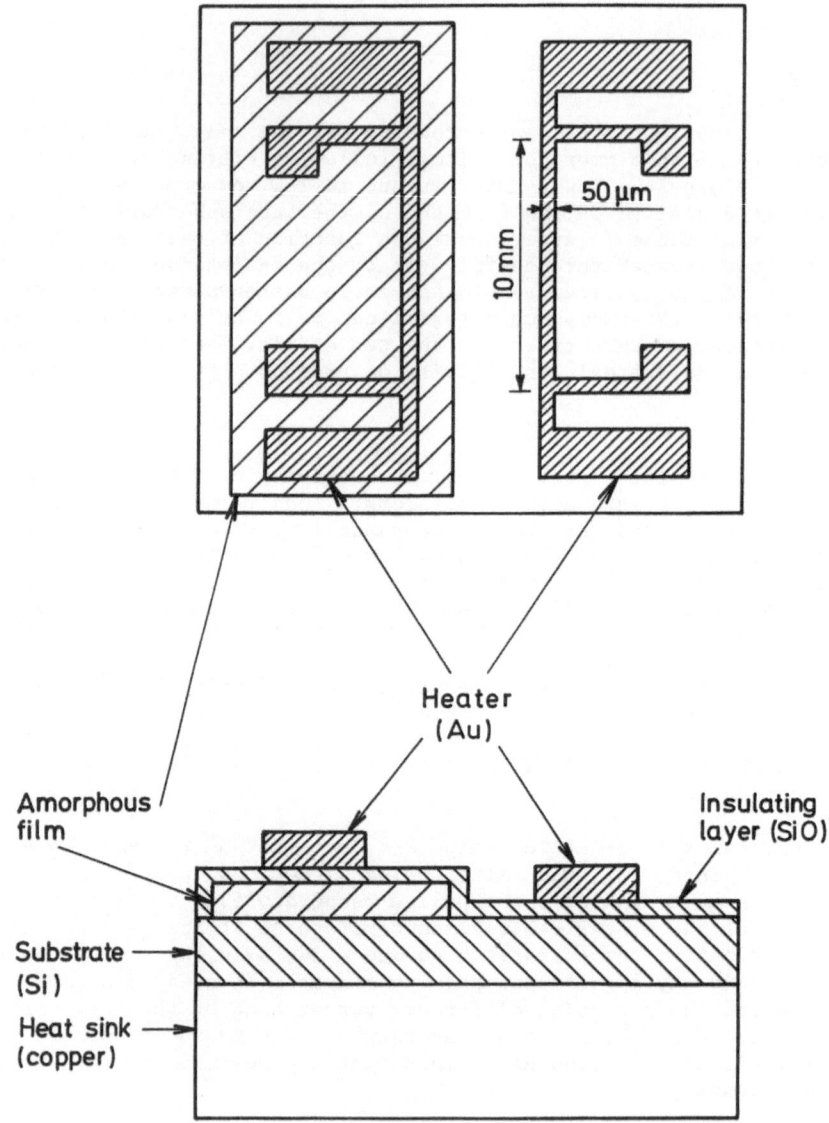

Figure 1.   Disposition of gold heating strips on the specimen.

The method is essentially of the steady state type but, because of the large power input, a gradual drift of temperature occurs as the substrate rises in temperature. However, there is no difficulty in taking this into account, since the time scale for this drift is much greater than that for the heating of the metallic strip.

The thermal conductivity of the film is found from the difference, ΔT, between the temperatures of the two heating strips when their power inputs are the same. Its value is given by $qd/A\Delta T$, where q is the power input to each strip, A is the area of contact with the film or substrate, and d is the thickness of the film. Since d is much smaller than the width or length of the heating strip, the isothermals in the film may be regarded as planar.

SPECIMEN PREPARATION

Our sample consists of an amorphous silicon film deposited by thermal evaporation on a polished single-crystal silicon substrate. The film covers only half of the substrate. A thin layer of silicon monoxide over the whole specimen provides electrical isolation.

Identical gold strips of 250 nm thickness, are laid down by thermal evaporation on the film and the bare substrate, using a lift-off photo-lithographic process (positive photoresist AZ1920). Electrical leads, for the introduction of electrical current and the measurement of potential difference, are attached to the arms shown in Figure 1.

The important thing to notice is that the configuration is identical for the two gold strips, with the exception of the amorphous film under one of them. Thus, for a given power input, the temperature difference depends solely on the thermal resistance of the portion of the film immediately below the relevant gold strip.

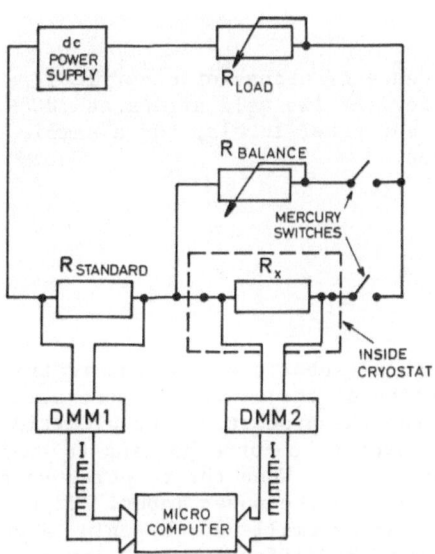

Figure 2.   Electrical Circuit.

The substrate is mounted on a heat sink within a continuous flow cryostat. For calibration purposes, its temperature is found by means of a rhodium-iron resistance thermometer.

In order to determine the electrical resistance as a function of temperature for the two gold strips, the potential drop is first measured using the lowest possible currents, so as to avoid any heating effect. This preliminary work is performed over the whole temperature range. Pure gold was selected as the material for the heating strips, since it has a continuous and reproducible variation of resistivity with temperature over the whole range. The electrical circuit is shown in Figure 2.

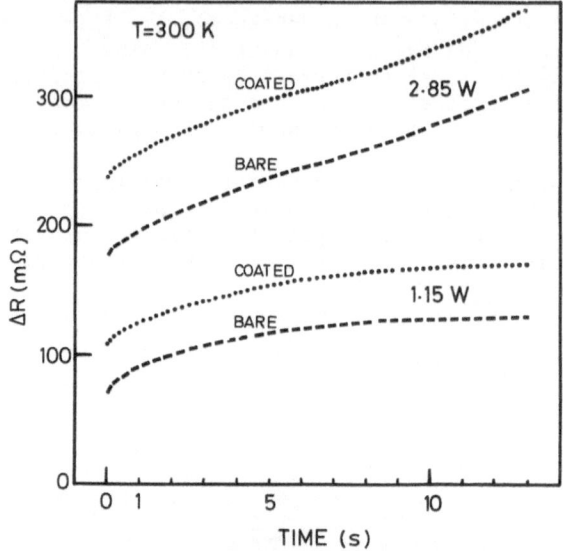

Figure 3. Dependence of change of electrical resistance on time for the two gold strips at 300 K, using different power levels, for a sample of amorphous silicon.

Then, at any particular substrate temperature the potential difference is determined as a function of the electric current, I, the power per unit length being obtained from the product of the electric current and I. The increase in resistance, over that corresponding to an infinitesimal current, indicates the temperature rise. When the temperature rise is not too large, it is directly proportional to the power input. Of course, the relative change of resistance is rather small but, by using a computer to determine the best fit to the results at different power levels, the ratio of the temperature increase to the power input can be accurately assessed.

One of the difficulties is that the power level is such that a gradual rise of temperature of the substrate is experienced. When the current is first switched on, the resistance of the sample changes rapidly as the temperature of the gold strip rises; thereafter, there is slower change as the substrate temperature also rises. The changes of potential drop were measured using a programmable digital voltmeter Fluke type 8848A, and the computer was programmed to remove the effect of the change in the substrate temperature.

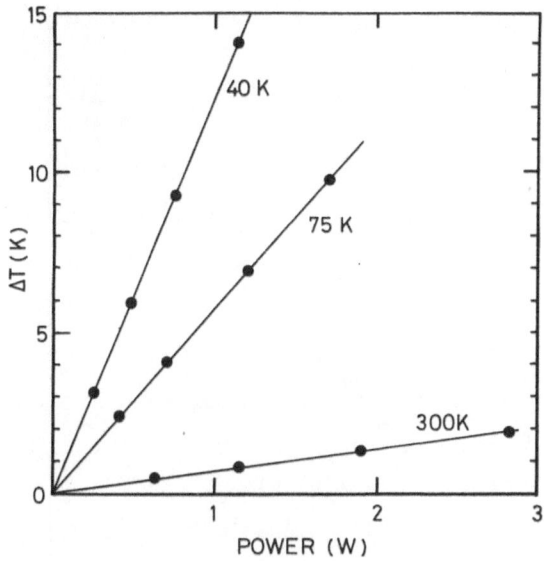

Figure 4. Variation of temperature difference between the strips with power at three different temperatures.

RESULTS

The experiments were performed on a sample of amorphous silicon of 1.2 μm thickness. As an example of the data that were obtained, Figure 3 shows plots of the change of resistance, ΔR, against time for the two gold strips at 300 K, using two different power levels. Figure 4 shows how the temperature difference between the strips was found to vary with power, at three different temperatures.

The variation of the measured thermal conductivity with temperature is shown in Figure 5. The data below 10 K are not thought to be particularly reliable because of the very small changes of resistance that had to be determined in this range. However, all the results below 100 K are consistent with a thermal conductivity that varies approximately linearly with temperature. Near room temperature the variation is less rapid and the thermal conductivity at 300 K is 4.2 W/mK.

Figure 5 also shows a theoretical curve for the minimum thermal conductivity of silicon, as predicted by Slack's theory[2]. The experimental results between about 50 K and 100 K are in fair agreement with the proposition that amorphous films should display the minimum thermal conductivity, but this is not the case at higher and lower temperatures.

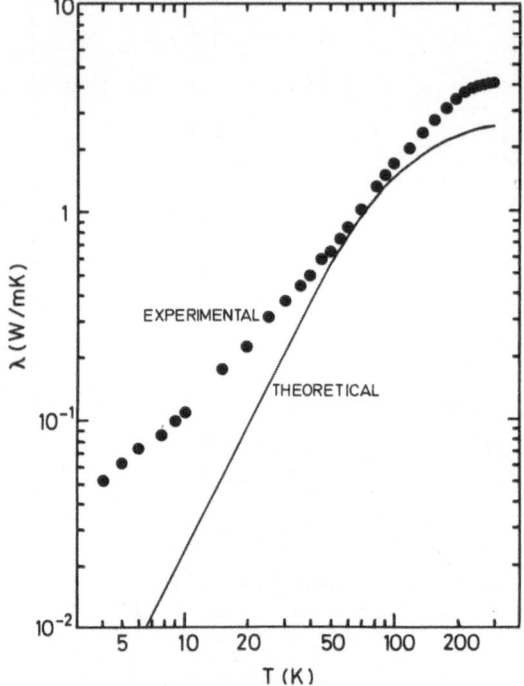

Figure 5. Temperature dependence of the thermal conductivity of amorphous silicon produced by thermal evaporation. The solid curve shows the theoretical variation of Slack's minimum thermal conductivity for silicon.

In particular, the experimental values for the thermal conductivity at low temperatures are much greater than those that are predicted by theory. However, it is noted that the experimental values of Graebner and Allen[3] for the thermal conductivity of amorphous germanium at low temperatures are also much larger than would be given from a calculation of the minimum thermal conductivity.

CONCLUSIONS

We have shown that it is possible to measure the thermal conductivity of amorphous films, over a wide temperature range, using a steady-state strip heating technique. The method has yielded values for amorphous silicon that are consistent at intermediate temperatures with theoretical predictions, using the minimum thermal conductivity concept. However, the agreement at higher and lower temperatures is poor. Experiments are now being carried out on amorphous silicon produced by DC magnetron sputtering, since it is possible that the data may be characteristic of the method of film preparation.

We also plan to carry out similar measurements on other amorphous semi-conductors and to study the effect of partial recrystallisation on the thermal conductivity.

ACKNOWLEDGEMENTS

The authors acknowledge the support given by the Australian Research Grants Scheme for this project.

REFERENCES

1.   Wieczorek, L., Paul, G.L. and Goldsmid, H.J., 1985, A CW laser technique for measuring the thermal conductivity of amorphous films, Proceedings of the Nineteenth Thermal Conductivity Conference, Cookeville, Tennessee.
2.   Slack, G.A., 1979, The thermal conductivity of nonmetallic crystals, Solid State Physics, 34:1.
3.   Graebner, J.E. and Allen, L.C., 1984, Thermal conductivity of amorphous germanium at low temperatures, Phys. Rev., 29: 5626.

PHONON IMAGES AND PHONON CONDUCTION

IN TRIGONAL CRYSTALS

A.K. McCurdy

Worcester Polytechnic Institute
Worcester, Massachusetts 01609

A.G. Every

University of the Witwatersrand
Johannesburg 2001, South Africa

ABSTRACT

   Monte Carlo phonon-focusing patterns provide graphic illustration of
the striking changes in phonon intensity arising from elastic anisotropy
and piezoelectric coupling. Patterns for each of the three modes show the
basic acoustic symmetry of the crystal and suggest crystallographic directions along which enhanced phonon conductivity can be expected in the boundary-scattering regime.

INTRODUCTION

   Phonon focusing in crystals depends upon the second-order elastic
constants [1]. In strongly piezoelectric crystals these constants can be
stiffened due to the electromechanical coupling between the stress field
and the accompanying electric field. The amount of elastic stiffening depends not only upon the piezoelectric and permittivity constants, but also
upon the direction of the wave vector for the propagating elastic wave [2].
By changing the ratios between the elastic constants elastic stiffening
modifies the shape of the phase and group-velocity surfaces and thus the
phonon focusing and the boundary-scattered phonon conductivity.

   Monte Carlo phonon focusing calculations have been reported by several authors [3-18]. Early results portrayed calculated intensities as a
function of the spherical polar angles θ and ϕ, first in matrix form [3-5],
then using 3-D graphics [6-11]. The most dramatic representation of phonon
intensities was obtained from the experimental results of Wolfe and co-
workers [10-14]. Experimental and subsequent calculated intensities were
displayed as black and white patterns on a television monitor. Later,
calculations displayed similar patterns but as negatives using a pen and
ink x-y plotter [15-18].

   Phonon-focusing patterns are presented for 3 elastically anisotropic
crystals whose low temperature phonon conductivity was discussed at the
19th ITCC [19]. These patterns were calculated using the polarization
vectors to determine the group velocities and reproduced on a laser printer.
The production of these patterns thus differs in two respects from those
reported previously for the same materials [12,13,16].

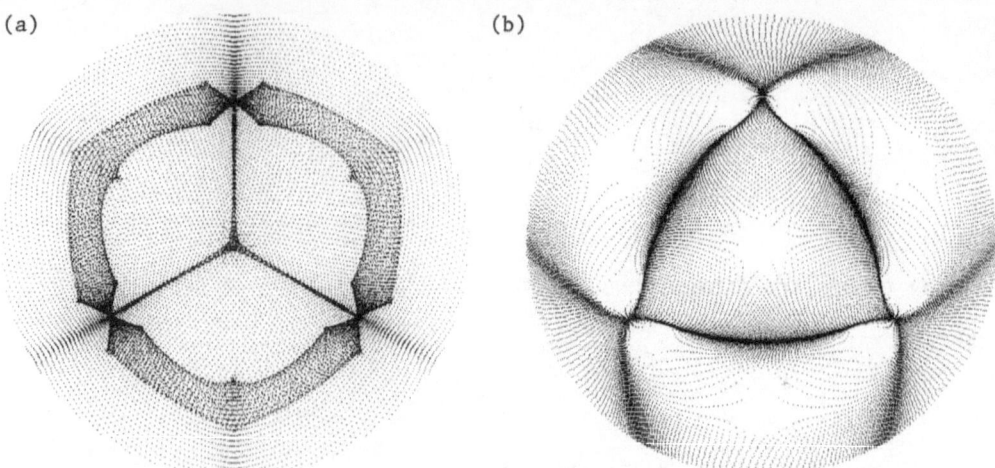

Fig. 1   Polar plot of the (a) ST and (b) FT phonon intensity patterns
         for LiNbO$_3$ with piezoelectricity ignored.   The intensity pattern
         for the L mode has little anisotropy.

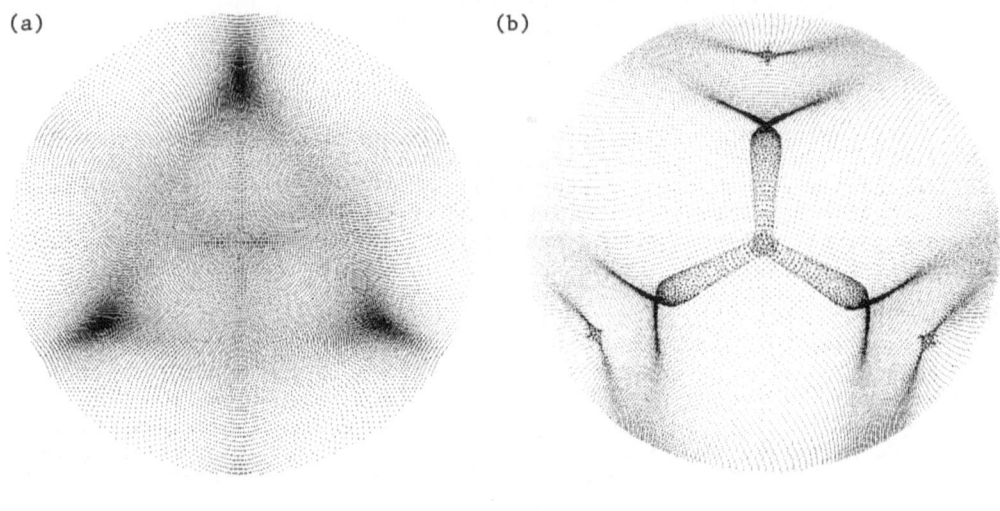

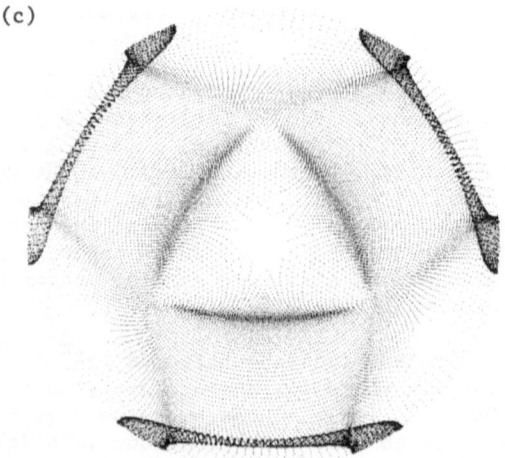

Fig. 2   Polar plot of the (a) L
         (b) ST and (c) FT phonon
         intensity patterns for
         LiNbO$_3$ with piezoelectric
         stiffening included.   The
         diameters of all patterns
         span 180 degrees.

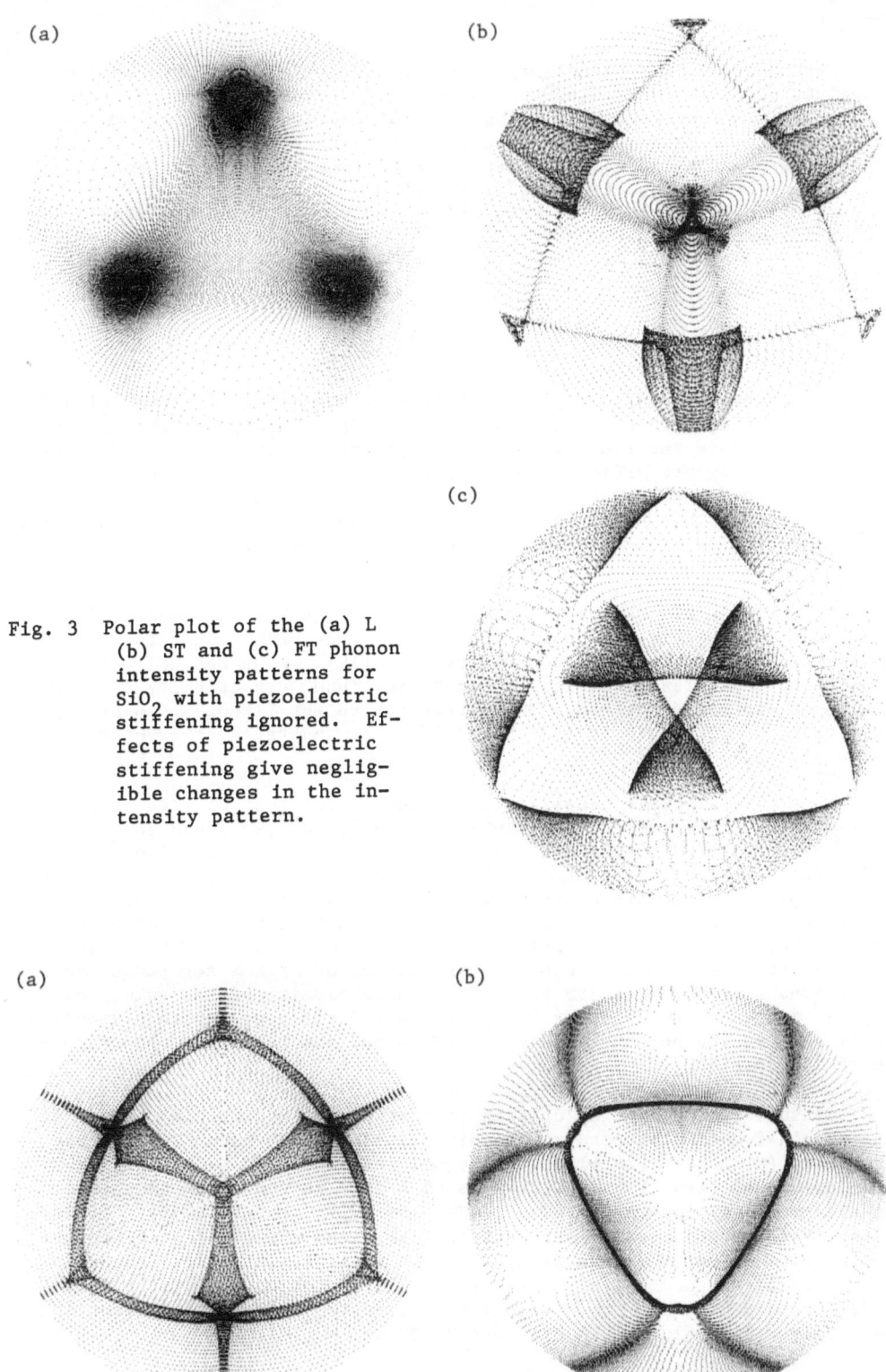

(a)

(b)

(c)

Fig. 3 Polar plot of the (a) L (b) ST and (c) FT phonon intensity patterns for SiO_2 with piezoelectric stiffening ignored. Effects of piezoelectric stiffening give negligible changes in the intensity pattern.

(a)

(b)

Fig. 4 Polar plot of the (a) ST and (b) FT phonon intensity patterns for Al_2O_3. The pattern for the L mode has little anisotropy.

RESULTS

Each Monte Carlo phonon image was produced on a laser printer and comprises approximately 20,630 group-velocity vectors. These ray vectors were generated systematically using a nearly equidensity of wave vectors reflecting the symmetry of the crystal. The center of each plot is the [001] axis and the radial distance is linear in angle with respect to this axis. To provide adequate visibility each dot is actually a small group of four X's. The darkness of each plot corresponds to the calculated phonon intensity.

DISCUSSION

In elastically anisotropic crystals an enhancement of energy flow is to be expected whenever the heat-flow axis is aligned along a highly focused direction. Since the boundary-scattered phonon conductivity, λ, is approximately inversely proportional to the square of the phonon velocity the greatest enhancements should occur for directions along which the slower transverse modes are highly focused. Furthermore, the greatest enhancement of heat flow along highly focused directions occurs in very long samples [20].

In strongly piezoelectric materials such as $LiNbO_3$ piezoelectric stiffening of the elastic constants increases the phase velocity and thus lowers the Casimir phonon conductivity, $<\lambda_c>$ [19]. Koos and Wolfe [13] have shown that piezoelectric stiffening in $LiNbO_3$ dramatically changes the phonon-focusing pattern. The patterns here for $LiNbO_3$ are based upon room temperature constants and thus differ from patterns at cryogenic temperatures.

When piezoelectric coupling is included directions of high phonon intensity for $LiNbO_3$ include the [100] (FT mode); [49°,90°] and equivalent axes (ST mode); [88°,210°] and equivalent axes, for example, [102°,90°] (FT mode). Values of the dimensionless phonon conductivity, $\lambda/<\lambda_c>$, for very long rods of square cross section were 1.346, 1.296 and 1.050, respectively, confirming an enhanced phonon conductivity. The value for $<\lambda_c>/T^3$ was 2204 D $Wm^{-1}K^{-4}$ where D is the side dimension. With piezoelectric coupling ignored, however, the values of $\lambda/<\lambda_c>$ for these same directions were 0.984, 1.355 and 0.868, respectively and the value for $<\lambda_c>/T^3$ increased to 2563D $Wm^{-1}K^{-4}$. Piezoelectric coupling thus dramatically changes the phonon-focusing patterns and the resulting phonon conductivity in $LiNbO_3$.

Piezoelectric coupling effects in quartz give negligible changes in the phonon-focusing patterns. The directions of high intensity include the [001] (ST mode); [22°,150°] and equivalent axes (FT mode); [53°,90°] and equivalent axes (L mode). Values of $\lambda/<\lambda_c>$ in very long samples of square cross section were 1.893, 1.430, and 1.145, respectively. It is clear that the greatest enhancement in λ is along the [001] axes because of the high focusing of the ST mode over a small solid angle. There is only a small enhancement along the [53°,90°] direction confirming that L mode focusing with its larger velocity contributes less to the phonon conductivity. The value of $<\lambda_c>/T^3$ was 2140D $Wm^{-1}K^{-4}$.

The phonon-focusing patterns in non-piezoelectric Al_2O_3 in some respects contrast with those of $LiNbO_3$ (piezoelectricity ignored). The direction of highest intensity is [55°,30°] and equivalent axes, for example [125°,90°]. The value of $\lambda/<\lambda_c>$ in very long samples of square

246

cross section was 2.213 where $<\lambda_c>/T^3$ was 915.4D $Wm^{-1}K^{-4}$. Values of $\lambda/<\lambda_c>$ along the [100], [010] and [001] axes were 0.989, 0.929, and 0.903, respectively [19].

Phonon-focusing patterns thus suggest crystallographic directions along which enhanced conductivity can occur and indicate for piezoelectric crystals when effects of piezoelectric coupling may be appreciable.

REFERENCES

[1] Maris, H.J., "Enhancement of Heat Pulses in Crystals due to Elastic Anisotropy", J. Acoust. Soc. Am. 50, 812-818 (1971).

[2] Auld, B.A., Acoustic Fields and Waves in Solids, Vol I, (J. Wiley, & Sons, New York, 1973) chapter 8, p. 265-314.

[3] Taylor, B., Maris, H.J., and Elbaum, C., "Focusing of Phonons in Crystalline Solids due to Elastic Anisotropy", Phys. Rev. B3, 1462-1472 (1971).

[4] Winternheimer, C.G., "Phonon Focusing and Phonon Conduction in Orthorhombic and Tetragonal Crystals in the Boundary-Scattering Regime", Ph.D. Thesis (Worcester Polytechnic Institute, 1975 (unpublished).

[5] Winternheimer, C.G., and McCurdy, A.K., "Phonon Focusing and Phonon Conduction in Orthorhombic and Tetragonal Crystals in the Boundary-Scattering Regime II", Phys. Rev. B18, 6576-6605 (1978).

[6] Rosch, F., and Weis, O., "Geometric Propagation of Acoustic Phonons in Monocrystals within Anisotropic Continuum Acoustics", Z. Phys. B25, 101-114 (1976).

[7] Rosch, F., and Weis, O., "Geometric Propagation of Acoustic Phonons in Monocrystals within Anisotropic Continuum Acoustics", Z. Phys. B25, 115-122 (1976).

[8] McCurdy, A.K., "Phonon Focusing and Phonon Conduction in Elastically Anisotropic Crystals", in Phonon Scattering in Condensed Matter, p. 341-344, ed., Maris, H.J., (Plenum Press, N.Y., 1980).

[9] McCurdy, A.K., "Phonon Conduction in Elastically Anisotropic Cubic Crystals", Phys. Rev. B26, 6971-6986 (1982).

[10] Northrop, G.A., and Wolfe, J.P., "Ballistic Phonon Imaging in Germanium", Phys. Rev. B22, 6196-6212 (1980).

[11] Northrop, G.A., Cotts, E.J., Anderson, A.C., and Wolfe, J.P., "Anisotropic Phonon-Dislocation Scattering in Deformed LiF", Phys. Rev. B27, 6395-6408 (1983).

[12] Every, A.G., Koos, G.L., and Wolfe, J.P., "Ballistic Phonon Imaging in Sapphire: Bulk Focusing and Critical-Core Channeling Effects", Phys. Rev. B29, 2190-2209 (1984).

[13] Koos, G.L., and Wolfe, J.P., "Piezoelectricity and Ballistic Heat Flow", Phys. Rev. B29, 6015-6017 (1984).

[14] Hurley, D.C., and Wolfe, J.P., "Phonon Focusing in Cubic Crystals", Phys. Rev. B32, 2568-2587 (1985).

[15] Every, A.G., and Stoddart, A.J., "Phonon Focusing in Cubic Crystals in Which Transverse Phase Velocities Exceed the Longitudinal Phase Velocity in Some Directions", Phys. Rev. B32, 1319-1322 (1985).

[16] Every, A.G., "Pseudosurface Wave Structures in Phonon Imaging", Phys. Rev. B33, 2719-2732 (1986).

[17] Every A.G., "Formation of Phonon-Focusing Caustics in Crystals", Phys. Rev. B34, 2852-2862 (1986).

[18] Every, A.G., and McCurdy, A.K., "Phonon Focusing in Piezoelectric Crystals", Phys. Rev. B36, 1432-1447 (1987).

[19] McCurdy, A.K., "Phonon Conduction in Elastically Anisotropic Trigonal Crystals", Thermal Conductivity 19, Proc. of the 19th International Thermal Conductivity Conference, ed., Yarbrough, D.W., (Plenum Press, in press).

[20] McCurdy, A.K., "Phonon Conduction in Elastically Anisotropic Cubic Crystals", Phys. Rev. B26, 6971-6986 (1982).

THERMAL CONDUCTIVITY OF WOOD-BASED PANELS[1]

Frederick A. Kamke

Assistant Professor
Department of Forest Products
Brooks Forest Products Center
Virginia Polytechnic Institute and State University
Blacksburg, VA 24061

ABSTRACT

Wood-based panels are used in a variety of structural and industrial
applications, in which their thermal properties may influence their overall
performance. Although the thermal properties of individual panel products
are relatively uniform, different panel types, and similar panels produced
by different manufacturers, have different thermal properties. These products are composite materials and their properties are different from those
of solid wood. Since most product standards that encompass these materials
are based on mechanical properties, there are few guidelines to aid in the
evaluation of the thermal properties of wood-based panels.

As an exploratory study, nine types of commercially produced wood-based
panels were tested for thermal conductivity using a guarded heat-flux
method. Panels were tested containing moisture and in the dry condition.
Increasing the moisture content or the specific gravity will increase
thermal conductivity. The relationship between thermal conductivity and
either moisture content or specific gravity appears to be linear. Since
many of these panels are not homogeneous through the thickness, the panel
thickness may influence the thermal conductivity. For a given specific
gravity, the descending order of thermal conductivity values for wood-based
products included in this study are: solid wood, plywood, particleboard,
and fiberboard. Although, particleboard exhibits considerable variability
that could upset this order.

INTRODUCTION

Wood is a hygroscopic, porous material. Its porosity is typically in
the range of 0.5 to 0.8. This naturally occurring anisotropic material
exhibits considerable variability in its physical properties. Most of the
variation can be accounted for by the variation of specific gravity.
----
[1]This research was funded in part by the American Plywood Association,
P. O. Box 11700, Tacoma, WA 98411. The author acknowledges the technical
assistance of William C. Thomas, Professor, Department of Mechanical
Engineering, Virginia Polytechnic Institute and State University.

Various composite products are manufactured from wood, such as particle-board, fiber board, and plywood. Operations involved in the manufacture of these products alter the wood structure, normally through densification. Additives, such as adhesives or sizing agents, may be blended with the wood elements. The orientation of the wood elements may also vary. The physical properties of the wood composite, including thermal conductivity, are, therefore, different than the properties of the solid wood from which it was made. The anisotropic and hygroscopic nature of wood results in a uniquely complex collection of thermal characteristics compared with those of other engineering materials. The manufacture of a wood-base composite further complicates the study of thermal properties.

In this study, nine types of wood-based panels were tested to compare their thermal conductivity. The effects of specific gravity, moisture content, and panel type were examined. The objective of this study was to supply the American Plywood Association (APA) with thermal conductivity data for a variety of wood-based panels using a standard test method (1). The data is to be used to evaluate alternative thermal conductivity test methods, to be adapted or developed by the APA.

The thermal conductivity of wood, as related to the anatomy of the structure, is explained in detail by Siau (10). The direction of heat flow relative to the wood cell orientation and the cell wall thickness have the largest effect on thermal conductivity. Thermal conductivity parallel to the grain (longitudinal direction) is approximately 2.5 times greater than in either the radial or tangential directions (Figure 1). While the cell orientation affects the thermal conductivity, it is the orientation of the molecular chains within the cell wall that have the greatest influence. The long-chain linear polymers (cellulose), which comprise the crystalline component of the cell wall, are arranged in bundles called microfibrils. These microfibrils are most closely aligned with the longitudinal axis of the cell. Apparently the transmission of thermal energy occurs more readily along the length of a microfibril than across a series of microfibrils. The cell wall substance, independent of the cell structure, has a thermal conductivity in the longitudinal direction roughly twice that of the radial or tangential directions (13).

The thermal conductivity of wood increases with an increase of temperature or moisture content (11). A plot of thermal conductivity of wood as a function of temperature and moisture content is shown in Figure 2. Note the inflection point at the freezing temperature of water and the discontinuity associated with the phase change above the fiber saturation point. Above the fiber saturation point (approximately 28 to 30 percent moisture content) the void space depicted in Figure 1 begins to fill with liquid water.

The thermal conductivity of wood-based composites, as for wood, are strongly dependent on specific gravity with little evidence of species influence (4, 9, 13). The largest data base of thermal conductivity values for wood-based composites was compiled by Lewis (9). This data base contains information on wood particle- and fiber-based panels. Empirical relationships for thermal conductivity as a function of specific gravity, temperature, and moisture content have been reported for certain wood-particle panels (8, 9, 14). Similar to solid wood, wood-based panels are anisotropic in regards to their thermal properties (7).

METHODS AND MATERIALS

The thermal conductivity of nine types of wood panels was measured using a guarded heat-flux method (1). The panel types and descriptions are listed in Table 1. All of the panels were produced commercially from either wood fibers, particles, flakes, veneers, or a combination of these wood elements.

The panels were tested in the condition as received. The dimensions were measured and a gross weight determined before the panels were placed in the guarded heat-flux apparatus (R-MATIC, Dynatech Corporation). The R-MATIC unit accepts a 61 cm (24 inch) square specimen. The MDF fiberboard, and HB fiberboard, panels measured 61 cm by 61 cm (24 inch by 24 inch). All the other panels were 30.5 cm by 30.5 cm (12 inch by 12 inch). The R-MATIC unit measures heat flux over a 645 cm^2 area (10 inch by 10 inch) in the center of each panel. Thus, the 30.5 cm square specimens were centered and surrounded on four edges by flexible foam insulation of equal thickness. This was necessary to minimize heat flow through the panel edges.

During the test one to two hours were required, depending on panel thickness and conductivity, for the panels to obtain apparent thermal equilibrium. A thermal steady-state was identified when five successive thermal resistance values agreed to within one percent, with no apparent trend, and the surface temperatures varied by no more than 0.11°C (0.2°F) when read at 5 minute intervals. Once thermal equilibrium was achieved, the thermal conductivity was recorded, and the sample removed from the R-MATIC unit. The panels were again weighed to determine if there was a net gain or loss of moisture. No significant weight changes were recorded. All of the panels (except HB fiberboard) were eventually dried to determine their to determine their moisture content. The HB fiberboard panels could not be dried without causing severe degradation. However, the moisture content of the HB fiberboard panels was probably about 2 to 4 percent.

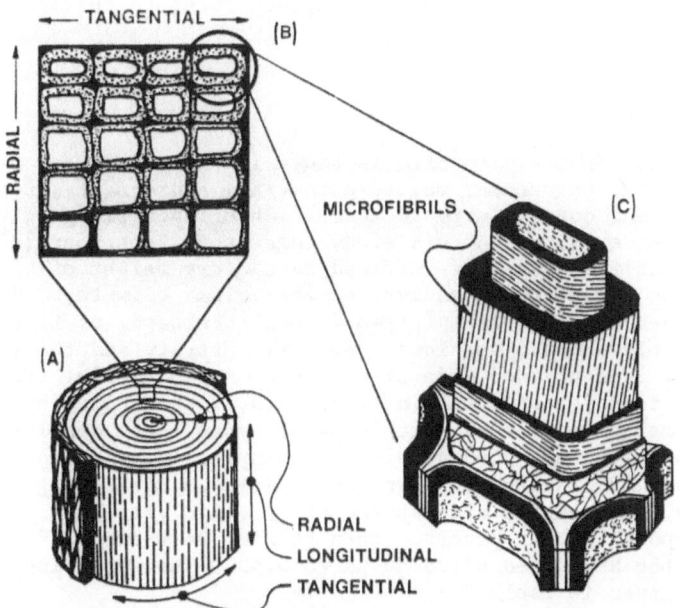

Figure 1.   Schematic of wood structure, (A) showing directional
            nomenclature with respect to the tree stem, (B) cross-section
            view showing organization of cells, and (C) orientation of
            microfibrils in the cell wall.

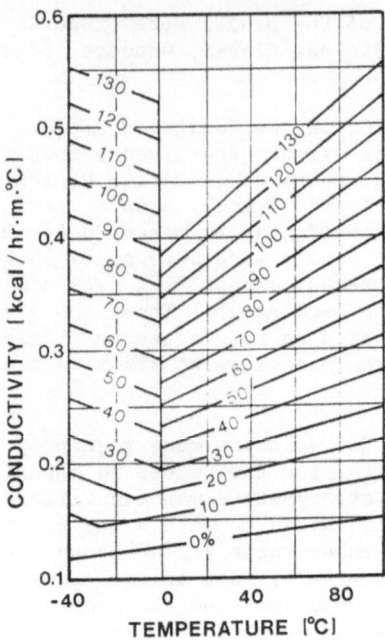

Figure 2. Thermal conductivity of birch (Betula spp.) as a function of
temperature and moisture content. Specific gravity is 0.515
(dry weight, wet volume basis) (11).

To determine if the moisture in the panels affected the thermal
conductivity six of the panels were oven dried, sealed in polyethylene
envelopes, and tested dry. The same test procedure was used for the dry
panels. The contribution of the polyethylene sheet to thermal resistance
was ignored.

RESULTS AND DISCUSSION

The thermal conductivity data is shown in Table 2, along with the panel
specific gravity, thickness, moisture content, and density at the time of
testing. Thermal conductivity as a function of specific gravity for all
the panel types examined in this study (except HB fiberboard) is plotted in
Figure 3. Specific gravity is defined as the dry weight of the material
divided by the weight of an equivalent wet-volume of water. The panels are
indicated based on panel type (plywood, particleboard, or fiberboard). In
general, the plywood had the lowest specific gravity and the lowest thermal
conductivity. The plywood made of Southeast Asian keruing (Panel A1),
however, had the highest specific gravity (0.77) and consequently the
highest thermal conductivity. Within panel types there was little variation
of the measured thermal conductivity. A simple linear regression of all
the panel data (including HB fiberboard) yields a misleading R-squared
value of 0.92, due to the wide spread of specific gravity between the HB
fiberboard and the other panels. When the two HB fiberboard panels are
eliminated, the R-squared value drops to 0.55. The resulting regression
equation is given in Table 3.

Figure 3 also shows predictive values of thermal conductivity as a
function of specific gravity and moisture content for solid wood (3, 5,
13). An average moisture content of 6 percent was used to construct the
graph since the moisture content of the panels used in this study averaged

Table 1. Wood-based panel types tested in this study.

Panel Reference	Panel[1] Type	Nominal Thickness (inch)	Species	Comment[2]
C	Ind. Particleboard	3/4	----	PF resin
D	OSB Particleboard	23/32	Aspen (Populus tremuloides)	PF resin
E	OSB Particleboard	7/16	Aspen (Populus tremuloides)	PF resin
A	HDO Plywood	25/32	Keruing (Dipterocarpus spp.)	13-ply
F	MDO Plywood	3/4	Douglas fir (Pseudotsuga menziesii)	7-ply
G	MDO Plywood	3/8	Douglas fir (Pseudotsuga menziesii)	3-ply
H	HBO Plywood	17/16	----	7-ply
I	HDO Plywood	15/32	Douglas fir (Pseudotsuga menziesii)	5-ply
J	Marine Plywood	3/8	Douglas fir (Pseudotsuga menziesii)	5-ply
K	MDF Fiberboard	1	----	----
L	HB Fiberboard	5/4, 3/2	----	1/8 inch tempered laminates

[1] HDO = high density overlay, MDO = medium density overlay, HBO = hardboard overlay, MDF = medium density fiberboard, HB = hardboard, OSB = oriented strandboard, Ind. = industrial

[2] PF = phenol formaldehyde

about 6 percent. The data depicted in Figure 3 indicates a consistent dependence of conductivity on specific gravity. The slope of the panel data with respect to specific gravity is nearly the same as reported for solid wood. Clearly the conductivity of the wood panels is less than the conductivity of solid wood for a given level of specific gravity.

The empirical equations for the relationships shown in Figure 3 are given in Table 3. A direct comparison of literature data is difficult due to differences in moisture content and swelling characteristics between various wood materials. The equations given in Table 3 have all been converted to a function of specific gravity.

Figures 4 and 5 show results of thermal conductivity measurements from this study in comparison to results reported by other investigators for particleboard and fiberboard, respectively. The greatest degree of variability is exhibited by the particleboard as indicated by the scatter shown in Figure 4. The results for fiberboard are quite consistent (Figure 5) and would appear to follow the relationship developed by Suzuki (12).

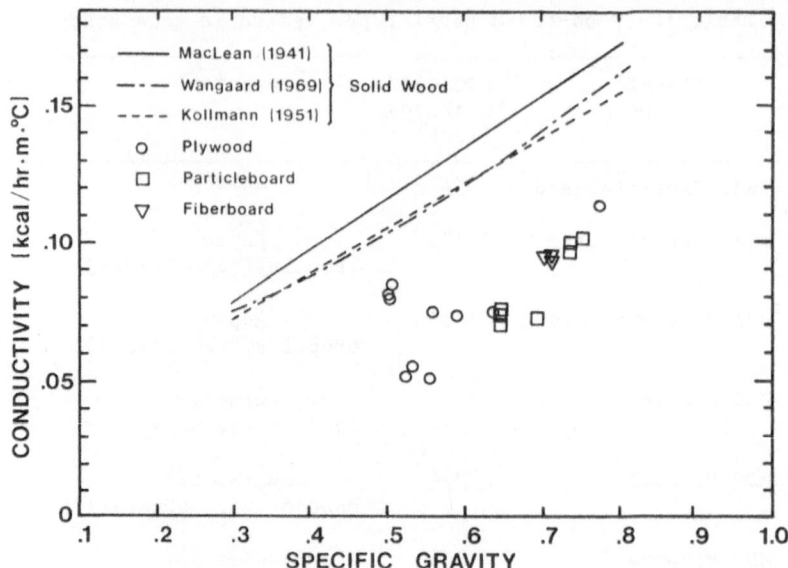

Figure 3. Thermal conductivity of wood-based panels included in this study as a function of specific gravity (dry weight, wet volume basis). Empirical relationships for solid wood are shown based on a moisture content of 6 percent.

A large variation in the particleboard data is expected. There is a wide variety of particle geometry and size that is suitable for particleboard manufacture. The amount of adhesive and other additives may also vary from less than 3 percent to over 12 percent of the dry panel weight. The particle size and shape will effect the void network between the wood elements, and therefore, should effect the conductivity.

Fiberboard is produced from individual wood cells or small bundles of cells that would have a relatively uniform size and shape in comparison with wood particles. Little or no adhesive is needed in fiberboard, usually less than 2 percent of the dry panel weight. As a result, the void network between the fibers should be uniform regardless of the panel's origin. Therefore, a small variation of thermal conductivity between different fiberboard panels would be expected.

Three panel types were tested that allowed a comparison of a possible effect of panel thickness on thermal conductivity with panels having similar construction. The OSB particleboard (Panels D and E) and HB fiberboard (Panel L) showed little or no effect of thickness on the measured thermal conductivity. This is the expected result for homogeneous materials. The HB fiberboard is homogeneous because of its laminar construction, with each laminate having identical properties. The OSB particleboard is not homogeneous due to the presence of a density gradient (high density surface layers and low density core). However, thermal conductivity has a nearly linear relation to specific gravity. This results in only a small difference of thermal conductivity when the average specific gravity is the same.

It is unclear why the Douglas fir plywood panels (Panels F and G) showed such a large difference of thermal conductivity. Plywood has a laminar construction. The only difference between these panels appeared to be the number of veneer layers and the panel thickness. Perhaps a difference in the number of gluelines or the veneer quality influenced the

Table 2. Summary of thermal conductivity data for the wood panels described in Table 1. Data was obtained by the guarded heat flux method (1). Average panel temperature was 24°C (75°F), with a temperature differential of 17° (30°F).

Panel Refer.	Thickness (cm)	Density[1] (g/cm^3)	Specific[2] Gravity	Moisture[3] Content (%)	Conduct. (kcal/ hr.m.°C)	Conduct. (btu.in/ hr.ft^2.°F)
A1	1.97	.83	.77	6.7	.11	.92
C1	1.92	.79	.74	6.7	.10	.81
C2	1.91	.78	.74	6.6	.10	.78
C3	1.90	.80	.75	6.6	.10	.82
D1	1.81	.67	.64	5	.07	.60
D2	1.81	.68	.65	4.6	.07	.60
D3	1.81	.67	.65	4.4	.08	.61
E1	1.19	.72	.69	4.3	.07	.58
E2	1.19	.68	.65	5	.07	.57
E3	1.19	.68	.65	5.6	.07	.57
F1	1.93	.54	.51	6.9	.09	.69
F2	1.92	.54	.50	6.8	.08	.66
F3	1.92	.54	.50	6.7	.08	.64
G1	.91	.59	.56	6.2	.05	.41
G2	.95	.57	.53	6.4	.06	.45
G3	.93	.56	.52	5.9	.05	.42
H1	2.69	.68	.64	6.7	.08	.61
I1	1.23	.59	.56	5.9	.08	.61
I2	1.21	.63	.59	6.3	.07	.60
J1	.95	.58	.53	8.6	.06	.45
K1	2.59	.76	.71	6.9	.10	.77
K2	2.58	.76	.71	6.6	.09	.77
K3	2.59	.75	.70	6.6	.09	.77
L1	3.11	1.42	----	----	.21	1.67
L2	3.76	1.42	----	----	.21	1.68

DRY PANELS:

Panel Refer.	Thickness (cm)	Density (g/cm^3)	Specific Gravity	Moisture Content (%)	Conduct. (kcal/ hr.m.°C)	Conduct. (btu.in/ hr.ft^2.°F)
C1	1.91	.74	.74	0	.09	.72
D1	1.78	.65	.65	0	.07	.57
E1	1.13	.72	.72	0	.07	.57
F1	1.90	.51	.51	0	.08	.61
K1	2.57	.72	.72	0	.08	.68
K2	2.54	.72	.72	0	.08	.69

[1] Density is based on weight and volume at time of testing.

[2] Specific gravity is based on dry weight and volume at time of testing.

[3] Based on dry weight of panel.

results. Plywood Panel F showed the greatest deviation from the thermal conductivity-specific gravity relationship given in Figure 3. An inspection of these panels showed a 6 to 1 ratio of visible knots in Panel F compared to Panel G. Knots distort the grain orientation such that a greater conductivity would be expected through the thickness of a laminate when knots are present.

The effect of moisture content on thermal conductivity of some selected wood panels is shown in Table 2. In all cases, the thermal conductivity of the dry panel is less than the thermal conductivity of the panel containing

Table 3. Summary of thermal conductivity relationships plotted in Figures 3, 4, and 5 based on specific gravity (G, dry weight, wet volume), moisture content (M, dry-basis fraction), and resin content (R, dry-basis fraction).

Material	Equation for Conductivity (kcal/hr.M.°C)	Source
Solid Wood	$(0.172 + 0.347\,M)\,G + 0.0205$	(5)
Solid Wood	$(0.159 + 0.235\,M)\,G + 0.0275\,M + 0.0187$	(2, 3)
Particleboard[1]	$0.0315 \exp(2.104\,G)$	(12)
Particleboard	$0.346\,(C_1 G + C_2) + \dfrac{0.0144}{C_2 G + 1}$	(9)

$$C_1 = 0.522\,M + \frac{0.248 + 0.186\,R}{1 + R}$$

$$C_2 = -0.956\,M - \frac{0.625 + 0.610\,R}{1 + R}$$

Fiberboard[1]	$0.113\,G + 0.0186$	(12)
Plywood,[2] Particleboard, Fiberboard	$0.14\,G - 0.010$	

[1]Assumes a moisture content of 6 percent in the air-dried condition reported by the author.

[2]Excludes HB fiberboard.

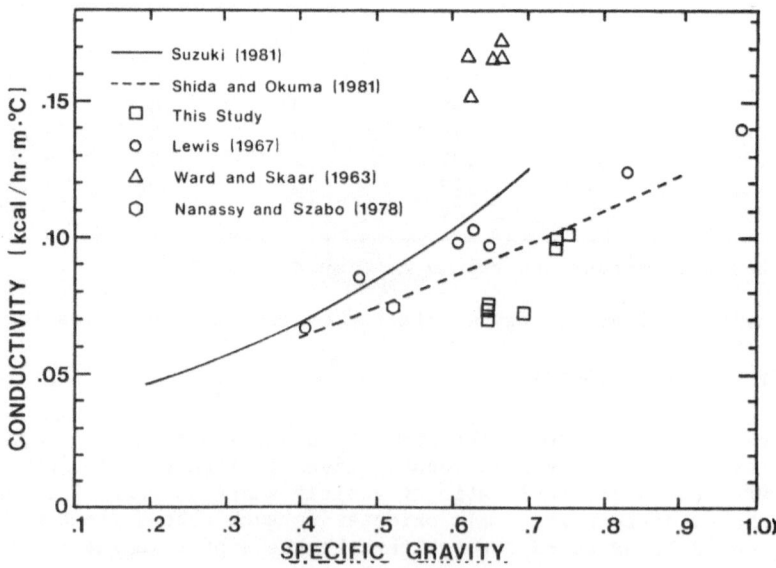

Figure 4. Thermal conductivity of particleboard panels as a function of specific gravity (dry weight, wet volume basis). All the data is adjusted to 6 percent moisture content based on MacLean's equation (5). The empirical relationships are given in Table 3.

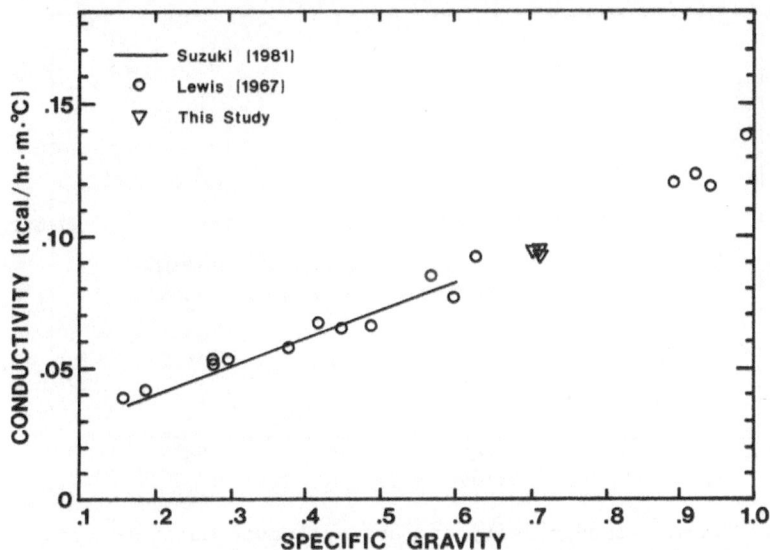

Figure 5. Thermal conductivity of fiberboard panels as a function of
specific gravity (dry weight, wet volume basis). All the data
is adjusted to 6 percent moisture content based on MacLean's
equation (5). The empirical relationship is given in Table 3.

moisture. This result is consistent with results reported by other
investigators (8, 9).

The effect of moisture content on obtaining thermal equilibrium in the
R-MATIC unit is shown in Figure 6. A MDF fiberboard panel, with an initial
moisture content of 6.9 percent, was placed in the R-MATIC unit and allowed
to come to apparent thermal equilibrium as described earlier. The panel
was removed, weighed, and returned to the R-MATIC unit. This procedure was
repeated six times over a period of 29 days. A net loss of 0.5 percent
moisture was recorded. The thermal conductivity also decreased over time,
with the decrease nearly proportional to the decrease of moisture content.
The relationship between thermal conductivity and average panel moisture
content is shown in Figure 6, where the thermal conductivity decreases with
a decrease in average panel moisture content.

At the end of 29 days the panel was removed from the R-MATIC unit.
Three small specimens were cut from the panel and sectioned into 5 layers
through the thickness to determine the moisture gradient. The results from
the three moisture gradient specimens were averaged and plotted in Figure
7. The moisture migrated from the warm side of the panel to the cool side.
No condensation was detected at the cool side. Because of the small amount
of air volume in the R-MATIC unit, the panel's sorption characteristics
control the humidity at the surfaces. The unit was not completely sealed,
therefore, a potential for a small change in total moisture content was
present during the 29 day period. Based on this limited test, it appears
that the average moisture content overshadows the effect of a moisture
content gradient on thermal conductivity. Since thermal conductivity shows
a near linear relationship with moisture content, as shown in Figure 3,
perhaps this is the expected result.

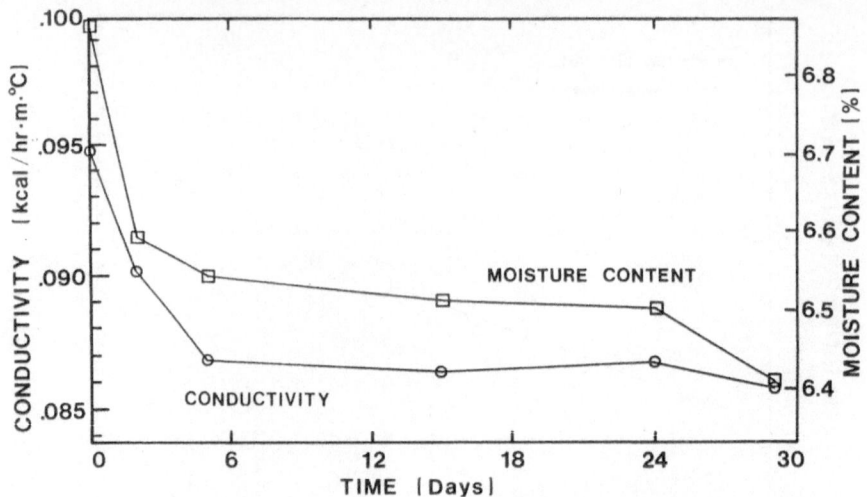

Figure 6.   Thermal conductivity and average panel moisture content for Panel K3 as a function of time during testing in the R-MATIC unit.

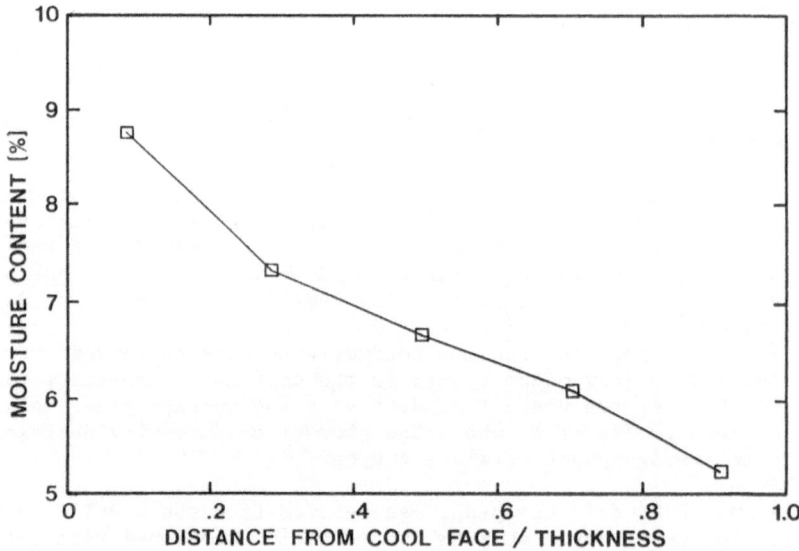

Figure 7.   Measured moisture content gradient for Panel K3 after 29 days of testing in the R-MATIC unit.

SUMMARY AND CONCLUSIONS

Nine types of commercial wood panels were tested for thermal conductivity using a guarded heat-flux method. Panels were tested with moisture and in the dry condition. Thermal conductivity is effected by both moisture content and specific gravity. Increasing the moisture content or the specific gravity will increase the thermal conductivity. A comparison between solid wood and wood-based panels, in terms of the relationship of thermal conductivity with specific gravity and moisture content, showed a definite similarity in trend but a difference in magnitude. The more homogeneous the construction of a panel, the less effect thickness will have on the measured thermal conductivity. A panel must be tested dry if true thermal equilibrium is desired. If tested with moisture, the panels must be carefully monitored so that a thermal conductivity value can be recorded when the apparent thermal equilibrium is first attained. Since the rate of moisture diffusion is very slow in comparison to the rate of thermal diffusion, the technique used in this project is a good approximation for determining thermal conductivity at low moisture contents. For purposes of comparison, however, the panels should be tested in a dry condition and sealed to prevent the adsorption of moisture.

LITERATURE CITED

1. ASTM. 1982. Part 18, ANSI/ASTM C 518-76, Standard test method for steady- state thermal transmission properties by means of the heat flow meter, in: "1982 Annual Book of ASTM Standards," American Society for Testing and Materials, Philadelphia, pp. 222-253.
2. Kollmann, F. and W. A. Cote, 1968. "Principles of Wood Science and Technology," Springer Verlag, New York, 592 p.
3. Kollmann, F. 1951. "Technologie des Holzes und der Holzwerkstoffe," Springer Verlag, Berlin, 1050 p.
4. Lewis, W. C., 1967. Thermal conductivity of wood-base fiber and particle panel materials. USDA For. Ser. Res. Pap. FPL 77. 12 p.
5. MacLean, J. D., 1941. Thermal conductivity of wood. Heating, Piping and Air Conditioning. 13:380-391.
6. Nanassy, A. J. and T. Szabo, 1978. Thermal properties of waferboards as determined by a transient method. Wood Sci. 11(1):17-22.
7. Schneider, A. and F. Engelhardt, 1977. Vergleichende Untersuchungen uber die Warmeleitfahigkeit von Holzspan-und Rindenplatten. Holz Roh Werk. 35:273-278.
8. Shida, S. and M. Okuma, 1980. Dependency of thermal conductivity of wood based materials on temperature and moisture content. Mokuzai Gakkaishi 26(2):112-117.
9. Shida, S. and M. Okuma, 1981. The effect of apparent specific gravity on thermal conductivity of particleboard. Mokuzai Gakkaishi 27(11):775-781.
10. Siau, J. F., 1984. "Transport Processes in Wood," Springer Verlag, New York, 245 p.
11. Steinhagen, H. P., 1977. Thermal conductive properties of wood, green or dry, from −40° to +100°C: a literature review. USDA For. Ser. Gen. Tech. Rep. FPL-9, 10 p.
12. Suzuki, M., 1981. Physical characteristics of multilayer particleboard. Mokuzai Gakkaishi 27(5):397-402.
13. Wangaard, F. F., 1969. Heat transmissivity of southern pine wood, plywood, fiberboard, and particleboard. Wood Sci. 2(1):54-60.
14. Ward, R. J. and C. Skaar, 1963. Specific heat and conductivity of particleboard as functions of temperature. For. Prod. J. 13(1):31-38.

PHONON SCATTERING AND THERMAL RESISTANCE DUE TO SPIN DISORDER

P. G. Klemens

Department of Physics and Institute of Materials Science
University of Connecticut
Storrs, Connecticut 06268

The interaction of phonons with spin sites in the disordered or paramagnetic state is estimated. Each spin site is a two-level system, capable of absorbing and reemitting phonons of matching frequency. The level spacing of each site is randomly distributed owing to the spin-spin interaction between the site and its nearest spin neighbors. This results in a reduction of the mean free path for a broad frequency band of lattice waves. Two parameters enter the theory. The spin-phonon interaction energy governs the strength of the interaction at each site. The spin-spin interaction energy determines the width of the band of affected phonons. The latter parameter is related to the ordering temperature. The thermal resistance can be estimated at temperatures which are high enough for complete disorder.

INTRODUCTION

In a solid containing a regular array of spin sites, phonons are absorbed and reradiated in transitions between spin levels. The phonon energy must equal the level spacing. In an array of oriented spins, all sites have the same level spacing, so that only phonons in a narrow frequency band are affected. This has only a small effect on the lattice thermal conductivity. However, if the level spacings of the spin sites are inhomogeneously broadened, a wider range of phonons is affected, and if the interaction with phonons is strong enough, the lattice thermal conductivity can be measurably reduced.

There are two sources of inhomogeneous broadening. The solid may be disordered, such as in a random solid solution, and each spin site may find itself in different surroundings. Alternatively, there may be appreciable interaction between neighboring spins, with a tendency for spins to become ordered at low temperatures. In the paramagnetic state, when spins are disordered, each spin site finds itself in a different surrounding, and the level spacing between the spin states varies from site to site. The corresponding broadening can be related to the spin-spin interaction and thus to the ordering temperature T_N.

PHONON RELAXATION RATE

Consider the direct spin-phonon interaction where a spin system makes a transition between two levels, separated in energy by energy E with the emission or absorption of a phonon of frequency ω where $E = \hbar\omega$. This one-phonon process is caused by a term $H' = Ae$ in the Hamiltonian linking the two spin states: e is the local strain due to the phonon and A is a parameter of the dimensions of energy. If one assumes the phonon gas to be in equilibrium and the spin distribution to depart from thermal equilibrium, the spin population tends toward equilibrium with the "direct" relaxation rate

$$\frac{1}{\tau_D} = 36\pi \left(\frac{\omega}{\omega_D}\right)^3 \frac{A^2}{(Mv^2)(\hbar\omega)} kT/\hbar \tag{1}$$

where ω_D is the Debye frequency, v the phonon velocity, M the average atomic mass, and k the Boltzmann constant. (1)

If instead of considering one spin center and summing over all normal modes one considers one normal mode and sums overall spin centers, and assumes the level distribution of the spin centers to be in thermal equilibrium, one obtains a phonon relaxation rate for phonon-spin interaction

$$\frac{1}{\tau_{PS}} = 2\pi \frac{A^2}{Mv^2\hbar} \frac{e^x - 1}{e^x + 1} g(\omega)\omega \tag{2}$$

where $x = \hbar\omega/kT$ and $g(\omega)d\omega$ is the number of spin centers per atom whose resonance frequency $\omega = E/\hbar$ lies in an interval $d\omega$ about ω. If x is small $(e^x - 1)/(e^x + 1) \simeq x/2$.

DISTRIBUTION OF LEVEL SPACING

Each spin sees a field which is zero in the average but fluctuates due to neighboring spins. The distribution $g(\omega)$ is thus centered at $\omega = 0$. Let each spin have z nearest neighbors and let the spin interaction energy be $\hbar\omega_o$. The distribution thus has a standard deviation $\hbar\omega_o z^{1/2}/4$. Since it makes no difference for phonon scattering whether ω is positive or negative, we can crudely write

$$g(\omega) = n/\Delta\omega \qquad \text{for } 0 < \omega < \Delta\omega$$
$$= 0 \qquad \text{otherwise} \tag{3}$$

where n is the number of spin sites per atom (n must be less than 1). Equating standard deviations,

$$\Delta\omega^2 = \frac{3}{4} z \omega_o^2 \tag{4}$$

Using a Bragg-Williams approximation, the ordering temperature T_N is related to ω_o by (2)

$$z \hbar \omega_o = kT_N \tag{5}$$

so that

$$\hbar\Delta\omega = (3/z)^{1/2} kT_N \tag{6}$$

For a simple cubic array of spin sites z=6 and $\hbar\Delta\omega = 0.7 \, kT_N$. We thus obtain from (2)

$$\frac{1}{\tau_{PS}} = n \frac{A^2 \pi}{\hbar\Delta\omega \, Mv^2} \simeq n \frac{1.4\pi}{Mv^2} \frac{A^2}{kT_N} \omega \qquad \omega < \Delta\omega \tag{7}$$

The intrinsic relaxation rate of phonons, due to anharmonic three-phonon interactions, at ordinary and high temperatures, is [3]

$$\frac{1}{\tau_i} = 2\gamma^2 \frac{kT}{Mv^2} \frac{\omega^2}{\omega_D} \tag{8}$$

where γ is the Grueneisen constant. This rate has the same frequency dependence as the spin relaxation rate, but the latter is effective only at frequency below $\Delta\omega$.

Since the spectral contribution to the specific heat per unit volume is $9Nk(\omega/\omega_D)^2$, and the thermal conductivity is

$$\lambda = \frac{1}{3} \int_0^{\omega_D} 9Nk (\omega/\omega_D)^2 \tau(\omega) v^2 d\omega \tag{9}$$

the intrinsic thermal conductivity, using (8), becomes

$$\lambda_i = 3Nk v^2 \tau_i(\omega_D) \tag{10}$$

with the integrand in (9) independent of ω.
Since the relaxation rates (7) and (8) must be added for $\omega < \omega_D$,

$$\lambda = \lambda_i \left[1 - \frac{\Delta\omega}{\omega_D} \frac{\tau_i}{\tau_i + \tau_{PS}} \right] \tag{11}$$

Comparing (7) and (8), and assuming that A is comparable to the strain dependence of the interaction energy, so that $A = \alpha k T_N$ where α is a numerical parameter, one finds

$$\lambda = \lambda_i \left[1 - \frac{\Delta\omega}{\omega_D} \left(1 + \frac{2\gamma^2}{1.4 n\alpha^2} \frac{T^2}{\theta T_N} \right)^{-1} \right] \tag{12}$$

The maximum reduction occurs in the limit of large α. Using $\hbar\Delta\omega = 0.7 kT_N$, this maximum is

$$(\delta\lambda/\lambda_i)_{max} = 0.7 T_N/\theta \tag{13}$$

However, as T is increased well above $(\theta/T_N)^{1/2}$, one expects a gradual reduction in $\delta\lambda/\lambda_i$ with increasing temperatures.

COMPARISON WITH EXPERIMENTS

Since the parameter $\alpha = A/kT_N$ is not known, it is better to compare the reduction in thermal conductivity to the maximum possible reduction given by (13), which depends only on the known Néel temperature. For UO_2, $T_N = 30$ K and $\theta = 380$ K. Thus, at temperatures near and above θ, the maximum depression of the thermal conductivity due to spin scattering should be about 5.5%. Since n=1/3, this would imply a large value of α, i.e. $\alpha/\gamma > 3$, and hence strong scattering of phonons by the spins.

The analysis of Moore and McElroy[4] leads to a much greater reduction (about 25%) at $T=\theta$, with $\delta\lambda/\lambda_i \propto 1/T$. The present estimate is quite inconsistent with the data, even if one were to allow for the effect of the assumption in their analysis that $1/\tau_{PS}$ is independent of ω. Since (13) depends on $\Delta\omega/\omega_D$, the most likely cause for error seems to be that (5) underestimates $\Delta\omega$.

Another source of the discrepancy may be that we have neglected spin-phonon interactions of the Raman type, i.e. a spin flip accompanied by the emission of a phonon ω_1 and the absorption of a phonon ω_2 such that $\omega_2 - \omega_1 = \pm\omega$, where $\hbar\omega$ is the level spacing. The phenomenological Hamiltonian would have the form $H' = Be^2$, and being second-order in phonon strain, H' is weaker than $H' = Ae$. At temperatures well above T_N, the phonon scattering would be very similar to point defect scattering with an effective mass-difference

$$\Delta M = B / v^2.$$

(14)

This apparent point defect scattering would be present even in a perfect mono-isotopic crystal above the disordering temperature, and since it affects primarily the high-frequency phonons, it would be additional to the direct process we have discussed. Even so, B would have to exceed kT_N by a substantial factor to cause reduction comparable to that attributed to spins in the case of UO_2.

In the case of MnF_2 and CoF_2, where T_N is above 70 and 40 K respectively, equation (13) may be applied to Slack's data(5) around the Debye temperature. The predicted depression $\delta\lambda/\lambda_L$ is then at most 10 to 15%, and thus consistent with the small difference observed between those crystals and $Zn F_2$ at ordinary temperatures.

In the case of Cr_2O_3, where T_N, = 308 K and $\Theta \simeq$ 700 K, (13) gives a maximum fractional reduction of about 30%. The observed relative depression of the thermal conducting(6) at 360 K, the highest measurement temperature, is comparable to that value.

Acknowledgment   The author wishes to thank Dr. R. K. Williams for drawing his attention to this problem and for helpful discussions.

References

(1)   See, for example, J. G. Castle, D. W. Feldman, P. G. Klemens and R. A. Weeks, Phys. Rev. 130, 577 (1963).
(2)   e.g., L. E. Reichl "A Modern Course in Statistical Physics", pp. 280-292. Texas Univ. Press, Austin, TX, 1980.
(3)   P. G. Klemens, in "Thermal Conductivity" ed. by R. P. Tye, Vol. 1, p. 49, Academic Press, London, 1969.
(4)   J. P. Moore and D. L. McElroy, J. Am. Ceramic Soc. 54, 40 (1971).
(5)   G. A. Slack, Phys. Rev. 122, 1451 (1961).
(6)   R. K. Williams, R. G. Graves and D. L. McElroy, J. Am. Ceramic Soc. 67, C151 (1984).

APPARENT THERMAL CONDUCTIVITY OF LOW DENSITY CONCRETE

OBTAINED WITH A RADIAL-HEAT-FLOW APPARATUS

R. D. Haynes
Remtech, Inc.
Huntsville, Alabama 35805

D. W. Yarbrough
Tennessee Technological University
Cookeville, Tennessee 38505

ABSTRACT

The thermal conductivites and diffusivities of six concretes containing perlite as an aggregate have been measured to within ±5% in the range 298 K to 322 K using an unguarded radial-heat-flow apparatus. The thermal conductivity was determined as a function of drying time, temperature, and density. The diffusivity was calculated by matching experimental time-temperature curves with computer generated curves. Computer simulations of the apparatus were also used to confirm radial heat flow for a central region of sufficient size for thermal measurements.

Concrete cylinders having densities from 1165 kg/m³ to 1516 kg/m³ had k_a values from 0.522 W/mK to 0.678 W/mK at 300 K. These thermal conductivity data are described by the following equation with an average deviation of 3.4%.

$$k(W/mK) = -0.0381 + 0.488 \times 10^{-3} \rho \ (kg/m^3)$$

The concrete specimens required from 100 to 150 days to reach constant k_a values. The concretes were found to have thermal diffusivities of approximately 1.5×10^7 m²/s. These concretes have thermal properties that make the material attractive for energy conservation applications.

INTRODUCTION

Concrete is one of the most widely used construction materials in the world because of its attractive physical properties and low cost. This material is composed of a binding agent (cement), mineral filler (aggregate), and water.[1] The aggregate comprises about 75 vol % of concretes and its properties heavily impact the overall properties of the concrete.[2] Many concrete formulations have cured densities greater than 1600 kg/m³ and apparent thermal conductivities, k_a, greater than 1.5 W/mK. These materials have low thermal resistances and do not have good energy conservation characteristics.

Lightweight concrete with corresponding reduced k_a can be produced by substituting a low density aggregate for the stone used in the heavier products. Expanded perlite can be used as an aggregate to produce concretes with densities in the range 1400 to 1900 kg/m³ or lower. The lightweight concretes formed with perlite as an aggregate have thermal resistance two to ten times that of the high density products. Unfortunately, the compressive strength of the low density concretes is generally less than that of high density concretes[3] and this forces a compromise between desired thermal properties and required strength.

Table 1.   Composition, Density, and Strength of Six Perlite Concrete Specimens.

Identification	Cement (Kg)	Water (Kg)	Perlite (Kg)	Density[b] (Kg/m³)	Compressive[c] Strength (MPa)
B[a]	4.299	1.935	0.993	1516	30.5
C	4.301	2.150	0.993	1396	28.5
D	4.400	2.640	1.540	1277	15.4
E	3.300	2.145	1.485	1165	11.4
F	3.780	2.268	1.349	1434	31.7
G	2.078	1.000	0.479	1420	23.6

[a] Specimen A was discarded
[b] Measured 28 days after specimen formation
[c] Compressive strength of 3 X 6 inch cylinder 28 days after specimen formation (Ref. 3)

LIGHTWEIGHT CONCRETE SPECIMENS

Lightweight concrete can be produced by using a low specific gravity aggregate, by aerating to form voids or closed cells, or to omit fine aggregate to form voids.[2,4] The concretes tested in this research were made from Portland cement, water, and expanded perlite having a bulk density of about 64 kg/m³. Table 1 contains the amount of each component that was used to produce specimens with aged densities from 1165 to 1516 kg/m³.

Thermal test specimens of the six concretes were cast into right circular cylinders with stainless-steel tubing along the center line as shown in Figure 1. The steel tubing was used as an electrical heater, and the outside of the cylinder was wrapped with a cooling coil to provide an isothermal boundary condition.

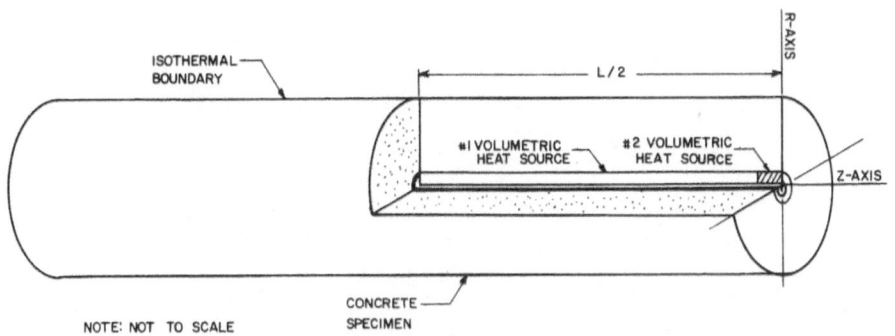

Figure 1.   Diagram of Concrete Test Specimen Showing Cast-in-Place Heater.

THERMAL MEASUREMENTS

The k_a of the six perlite concrete specimens were determined for a limited temperature range by radially dissipating measured D. C. power from the cast-in-place steel tubes. Figure 2 shows the circuit that was used. D. C. current through the steel tubing was determined from the voltage drop across the shunt resistor R_s, shown in Figure 2, measured with a digital multimeter[5] that was periodically compared with an L & N Type K-5 potentiometer.[6] Resistor R_s was compared with a standard resistor[7] to establish its resistance. A stable D. C. power supply[8] was used to provide constant current to the test specimen.

Figure 3 shows the dimensions of the cylindrical concrete test specimens and locates the temperature sensors. Three type-T thermocouples were bonded to the steel tube near the center of the cylinder to obtain the temperature of the hot boundary. Three type-T thermocouples were bonded to the outside surface of the concrete cylinder to obtain the cold boundary temperature. In both cases, the thermocouple output was converted to temperature using NBS Monograph 125.[9] The arithmetic average of the three interior temperatures was taken as the inside surface temperature while the average of the three centrally located exterior thermocouples provided the outside surface temperature. The thermocouple-derived temperatures were checked against a set of precision mercury-in-glass thermometers[10] prior to the thermal testing.

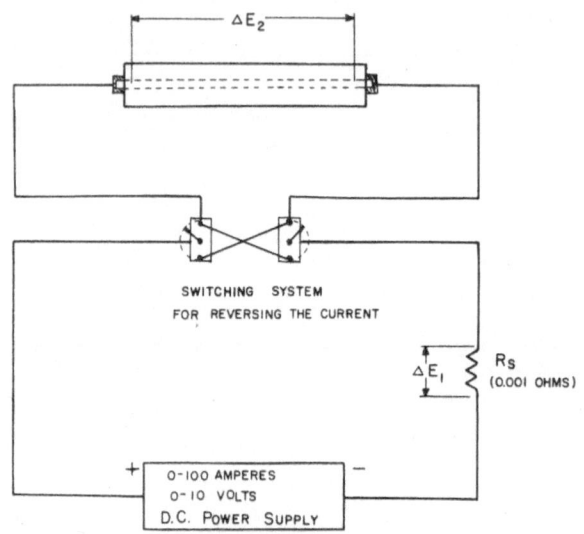

Figure 2. Electrical System Used to Measure k_a.

Holes were drilled in the ends of the stainless-steel heater tubes to increase the local electrical resistance. This step was taken to provide some thermal guarding at the ends of the heater and reduce axial heat flow. A series of HEATING5 thermal simulations was performed to demonstrate that the cylindrical system would have radial-only heat flow in the central region where temperatures were measured.[11] Figure 4 is a typical HEATING5 result that shows less than 1% error due to axial heat flow if temperature measurements are made within 14.6 cm of the center. The presence of the passive end guards increases the potential measurement region. The HEATING5 analysis provided justification for using the one-dimensional form of Fourier's Law to calculate k_a from power and temperature measurements.

$$k_a = EI\ln(r_o/r_i)/2\pi L(T_i - T_o) \tag{1}$$

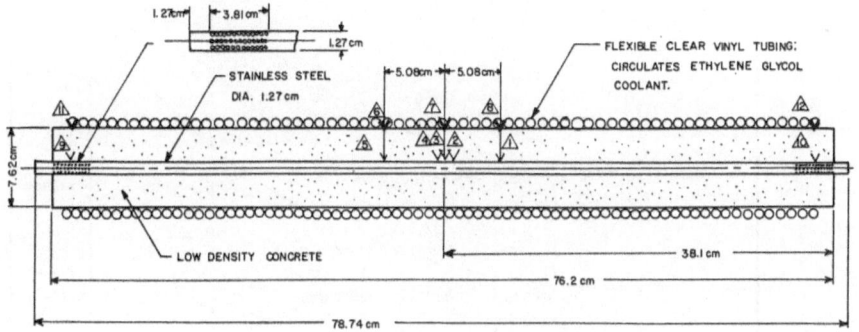

Figure 3. Diagram of Cylindrical Test Specimen Showing
Dimensions and Thermocouple Locations

The concrete test specimens were assembled for thermal testing after a 28-day curing period. The specimens, however, continued to lose water for weeks after the initial curing period. Cubes about 5.0 cm on each side formed from each batch of concrete were maintained in the same environment as the cylinders. The weight of the cubes was monitored and compared with oven-dry weight to assess the moisture content of the cylinders. This technique provided a way to determine k_a as a function of moisture content.

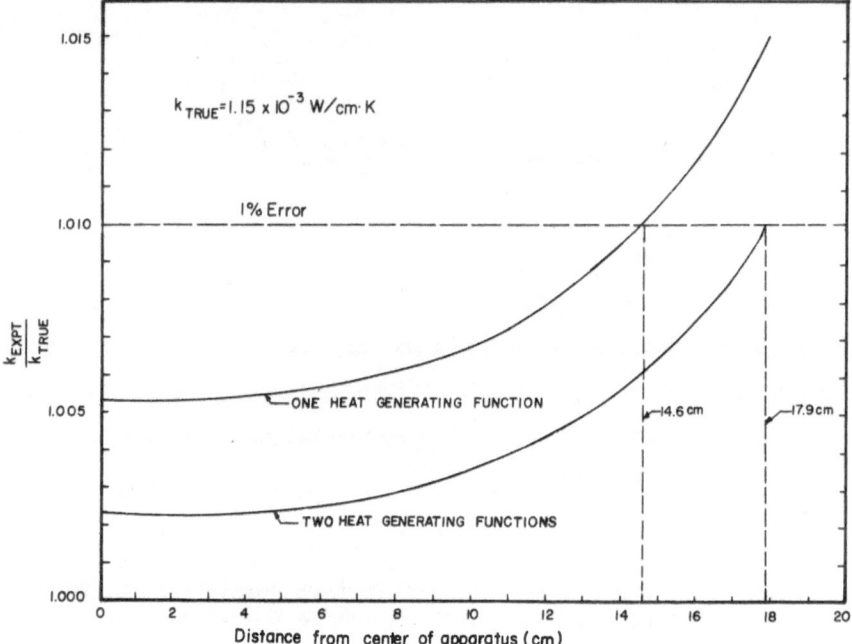

Figure 4. Results of HEATING5 Analysis Showing Error Due to Axial Heat Flow for Single Tube Measurements (One Heat-generating Function) and Tube with Passive End Guard (Two Heat-generating Functions).

The k_a of the test specimens decreased with time as moisture content decreased. Thermal data for each cylinder was used to project the k_a when the concrete reached moisture equilibrium with the nominally 50% R.H. laboratory conditions. The Method of Least Squares was used to evaluate the parameters k_∞, b, and c in Equation 2. The parameter k_∞ was then taken as the equilibrium value for k_a. The total measurement times exceeded 3 months for each specimen.

$$k_a = k_\infty + be^{-ct} \tag{2}$$

A determinate error analysis was completed for the apparatus used to measure k_a.[12] This analysis shows the error to be between 5 and 6% for the measurement conditions that were used.

Thermal diffusivities for the concrete specimens were obtained from transient thermal data generated by stepping the power to the central heater from zero to a steady value. The response of the heater temperature that was initially surrounded by isothermal material was recorded and compared with temperatures obtained from HEATING5 simulations to obtain α.

DISCUSSION OF RESULTS

Figure 5 shows the k_a for each of the six perlite concretes as a function of moisture content. These data were used to obtain $\Delta k/\Delta m$, the change in k_a due to a 1.0 weight % change in moisture content. The results for $\Delta k/\Delta m$ are shown in the figure.

Figure 6 shows k_a as a function of temperature for the six perlite concretes. The data show a slight increase in k_a with temperature and were used to derive the temperature coefficients dk/dt included in the figure.

The results for the thermal diffusivity experiments are shown in Table 2. The α values were obtained by matching experimental time-temperature data with time-temperature data calculated using the previously measured k_a and a specific heat of 3000 J/kgK. Table 2 also summarizes the data for the six perlite concretes used in the thermal simulations.

Figure 7 compares k_a data obtained in this research with k_a data for perlite concretes obtained by Wilson[13] in a longitudinal-heat-flow apparatus. The figure shows a curve in k_a versus density constructed from perviously published data reviewed by Haynes[12] and Wilson[13] and a curve published by Tye and Spinney.[13] The agreement between Haynes and Wilson is excellent. The agreement between the perlite concrete curve and the data from Tye and Spinney is reasonable, at least, in the density range below 1500 kg/m³. The solid curve in Figure 7 clearly shows k_a below the present values. A direct comparison must be viewed cautiously, however, because a variety of aggregates was used to produce the concretes represented in the literature review.

Table 2. Thermal Properties Used to Match Experimental and Calculated Time-Temperature Curves.

Specimen	ρ	k_a	C_p	α
	(kg/m³)	(W/mK)	(J/kgK)	(m²/s)
B	1516	0.669	3000	1.5×10^{-7}
C	1396	0.656	3000	1.6×10^{-7}
D	1277	0.578	3000	1.5×10^{-7}
E	1165	0.522	3000	1.5×10^{-7}
F	1434	0.676	3000	1.6×10^{-7}
G	1420	0.678	3000	1.6×10^{-7}

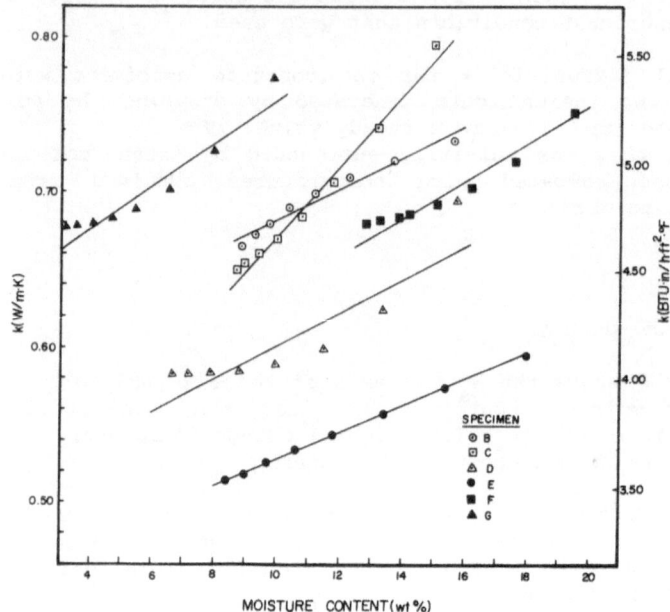

Figure 5. The Effect of Moisture on k_a for Six Perlite Concrete
Specimens. $\Delta k/\Delta m$: (b) 0.010 (c) 0.023 (d) 0.011 (e) 0.0086
(f) 0.012 (g) 0.014

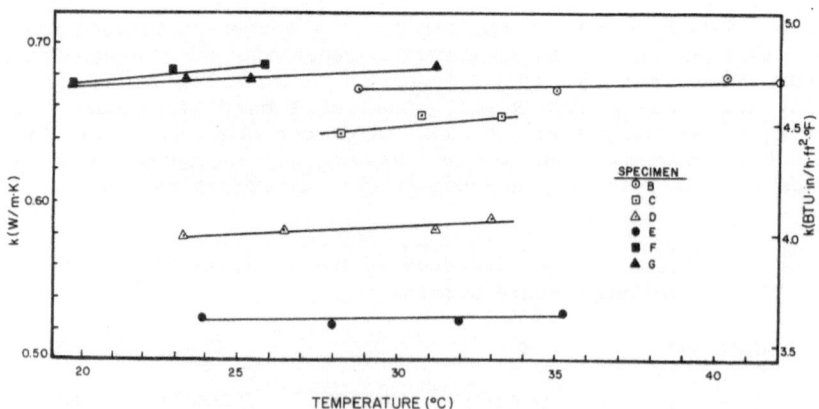

Figure 6. k_a as a Function of Temperature for Six Perlite
Concrete Specimens.

272

SUMMARY

The k_a for perlite concrete near 25°C can be described to about ±5% by Equation 3.

$$k_a \ (W/mK) = -0.0381 + 0.488 \times 10^{-3} \ \rho \ (kg/m^3) \tag{3}$$

The thermal conductivity of perlite concretes increases with moisture content. The average value for $\Delta k/\Delta m$ was found to be 1.3×10^{-2}. Values for dk/dT were found to range from 3.2×10^{-4} to 2.6×10^{-3} in the temperature interval studied. The increase in k_a with T was small but measurable. An average value of 1.55×10^{-7} m^2/s was established for the thermal diffusivity of the perlite concrete. For perlite concrete in the density range of 1200 to 1500 kg/m³ it is possible to obtain thermal conductivities one third that of normal structural concrete.

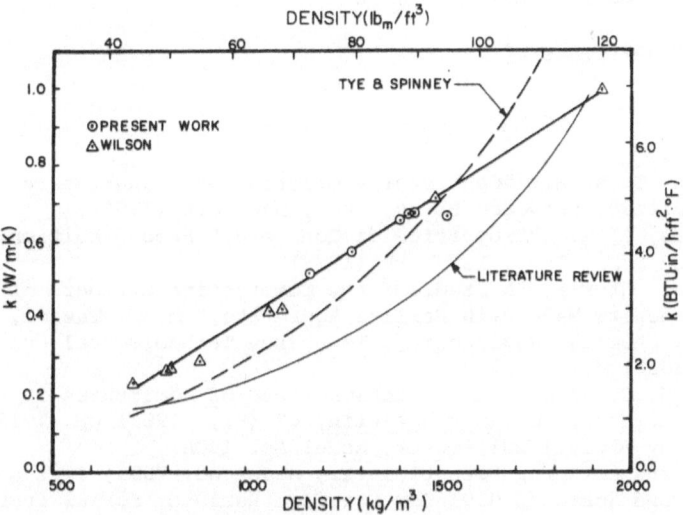

Figure 7. Thermal Conductivity of Concretes as a Function of Density.

ACKNOWLEDGMENT

This project was supported by the Tennessee Technological University Center for Manufacturing Research and Technology Utilization and by Chemrock Corporation, Nashville, Tennessee. Appreciation is extended to both organizations.

273

NOMENCLATURE

C_p Specific Heat, J/kgK

E   Voltage Difference, Volts

I   Current, Amperes

k_a or k   Apparent Thermal Conductivity, W/mK

k_∞ Equilibrium Value for k_a, W/mK

L   Length, cm

r_i, r_o Inside Radius, Outside Radius, cm

T   Temperature

t   Time

α Thermal Diffusivity, m^2/s

ρ Density, kg/m^3

REFERENCES

1. Akroyd, T. N. W., "Concrete: Properties and Manufacture," First Edition, Pergamon Press, Inc., New York (1962).
2. Neville, A. M., "Properties of Concrete," Second Edition, John Wiley and Sons, Inc., New York (1973).
3. Michael, Ntoria, "A Study of the Compressive Strength of Lightweight Concrete Made with Perlite Aggregate," M. S. Thesis, Department of Chemical Engineering, Tennessee Technological University (1986).
4. Lankard, D. R. and L. E. Hackman, "Use of Admixtures in Refractory Concretes," Ceramic Bulletin, 62 (9), (1983) pp. 1019-1035.
5. Keithley Digital Multimeter, Model No. 160B.
6. Leeds and Northrup Potentiometer, Model No. 7555, Type K-5.
7. Leeds and Northrup 0.01 Ohm Standard Resistor Serial Number 195081.
8. Hewlett-Packard Model 6260B D. C. Power Source.
9. Powell, R. L., W. J. Hall, C. H. Hyink, Jr. and L. L. Sparks, "Thermocouple Reference Tables Based on IPTS-68," National Bureau of Standards Monograph 125, U. S. Government Printing Office, Washington, D. C. (1974).
10. Standard Thermometers Manufactured by H-B Instrument Company, Philadephia, Pennsylvania.
11. Turner, W. D., D. C. Elrod, and I. I. Simon-Tov, "HEATING5: An IBM 360 Heat Conduction Program," ORNL/CSD/TM-15, Oak Ridge National Laboratory, Oak Ridge, Tennessee.
12. Haynes, R. D., "Thermal Conductivity of Low-Density Concretes Using a Radial-Heat-Flow-Apparatus," M. S. Thesis, Department of Chemical Engineering, Tennessee Technological University (1986).
13. Wilson, M. A., "Thermal Resistance of Lightweight Concrete," M. S. Thesis, Department of Chemical Engineering, Tennessee Technological University (1984).
14. Tye, R. P. and S. C. Spinney, "Thermal Conductivity of Concretes: Measurement Problems and the Effect of Moisture," Institute International du Froid, Commission BI, Washington, D. C. (1976).

SESSION 6

METHODS

THERMAL CONDUCTIVITY MEASUREMENTS WITH OPTICAL FIBER SENSORS

B. J. White, J. P. Davis,
and L. C. Bobb

Naval Air Development Center
Warminster, PA 18974

D. C. Larson

Department of Physics
Drexel University
Philadelphia, PA 19104

ABSTRACT

The thermal conductivity of a liquid can be determined through the use of a short segment of conductively coated optical fiber, which serves as both the heating element and the thermometer. Experiments were performed in which a fiber was heated while immersed in ethylene glycol, glycerol, and water. Values of the thermal conductivity were calculated from data which show the fiber temperature as a function of time after the application of voltage to the fiber's conductive coating.

INTRODUCTION

The measurement of the thermal conductivities of fluids has been accomplished through the use of the transient hot-wire technique.[1] In this method, a thin wire surrounded by a medium of much lower thermal conductivity is heated via the passage of electrical current through the wire. The resulting temperature increase of the wire, which depends upon the thermal conductivity of the surrounding fluid, is recorded as a function of the time. In this paper, we describe a variation of the transient hot-wire technique, in which a short segment of conductively coated optical fiber in an all-fiber Mach-Zehnder interferometer is heated resistively by passing current through the conductive coating.[2] The increase in the fiber temperature results in both a change in the refractive index of the fiber core and an axial strain in the fiber and is observed as a shifting of the interference pattern at the output of the interferometer. This shifting of the interference fringes may be directly correlated to the temperature change, and the thermal conductivity determined from the rate of change of the temperature.

EXPERIMENTAL SETUP

A cross-sectional view of the ITT Type T-1601 single-mode optical fiber used in this study is shown in Figure 1. It consists of a silica core, B_2O_3-doped silica cladding, and silica substrate, and has a total diameter of 80 μm. For these experiments, a thin film of gold was sputtered onto the surface of a short length of the silica fiber before the fiber was inserted into the interferometer.

The Mach-Zehnder interferometer, shown in Figure 2, was fabricated from two 2x2 ITT fused fiber couplers. Two fibers connected to one of the couplers had been spliced to two of the fibers emerging from the other to obtain the desired configuration. Light from a helium-neon laser was focused onto the core of the input fiber and was split evenly between the two arms of the interferometer at the first coupler. After traversing the two paths, the light was mixed at the second coupler and collected by a photodiode at the output. The phase of the light exiting each arm of the interferometer is affected by strains and/or refractive index changes in the fiber. If the fiber in one arm is heated relative to that in the other arm, a change in the relative phase of the light is observed as a displacement of interference fringes at the output. The interferometer output was recorded on a Gould ES 1000 recorder. In this manner, it was possible to measure the actual phase shift (by counting the number of fringes produced) per unit temperature change caused by the application of current to the conductive coating.

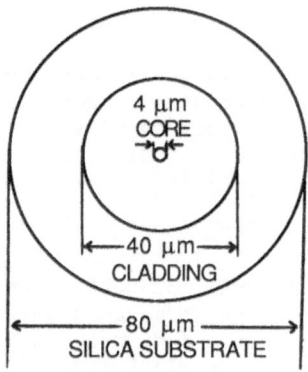

Figure 1.   Cross-sectional view of ITT single-mode optical fiber.

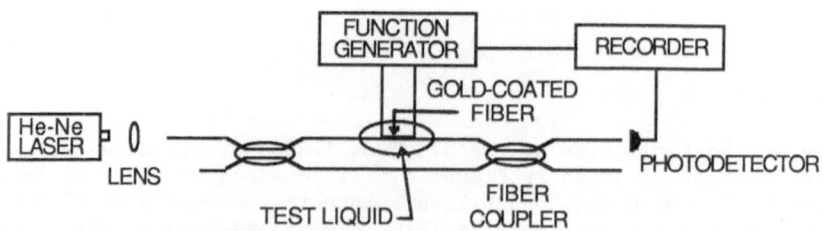

Figure 2.   Optical-fiber Mach-Zehnder interferometer.

DETERMINATION OF FIBER TEMPERATURE SENSITIVITY

Theoretical

The temperature sensitivity of the fibers was determined both theoretically and experimentally. The phase of a wave propagating in a fiber of length L is given by

$$\phi = \frac{2\pi nL}{\lambda} \tag{1}$$

in which n is the effective refractive index and λ is the wavelength of the light in free space. In the single-mode fiber used in this study, 80% of the light is confined to the core. Therefore, the effective refractive index in Eq. (1) may be replaced by the core refractive index. A change in fiber temperature ΔT results in a phase shift $\Delta\phi$ of the light in the fiber because of the change in fiber length due to thermal expansion or contraction, the temperature-induced change in the refractive index of the fiber core, and the photoelastic effect:

$$\frac{\Delta\phi}{\phi} = \frac{\Delta L}{L} + \frac{\Delta n}{n} = \varepsilon_z + \frac{1}{n}\left(\frac{\partial n}{\partial T}\right)_\rho \Delta T + \left(\frac{\Delta n}{n}\right)_T \tag{2}$$

in which ε_z is the axial strain in the core and ρ is the core density. The last term is related to the photoelastic effect, whereby the strain changes the refractive index, and may be expressed in terms of the strains and the elastooptic or Pockels coefficients. An expression for the temperature sensitivity $\frac{\Delta\phi}{\phi\Delta T}$ is thus derived:

$$\frac{\Delta\phi}{\phi\Delta T} = \frac{1}{n}\left(\frac{\partial n}{\partial T}\right)_\rho + \frac{1}{\Delta T}\left\{\varepsilon_z - \frac{n^2}{2}\left[(P_{11} + P_{12})\varepsilon_r + P_{12}\,\varepsilon_z\right]\right\} \tag{3}$$

P_{11} and P_{12} are the Pockels coefficients of the core and ε_r is the radial strain in the core. For this analysis, we considered the fiber to consist of four concentric layers, i: the core, the cladding, the substrate, and the very thin (~ 0.1 µm) gold coating. If we assume uniform fiber temperature T, the stresses σ^i for each fiber layer are given by[3]

$$\begin{bmatrix} \sigma_r^i \\ \sigma_\theta^i \\ \sigma_z^i \end{bmatrix} = \begin{bmatrix} \lambda_i + 2\mu_i & \lambda_i & \lambda_i \\ \lambda_i & \lambda_i + 2\mu_i & \lambda_i \\ \lambda_i & \lambda_i & \lambda_i + 2\mu_i \end{bmatrix} \begin{bmatrix} \varepsilon_r^i - \gamma_i T \\ \varepsilon_\theta^i - \gamma_i T \\ \varepsilon_z^i - \gamma_i T \end{bmatrix} \tag{4}$$

in which γ_i is the coefficient of thermal expansion of the ith layer and λ_i and μ_i are the Lamé parameters which are related to the Young's moduli E_i and the Poisson's ratios ν_i by:

$$\lambda_i = \frac{\nu_i E_i}{(1 + \nu_i)(1 - 2\nu_i)} \quad \text{and} \quad \mu_i = \frac{E_i}{2(1 + \nu_i)} \tag{5}$$

Employing the methods of Schuetz et al.[4] and inserting the necessary boundary conditions, we obtain eight equations with eight unknowns, which, when solved, yield the strain information necessary to solve Eq. (3). When

uniform fiber temperature is assumed, the calculated temperature sensitivity
of the fiber is

$$\frac{\Delta\phi}{\phi\Delta T} = 7.08 \times 10^{-6}/K \tag{6}$$

which is easily manipulated to obtain an experimentally more practical quantity:

$$\frac{\Delta\phi}{L\Delta T} = 102.5 \frac{rad}{m \cdot K} = 16.3 \frac{fringes}{m \cdot K} \tag{7}$$

Experimental

Changes in the fiber temperature were determined experimentally by
applying different amounts of current to the conductive portion of the fiber
and measuring the voltage drop across it. The resistances were calculated
and the temperature changes found using the relation

$$\Delta T = \frac{\Delta R}{Ry_T} \tag{8}$$

in which ΔR is the resistance change corresponding to a temperature change
ΔT, R is the room-temperature resistance, and y_T is the temperature coefficient of resistivity which, for bulk gold, is equal to 0.0034/K. Studies
of the thickness dependence of y_T indicate that its value is reduced for
thin films.[5] A value of y_T for the coating on the fiber was determined
experimentally by placing the gold-coated portion of the fiber in a furnace
and monitoring the resistance as the temperature was increased. The
resistance and temperature data were then inserted in Eq. (8), which was
solved for y_T. This procedure yielded a value of 0.0027/K for the temperature coefficient of resistivity of the fiber coating. When this value
is used to calculate ΔT for different current steps applied to the fiber
coating and these temperature changes are correlated with the corresponding
number of fringes and fiber length, an experimental value of the temperature
sensitivity is determined. Because of approximations in the theoretical
model described above and uncertainties in the values of some of the parameters used in that determination, the value of the temperature sensitivity
obtained in Eq. (7) does not agree precisely with that obtained experimentally. Both our research and that performed in a similar study by Lagakos
et al.[6] indicate a value of the temperature sensitivity 2.7 to 2.9% lower
than the theoretical value. Hence, the temperature changes reported in this
paper were calculated from the number of fringes produced, the length of
fiber being heated, and the experimental temperature sensitivity

$$\frac{\Delta\phi}{L\Delta T}\bigg|_{exp} = 15.9 \frac{fringes}{m \cdot K} \tag{9}$$

The temperature changes determined by dividing the number of fringes produced for each current step by this value of the temperature sensitivity
were found to agree well with those calculated by inserting the corresponding resistance changes into Eq. (8).

A complication that frequently arises in the determination of thermal conductivity stems from the onset of convective losses from the heat source. Studies by Vest and Lawson[7] and Parsons and Mulligan[8] indicate that there is a delay time before the onset of significant convection near a suddenly heated horizontal wire. During this time interval, heat transfer from the wire to the surrounding medium is primarily by conduction. The time delay t^* before convection is well-established can be estimated from the relation[7]

$$t^* = 43\left(\frac{Qg\gamma\sqrt{\alpha}}{k\nu}\right)^{-\frac{2}{3}} \tag{10}$$

in which Q is the power per unit length, g the gravitational acceleration, γ the thermal expansion coefficient of the surrounding medium, α its thermal diffusivity, k its thermal conductivity, and ν its kinematic viscosity.

The results of the experiments by Vest and Lawson in which they observed interference fringes emanating from a heated tungsten wire, agree fairly well with this expression for the delay time. In their experiments, they define the delay time as "the first time at which an appreciable lack of symmetry in the outermost isotherm was observed." In our studies[9] of the heating of conductively coated optical fibers, we found that the delay time before the onset of convection was somewhat less than that suggested by Eq. (10). This reduced delay time is obtained because the heat loss is determined right at the fiber surface, rather than at some later time when the outermost isotherm has moved some distance from the wire. For our purposes, we define the onset of convection as occurring when the curve describing the fiber temperature as a function of time begins to deviate from Carslaw and Jaeger's conduction model[10] described below. As a first approximation, though, we used the above relation to determine whether the heat loss from a fiber in each liquid was likely to be conductive for the duration of the experiment. In the case of glycerol, when a power of 1.01 W is applied to a 3.0-cm-long fiber, a delay time of 157 s is calculated. This is considerably greater than the time (less than 10 s) for the experimental production of fringes. We may therefore conclude that most of the heat loss from the fiber in this case is by conduction. (It should be noted that, though the fiber continues to increase in temperature in cases where the heat loss is solely by conduction, the temperature change per unit time becomes smaller, such that, ultimately, it becomes impossible to distinguish the interference fringes from random drift in the interferometer.)

For ethylene glycol, the calculated t^* is 14 s when 0.512 W is applied to a 4.0-cm-long fiber. Since observable fringes cease to be produced after 2 to 3 s, we can assume that most of the heat loss from the fiber in ethylene glycol is also by conduction.

The results of experiments performed with water as the surrounding medium suggest that, at higher powers (which result in correspondingly lower delay times), heat is lost by both conduction and convection. A delay time of 4 s was obtained for a 4.0-cm-long fiber to which 1.100 W was applied. This is slightly larger than the 2 s or less required for fringe production, but, as mentioned above, reduced delay times are observed with the all-fiber interferometric method.

EXPERIMENTAL RESULTS

A series of experiments was performed in which square-wave voltages of different magnitude were applied across gold-coated optical fibers immersed

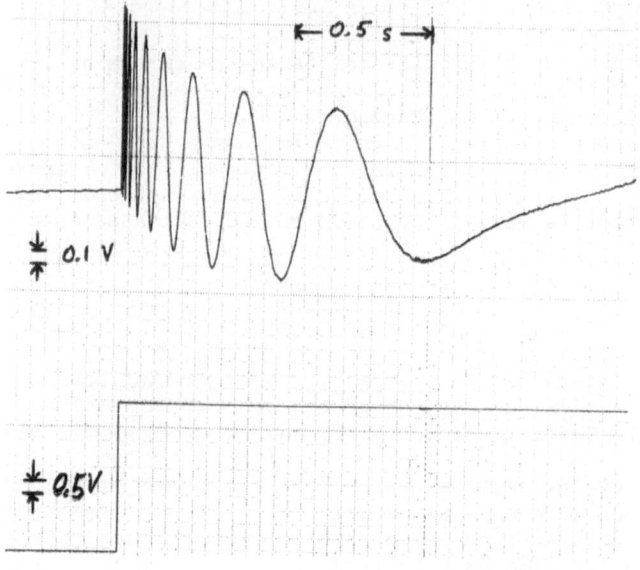

Figure 3. Interference fringes (upper trace) generated
when a power of 0.968 W (lower trace) was
applied across a 4.0-cm-long fiber in water.

in ethylene glycol, glycerol, and water. The change in the fiber temperature as a function of time was determined by monitoring the number and
spacing of the interference fringes produced. A sample recording of fringes
generated after the application of 0.968 W to a 4.0-cm-long optical fiber in
water is shown in Figure 3. It can be seen that the fringes spread out with
time (horizontal axis) as the fiber is heated. The temperature change can
be determined at any time by dividing the number of fringes produced until
that time by the value in Eq. (9) and by the length of fiber being heated.

The experimental data were compared with the solution of the conduction
equation given by Carslaw and Jaeger[10] for a cylinder of radius a and infinite conductivity surrounded by an infinite medium and heated at the rate Q
per unit length per unit time t. The solution for large τ is approximately

$$\Delta T = \frac{Q}{4\pi k}\left\{\left[\ln \frac{4\tau}{C}\right]\left[1 + \frac{\alpha_1 - 2}{2\alpha_1\tau}\right] + \frac{1}{2\tau}\right\} \qquad (11)$$

in which $\tau = kt/\rho c_p a^2$, $C = 1.7811$, and $\alpha_1 = 2(\rho c_p)_{fluid}/(\rho c_p)_{fiber}$, where ρ is the
density and c_p is the heat capacity. For sufficiently large values of the
time, the last two terms in Eq. (11) become negligible so that a curve of ΔT
versus $\ln \tau$ has a linear asymptote of slope Q/k, and thus, if Q is known, k
may be determined.

For each liquid, the temperature change was determined as a function of
time for each of several power levels applied to the fiber's conductive
coating. These time-dependent changes in temperature were then normalized
by dividing each by its respective power per unit length, Q. Since τ is
dependent upon k, we plotted $\Delta T/Q$ versus $\ln t$, where t is measured in
seconds. These curves were consistently reproducible for the different
power levels attempted. This is evidenced in Figure 4, which shows data
sets obtained at three different powers while a 4.0-cm-long segment of
optical fiber was immersed in ethylene glycol.

A curve derived from the theoretical conduction model is shown along with its linear asymptote for large t and the ethylene glycol data of Figure 4. The final slope of the theoretical curve may be made approximately equal to that of the experimental data by adjusting the value of k in Eq. (11). A value of 2.80×10^{-3} W/cm·K was obtained in this manner for the thermal conductivity of ethylene glycol. This value is somewhat higher than the accepted thermal conductivity[11] at 300 K of 2.576×10^{-3} W/cm·K, but still within the range of values obtained by other investigators. The values of k we obtained at different power levels were consistently high and may be indicative of the presence of an impurity, such as water, in the sample.

Data recorded after the sudden application of two different powers to a 3.0-cm-long fiber immersed in glycerol are plotted in Figure 5 along with

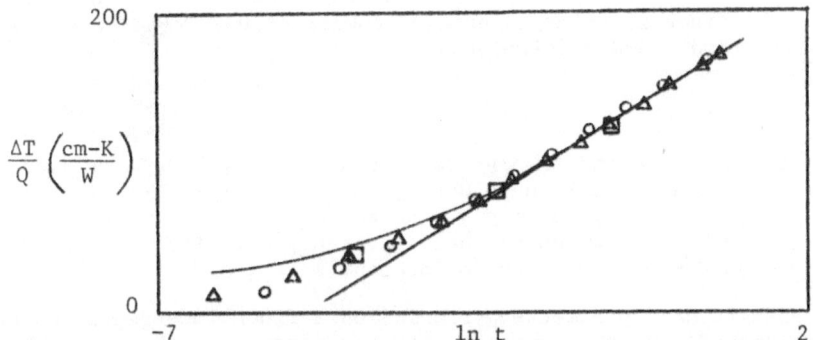

Figure 4. Plot of the normalized temperature change as a function of the natural logarithm of the time for a 4.0-cm-long fiber in ethylene glycol. The symbols represent the experimental data points obtained at different powers: squares -- 0.159 W, circles -- 0.436 W, triangles -- 0.512 W. A value of $k=2.80 \times 10^{-3}$ W/cm·K was used in fitting the theoretical curve and its linear asymptote (solid curves) to the data.

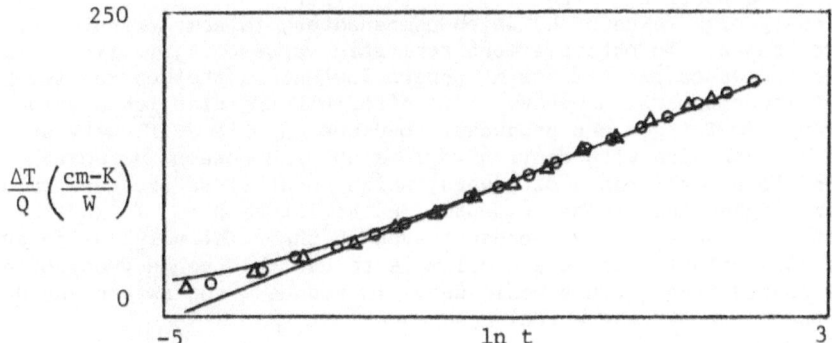

Figure 5. Plot of the normalized temperature change as a function of the natural logarithm of the time for a 3.0-cm-long fiber in glycerol. The triangles and circles represent the data points at applied powers of 0.515 W and 0.654 W, respectively. The theoretical curve and its linear asymptote (solid curves) were obtained for $k=2.93 \times 10^{-3}$ W/cm·K.

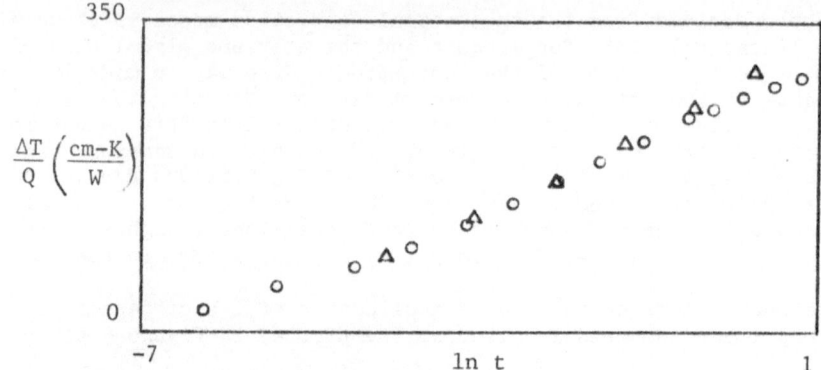

Figure 6. Plot of the normalized temperature change as a function
of the natural logarithm of the time for a 4.0-cm-long
fiber in water at applied powers of 0.312 W (triangles)
and 1.100 W (circles).

the curve derived from the theoretical conduction model and the linear
asymptote. The theoretical curve was again made to fit the data points by
suitable selection of a value for k. As shown, the best fit was obtained
for $k=2.93 \times 10^{-3}$ W/cm·K, which compares favorably with the value at 300 K of
2.880×10^{-3} W/cm·K recommended by Touloukian.[11]

The results of experiments performed on a fiber immersed in water
yielded values of k which increased with an increase in the magnitude of the
power to the fiber. This change in the thermal conductivity was in excess
of what the temperature dependence of k would suggest and is indicative of
an additional mechanism of heat loss. The increase in k with power may be
the result of the onset of a significant amount of convection at higher
power levels. This heat loss, when combined with that lost conductively, is
manifested as a lower overall temperature change than that which would be
expected if conduction alone were acting to transfer heat from the fiber.
The presence of convection also decreases the slope of the $\Delta T/Q$ vs. ln t
curve as the fiber approaches thermal equilibrium. It is possible, however,
to obtain reasonably good agreement between Carslaw and Jaeger's linear
asymptote and a portion of the slope of the experimental curve by selecting
an erroneously high value of k, which compensates, to some extent, for the
convective losses. To obtain a more realistic value of k, it is necessary
to perform the water experiments at powers low enough that convection hasn't
begun. At lower powers, however, it is difficult to establish a value for k
because very few fringes are produced. Because of this difficulty at low
powers and the problem with convective loss at high powers, we were
constrained to a small range of powers, which nonetheless produced values of
k that were higher than those presented in the literature. It is likely,
then, that some convection was present even at the relatively low powers
selected. One solution to this problem is to use a longer segment of conductively coated fiber. This would serve to reduce Q and extend the delay
time for the onset of convection.

The results of two experiments performed at different powers while the
fiber was immersed in water are shown in Figure 6. At a power of 0.312 W,
the slope of the curve remains linear with ln t, and a value for the thermal
conductivity of 6.53×10^{-3} W/cm·K was found. The data taken at 1.100 W exhibit a smaller slope during the linear portion of the curve, and good
agreement between the data and the theoretical model was obtained with a k
of 6.90×10^{-3} W/cm·K. The experimentally determined values of k at power
levels ranging from 0.216 W to 1.100 W are plotted in Figure 7. The first

Table 1. Summary of Results of Thermal Conductivity Measurements

Liquid	Experimental k (mW/cm·K)	Accepted[11] k (mW/cm·K)	Range of accepted values	Deviation of present experiment from accepted value of k
Ethylene glycol	2.80	2.576	-7% to +10%	+8.7%
Glycerol	2.93	2.880	-2% to +3%	+1.7%
Water	6.21	6.084	-2% to +3%	+2.1%

four points fall roughly on a line and when a least-squares fit is extrapolated to zero power, for which the delay time before the onset of convection is infinite, a value of k of 6.21×10^{-3} W/cm·K is obtained. This exceeds the accepted value[11] at 300 K of 6.084×10^{-3} W/cm·K by only 2.1%. This method, then, appears to be feasible for determining the thermal conductivity of liquids for which convection has begun.

A table summarizing our results and comparing them with Touloukian's recommended values of the thermal conductivity is presented above. The dispersion in the values at 300 K found by the various investigators cited in the Touloukian work is indicated for each liquid. As shown, all of the values of k obtained in the present study are within the acceptable ranges.

CONCLUSIONS

We have demonstrated that thermal conductivities of liquids may be determined through the use of an optical-fiber interferometric technique in which the optical fiber acts as both the heating element and the thermometer. Good agreement was obtained between our experimental results and published values of the thermal conductivities of glycerol and ethylene glycol. The experiments performed with the fiber in water yielded progressively higher values of k as the power to the fiber was increased and indicate that care should be taken when employing this technique to avoid or to compensate for the introduction of convective motion in the liquid.

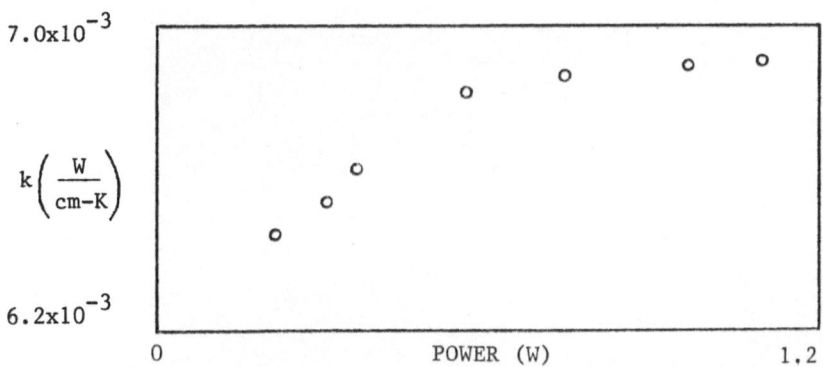

Figure 7. Experimentally determined values of the thermal conductivity of water as a function of power to the fiber.

REFERENCES

1. N. Fox, N. W. Gaggini, and R. Wangsani, "Measurement of the thermal conductivity of liquids using a transient hot wire technique," Am. J.
   Phys. 55, 272 (1987).

2. B. J. White, J. P. Davis, L. C. Bobb, H. D. Krumboltz, and D. C. Larson,
   "Optical-fiber thermal modulator," J. Lightwave Tech. LT-5 (Sept.
   1987).

3. J. F. Nye, **Physical Properties of Crystals**, Oxford: Oxford University
   Press, 1976.

4. L. S. Schuetz, J. H. Cole, J. Jarzynski, N. Lagakos, and J. A. Bucaro,
   "Dynamic thermal response of single-mode optical fiber for interferometric sensors," Appl. Opt. 22, 478 (1983).

5. K. L. Chopra and L. C. Bobb, "Electrical resistivity studies on polycrystalline and epitaxially grown gold films," Acta. Met. 12, 807
   (1964).

6. N. Lagakos, J. A. Bucaro, and J. Jarzynski, "Temperature-induced optical
   phase shifts in fibers," Appl. Opt. 20, 2305 (1981).

7. C. M. Vest and M. L. Lawson, "Onset of convection near a suddenly heated
   horizontal wire," Int. J. Heat Mass Transfer 15, 1281 (1972).

8. J. R. Parsons, Jr. and J. C. Mulligan, "Onset of natural convection from
   a suddenly heated horizontal cylinder," J. Heat Transfer 102, 636
   (1980).

9. B. J. White, J. P. Davis, L. C. Bobb, and D. C. Larson, "Heat transfer
   from resistively heated optical fibers," to be published.

10. H. S. Carslaw and J. C. Jaeger, **Conduction of Heat in Solids**, Oxford:
    Oxford University Press, 1959.

11. Y. S. Touloukian, "Thermal conductivity -- nonmetallic liquids and
    gases," Vol. 3 of **Thermophysical Properties of Matter**, The TPRC Data
    Series, New York: IFI/Plenum, 1970.

EFFECT OF EXPERIMENTAL VARIABLES ON FLASH THERMAL DIFFUSIVITY DATA ANALYSIS

James N. Sweet

Sandia National Laboratories
P. O. Box 5800
Albuquerque, NM 87185

ABSTRACT
         Flash thermal diffusivity data is usually analyzed with a
thermal model which assumes axial heat conduction and uniform
illumination of the flashed surface. For high accuracy data reduction,
it becomes important to bound the errors caused by radial heat flow and
by non-uniform laser beam profiles. These effects are examined
analytically for a case in which the incident laser beam is confined to a
radius smaller than the sample radius. The dependence of the output of
an averaging detector on the magnitude of the radial heat transfer
coefficient is presented and the linear dependence of radial and axial
loss sensitivity coefficients is discussed. From this discussion, we
conclude that inclusion of radial loss effects in analysis of the thermal
response of multilayer structures is not important unless the radial loss
factor is very large. Analytical results are presented for the
temperature vs. time response of a two layer composite sample with
interfacial thermal resistance and high thermal losses at the sample
faces. The use of these results to reduce data for two multilayer
samples is presented to show the utility of new data reduction
techniques.

This work performed at Sandia National Laboratories supported by the U.S.
Department of Energy under Contract Number DE-AC04-76-DP00789.

INTRODUCTION

The flash or pulse method of measuring the thermal diffusivity of liquid and solid materials is now widely used, although commercial equipment for making this measurement is just beginning to become available. The method has been discussed extensively in the literature and reviewed relatively recently by Taylor[1] and Taylor and Maglic[2]. Despite the relative maturity of the technique, there are still questions about its accuracy and its sensitivity to departures of the actual experimental conditions from those postulated in the theory used to reduce the temperature vs. time data. The purposes of this paper are to reexamine the effects of some experimental variables on the accuracy of flash diffusivity measurements and to present examples of the use of the method in determination of the diffusivity or thermal contact resistance of one layer in a multilayered composite sample.

In a critical examination of the flash method, Taylor[3] identified three sources of non-measurement error: (1)finite pulse time effects, (2) heat losses or gains from the sample surfaces during the measurement, and (3) non-uniform heating by the incident laser pulse. Finite pulse time effects become important when the half rise time of the sample back-surface temperature, $t_{1/2}$, is significant relative to the pulse time of the laser, τ_p. Methods of correcting for finite pulse time effects are well documented and have been implemented in numerous calculations and computer programs[1,2,4]. The major uncertainty arises from imprecise knowledge of the laser pulse shape or from random shifts in the pulse shape or pulse position from flash to flash. If $t_{1/2} > 2\tau_p$, the exact shape of the pulse is unimportant, and various approximate shapes can be used[4].

Heat losses are usually taken into account in an approximate way by assuming that a linear or Newton's law of cooling heat transfer condition applies at the sample surfaces[5]. A schematic diagram of a laser pulse measurement with a cylindrical sample is shown in Fig. 1. The axial and radial heat transfer coefficients, h_z and h_r, are presumed to characterize both radiative losses from the sample to the surrounding furnace structure as well as conductive losses to sample supports. In the radiative case, the usual $h \propto (T^4 - T_0^4)$ radiation law is linearized, using the assumption that the deviation of the sample temperature, T, from the ambient temperature, T_0, satisfies the condition, $|T - T_0| \ll T_0$. In a recent study by Degiovanni, Sinicki, and Laurent[6], numerical heat transfer analysis techniques were used to show that the linear approximation of radiative losses did not introduce significant error(<1%), for any reasonable values of T_0 and maximum front surface temperature, T_{max}.

288

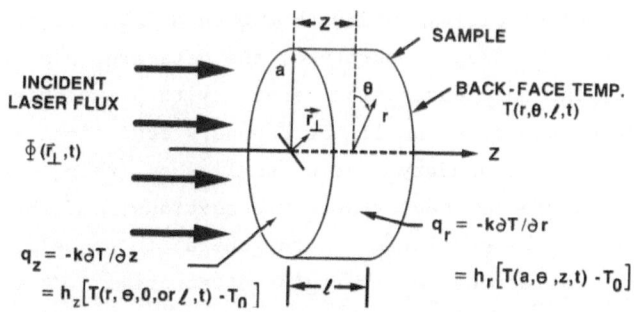

Fig. 1   Sample geometry for flash thermal diffusivity measurements.
The incident laser flux, $\Phi(\vec{r}_\perp, t)$, may vary in the plane of the
sample face, as described by the in-plane vector $\vec{r}_\perp$. The heat
transfer from the front and back face is described by a heat
transfer coefficient h_z and the heat transfer from the radial
boundary is described by a heat transfer coefficient h_r.

The effects produced by a non-uniform laser beam profile are harder to
quantify. Non-uniformities can be caused not only by illumination
variations over the sample surface, but also by some of the incident energy
striking the sample support structure. It has been noted by ourselves and
others[3] that significant errors can be introduced by laser misalignment
even though the temperature vs. time data agree well with a theoretically
calculated curve. As we shall discuss in the next section, beam non-
uniformities over the sample surface will not cause errors if a detector is
used which responds linearly to the average temperature of the sample back
surface. Another potential source of error is misalignment of the detector
viewing optics such that the detector is not focused solely on the sample
back surface.

Beam Shape and Detector View Effects

Ideally, we desire an experimental configuration in which variations
in incident beam intensity over the sample front surface do not produce
significant variations in the experimental data. The cylindrical symmetry
of the flash diffusivity sample, as shown in Fig. 1, plays an important
role in minimizing the sensitivity to flux inhomogeneities. Beedam and
Dalrymple[7] and McKay and Schriempf[8] have theoretically analyzed the
effects produced by radial beam inhomogeneities for the situation in which
a point temperature detector such as a thermocouple is utilized. They find

that significant errors can be introduced in the diffusivity determination
unless information about the laser beam shape is used in the temperature
calculation, even for cases in which the non-uniformity is moderate.
Recently, Degiovanni et al.[6] showed that when a detector is used which
measures the average surface temperature, the data are invariant to radial
beam variations. The only assumption necessary to prove this result from
the heat conduction equation and linear boundary conditions is that of zero
heat flux at the radial boundary. As we shall show, this is not a critical
assumption for usual radial heat transfer conditions, and thus the
Degiovanni result can be considered quite general in scope for incident
beams with radial variations and averaging detectors.

It is easy to show from the linear heat conduction equation that
azimuthal as well as radial variations in the incident beam do not effect
the results if a temperature averaging detector is used. Using the
geometry shown in Fig. 1, we consider the time and position dependent
temperature, $T(\vec{r},t) = T(r,\theta,z,t)$ in the sample after the arrival of the
laser flash. It can be shown that the temperature produced by a laser
pulse with a finite pulse length and specified pulse shape can be derived
from the response to an instantaneous pulse[4,5], so we shall consider the
laser pulse to be a zero time-width impulse. The mathematical form for the
incident pulse is thus taken as,

$$\Phi(\vec{r}_\perp,t) = \Phi_0 \bar{\tau} \delta(t) f(\vec{r}_\perp), \tag{1}$$

where Φ_0 = average pulse energy flux, $\bar{\tau}$ = pulse length, and the product
$\Phi_0\bar{\tau}$ = pulse energy absorbed by the sample. $\delta(t)$ = unit impulse or Dirac
delta function, and $f(\vec{r}_\perp)$ is the pulse energy distribution function over
the front surface of the sample as characterized by the in-plane radius
vector, $\vec{r}_\perp$. $f(\vec{r}_\perp)$ is assumed to a be normalized over the sample surface,
$\int f(\vec{r}_\perp)d^2\vec{r}_\perp = 1$, so that the pulse energy is independent of f. The solution
for the temperature can then be written as,

$$T(r,\theta,z,t) = T_\infty \sum_{m=1}^{\infty} \sum_{n=1}^{\infty} \sum_{k=0}^{\infty} \left[\frac{\exp[-t/\tau_{mn}] \, R_k(y_m r/a) \, Z(x_n z/1) \, \Theta(k\theta)}{N_R(y_m,k) \, N_Z(x_n) \, N_\Theta(k)} \right] \tag{2}$$

$$* \int_{\pi}^{\pi} d\theta' \int_0^{a} r' dr' \left\{ f(\vec{r}_\perp')\Theta(k\theta')R_k(y_m r'/a) \right\}$$

The various expressions in Eq.(2) are fully defined in Appendix A. R, Z,
and Θ are the radial, axial, and azimuthal eigenfunctions, respectively and
the corresponding N functions in the denominator of Eq.(2) are the
normalizing factors for these functions. The effect of the incident pulse

290

intensity function is contained in the double integral on the second line
of Eq. (2).

It is easy to see that any θ variation in $f(\vec{r}_\perp)$ does not affect the
form of the back-surface temperature if an averaging detector is used. The
average back-surface temperature is given by,

$$<T(t)>_{BS} = (1/\pi a^2) \int_{-\pi}^{\pi} d\theta \int_0^a rdrT(r,\theta,\ell,t) \tag{3}$$

Since $\Theta(\theta)$ in Eq. (2) is $\cos(k\theta)$ (see App. A), the θ integral in Eq. (3)
is zero for $k \neq 0$, so only the $k = 0$ term in Eq. (2) is left after the
average is performed. Thus, any angular dependence in the intensity
function, f, is washed out when the average temperature is measured. This
conclusion depends critically on the assumption that the detector
centerline is coincident with the sample centerline.

Any effect of pulse non-uniformity must arise from the radial integral
in Eq. (2). As stated above, Degiovanni et al.[6] show that radial
averaging removes these inhomogenieties if the radial heat flux on the
sample circumference, $k\partial T(a,\theta,z,t)/\partial r|_{r=a} = 0$. If this quantity is non-
zero, it can be shown that the Degiovanni result is still valid if radial
losses are small but that it becomes increasingly less valid as h_r
increases. Rather than presenting detailed analysis to support this
conclusion, we show results of a calculation for the back-face temperature
vs. reduced temperature, $t_r = \alpha t/\ell^2$, for a one layer sample with various
degrees of restriction of the incident beam into a radius $a_\ell \leq a$. Fig.
2(a) shows the average back-face temperature relative to the maximum value
of that temperature for a sample with an axial loss parameter, $b_z = 0.2$, a
thickness/radius ratio of 1/3, and equal radial and axial heat transfer
coefficients, $h_r = h_z$, with illumination to sample radius ratios, $a_\ell/a =$
0.2, .4, .6, .8, and 1.0. The laser beam has a constant intensity for
$r < a_\ell$ and zero intensity for $r > a_\ell$. We see that restricting the laser
beam to a radius smaller than that of the sample has only a small effect on
the temperature history, in agreement with the Degiovanni prediction. In
addition, the leading edge data ($t_r < 0.4$) are practically invariant to the
illumination radius, indicating that leading edge data reduction methods
are not influenced much by radial asymetries. In contrast to this
behavior, Fig. 2(b) shows the result of a calculation for a similar sample
but with enhanced radial heat transfer such that $h_r/h_z = 10$. In this case,
fairly substantial changes occur in the average back-face temperature.
Even in this case, however, data analysis techniques which concentrate on
early time data should still yield relatively accurate results.

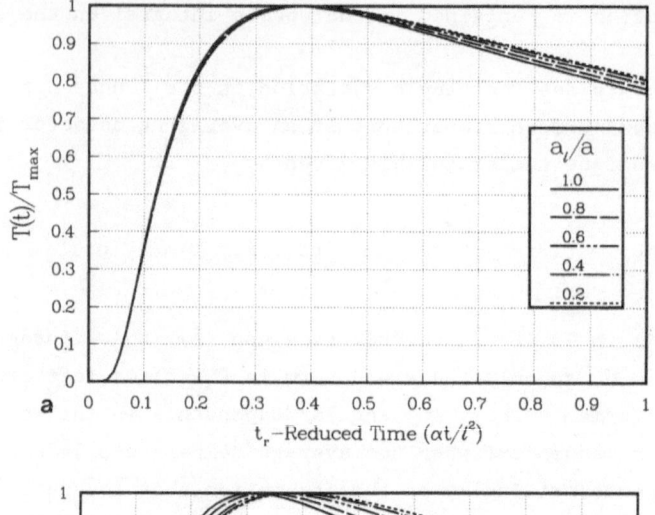

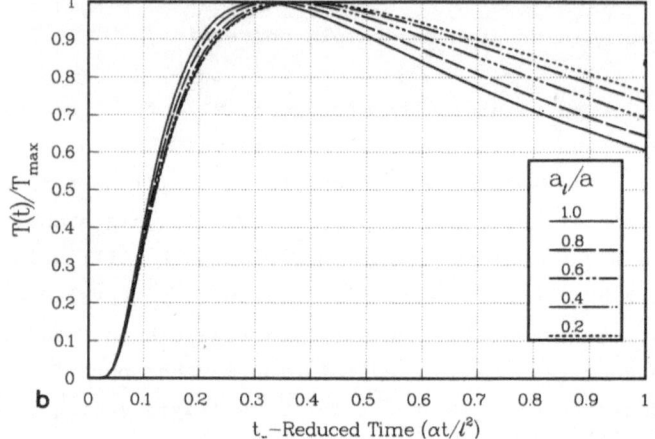

Fig. 2(a)   Theoretical temperature excursions for a homogeneous or
single layer sample with the incident laser flux constant
in magnitude but confined to a radius $a_\ell \leq a$, where
a=sample radius.   For this calculation, $h_r = h_z$, $b_z = h_z \ell/k$
= 0.2, and ℓ/a = 0.3.

(b)   A calculation similar to that in Fig. 2(a) except that $h_r = 10h_z$.

It is interesting to note that the effect produced by an aperture in
the incident beam is the same as the effect produced when a similar
aperture with a radius $a_d \leq a$ is placed in the field of view of the back-
face temperature detector.   From Eqs.(2) and (3), it can be shown that with
the appropriate averages, the radial or m sum in Eq.(2) becomes,

$$\sum_m \rightarrow \sum_m \left\{ \frac{4a^2 J_1(y_m a_\ell/a) J_1(y_m a_d/a)}{a_\ell a_d J_0^2(y_m)(y_m^2 + b_r^2)} \right\} \qquad (4)$$

Eq.(4) shows the complete symmetry between the detector and incident laser radii, a_d and a_ℓ. In Eq.(4) J_0 and J_1 are ordinary Bessel functions.

A remaining question relates to the variation of the back face temperature with variations in the radial losses. Although the radial and axial heat transfer coefficients, h_r and h_z, enter separately in the theory, Eq.(2), it is not necessary to know both of them, and in fact, they cannot be determined separately in a flash diffusivity experiment. The insensitivity of the temperature to variations in h_r was first pointed out by Cape and Lehman[9]. These authors used various approximations to show that the radial and axial loss parameters, b_z and b_r, entered the theory as the linear combination, $b = b_z + (\ell/a)^2 b_r$. These loss factors are related to h_r and h_z through the relations, $b_z = h_z \ell/k$ and $b_r = h_r a/k$, respectively, where k = thermal conductivity of the sample. Many calculations indicate that b_r and b_z are linearly dependent parameters; variations in either affect the average back-face temperature in exactly the same way. This is shown in Figs. 3(a) and 3(b) for calculations associated with a two-layer sample composed of a front layer of graphite and a back layer of CP explosive. Fig 3(a) shows the normalized sensitivity coefficients $\beta \partial T/\partial \beta$[10], vs time for various parameters $\beta = \alpha_2$, T_∞, b_2, and b_r, where b_2 is the loss parameter of the second layer (see App. B). These coefficients are useful for showing how a given fractional parameter change, $\delta\beta/\beta$, affects the temperature. Similar results have been shown by Koski[4] for a single layer sample. The similar shapes of the sensitivity curves for b_2 and b_r in Fig. 3(a) indicate that these parameters are close to being linearly dependent. This means that the parameters cannot be determined independently in a least-squares data fitting process[10]. The other parameters, α_2 and T_∞, have associated sensitivity coefficients which are not linearly dependent with each other or with b_2. Fig. 3(b) shows the results or a temperature calculation which demonstrates the linear dependence of b_r and b_2. The solid curve is for b_2 = 1.33 and h_r/h_z = 1, while the dashed curve has b_2 = 1.20 and h_r/h_z = 10. It is evident that both sets of parameters yield nearly identical results.

The above results indicate that the accuracy of flash diffusivity property measurements should not be critically affected by uncertainties in the laser intensity profile unless the radial losses are very high. It is also evident that uncertainties about the magnitude of the radial loss factor should not affect the value of a thermal diffusivity or contact resistance determined by least-squares data fitting techniques. A potential cause of experimental results being at variance with these predictions is the deviation of an actual infrared detector from an ideal thermal averager. If the sample is thin and the pulse asymmetry is large, then there could be a significant difference in temperature across the

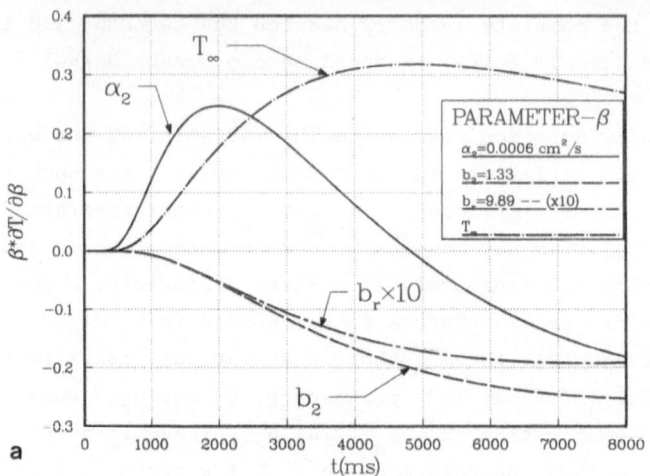

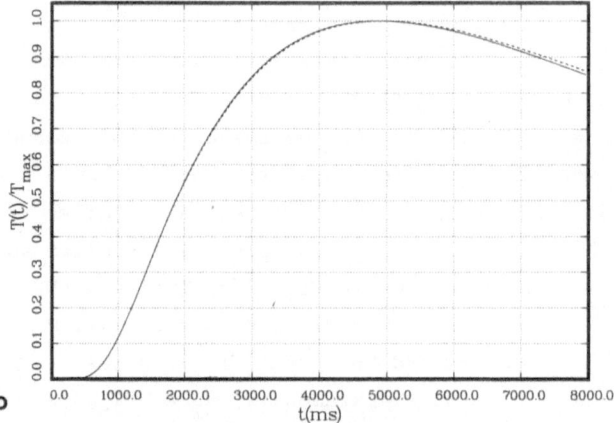

Fig. 3(a) Sensitivity coefficients for a two layer sample described
in the text. The front layer is graphite with α_1 = 0.89
cm^2/s and ℓ_1 = 0.1 cm. The radial loss sensitivity
coefficient has been multiplied by a factor of 10 for ease
in comparison with the second layer axial loss coefficient.

   (b) Transient calculation using Fig. 3(a) parameters. The
solid curve corresponds to the Fig 3(a) conditions, while
the dashed curve is for h_r/h_z = 10 and b_2 = $h_z\ell_2/k_2$ = 1.20.

sample back surface, especially early in the transient. If this
temperature difference is large enough to produce a nonlinear response in
the detector, the hottest parts of the sample will be too heavily weighted
in the measurement, possibly resulting in significant errors. Thus, the

results in this section must be accompanied with the assumption that radial variations in the pulse shape do not produce back-surface temperature variations large enough to cause the detector to depart from linearity.

Two Layer Sample Property Measurements

There has been increasing interest in using flash diffusivity to measure the thermal diffusivity of one layer in a planar multilayer structure with the restriction that the thermal properties of the other layers in the structure are known. Taylor has reviewed past work in this area[1,2], much of which was originated by him. Koski has recently extended the analysis of layered sample data by presenting the sensitivity coefficients for a number of cases[11], utilizing the analytical solutions developed by Lee[12] for the temperature vs. time history of two and three layer samples with no thermal losses. For measurements on thick, low conductivity samples, the magnitude of a layer loss parameter, $b_i = h\ell_i/k_i$, can be appreciable, even at low temperatures. In this light, we have extended Lee's two layer analysis to include both thermal losses at the boundaries and a thermal contact resistance between the layers. The solution is given in Appendix B in a form suitable for programming. This solution had been previously derived by Pratt[13], but left in a complicated form which proved difficult to compare directly with Lee's solution in the no-loss limit. For that reason, we have reanalyzed the loss case, following Lee's notation as much as possible.

A.   AXM9Q Graphite-CP Explosive

Measurements on explosives will usually require a high conductivity front layer to reduce the maximum temperature seen by the explosive. This type of measurement has been discussed previously by Shoemaker, Stark, and Taylor[14] and by Shoemaker, Stark, Kosigoe, and Taylor[15]. Our samples consisted of a front layer of Poco AXM9Q graphite ≈ 0.1 cm thick pressed onto a disk of CP explosive powder of approximately the same thickness. Theoretical curves characterizing one of our samples have been presented in Fig. 3(b). A comparison between a least squares fit and data derived from primary transient recorder data is shown in Fig. 4(solid curve). The data in Fig. 4 were derived from the primary data by low-pass filtering and then averaging of adjacent channels to reduce the amount of data to a manageable size. The least squares parameters were $\alpha_{CP} = 5.9 \times 10^{-4} cm^2/s$ and $b_2 = 1.32$. Also shown for interest are the theoretical single layer solutions found by our single layer analysis program described previously by Koski[4]. The single layer curve which best matches the leading edge data, t < 5000 ms, was found using our PULSE analysis, based on the original work by Clark and

Taylor[16](dashed line). It predicted a diffusivity, $\alpha_{CP} = 5 \times 10^{-4} cm^2/s$ and loss parameter $b_2 = 0.12$. The dot-dash curve came from our COWAN program, based on the trailing edge analysis of Cowan[17], yielding $\alpha_{CP} = 4 \times 10^{-4} cm^2/s$ and $b_2 = 1.36$. The PULSE curve predicts a CP diffusivity which is within $\approx 20\%$ of the least-squares result, while the COWAN curve predicts a loss parameter which is reasonably close to the correct value. The CP diffusivity predicted by a leading edge analysis and CP loss parameter predicted by a trailing edge analysis provide good starting values for a non-linear least squares solution. Shoemaker[14] has presented a similar comparison of single-layer vs. two-layer temperature profiles in his study on explosives.

Several points of interest may be noted in regard to the CP data. First, the measured thermal diffusivity for the data shown in Fig. 4 is a very low value, $0.00059 cm^2/s$, corresponding to a conductivity, $k_{CP} = 0.093$ W/m-K. This is a conductivity more typical of a powder than a pressed

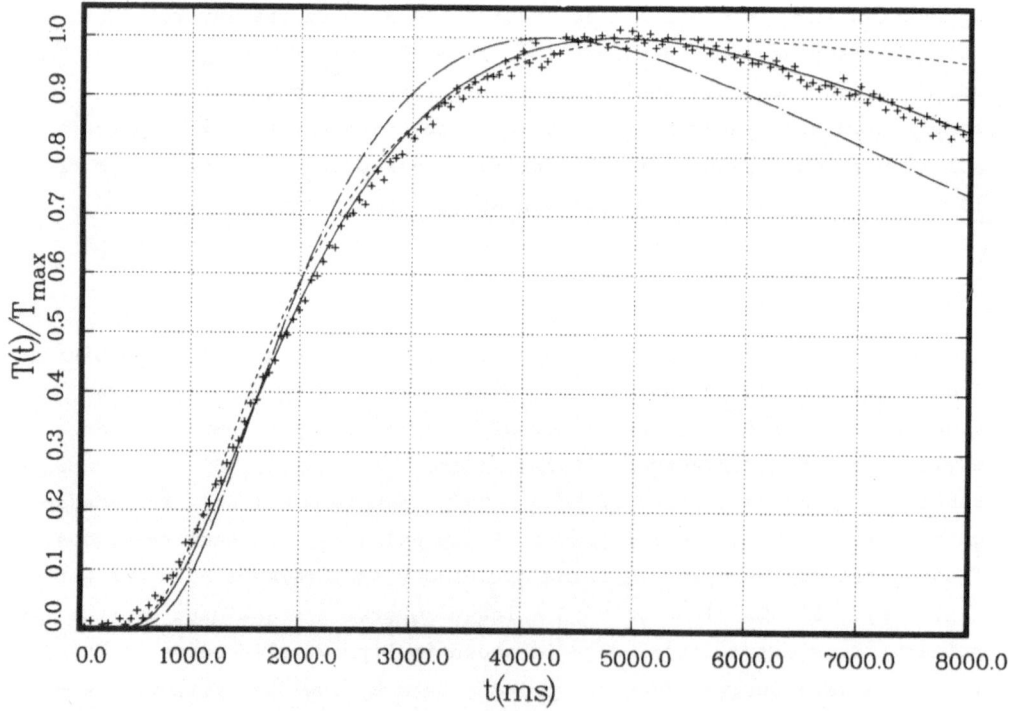

Fig. 4. Data(+ signs) and curve fits to CP explosive data at T=23.6°C. Solid curve-least squares curve-fit giving $\alpha_2 = \alpha_{CP} = 0.00059 \ cm^2/s$ and $b_2 = 1.32$, with $h_r/h_z = 50$. Dash curve-fit using a single layer leading edge or Clark & Taylor analysis, yielding $\alpha_2 = 0.00050 \ cm^2/s$. Dot-dash curve-fit using a single layer Cowan's or trailing edge analysis, yielding $\alpha_2 = 0.00040 \ cm^2/s$.

material, but conductivities almost this low were found by Shoemaker et al. for some pressed powder samples[14,15]. The data are quite insensitive to the unknown contact resistance, a desirable feature. We used a nominal value of 0.1 cm^2-K/W for the contact resistance. Changing this by an order of magnitude to 1 cm^2-K/W produced a change in the diffusivity to 6.013×10^{-4} cm^2/s, about 2%.

Another point of interest is that the shot-to-shot "random" variation in the CP thermal diffusivity is larger in magnitude than the uncertainty in that parameter estimated from a single least-squares data analysis. For the sample in Fig. 4, the 95% confidence limits for α_{CP} derived from the diagonal elements of the covariance matrix and the sum of residual squares was $(5.884 \pm .099) \times 10^{-4}$ cm^2/s, corresponding to an uncertainty $\approx \pm 1.7\%$. Other measurements on this sample at the same temperature, however, led to α values outside this range; the difference between the largest and average α was 0.17×10^{-4} cm^2/s, or about 2.9%. These apparently random shot-to-shot variations presently provide the major limitation on overall system accuracy, assuming that the critical sample parameters, such as the CP thickness, are known to an accuracy sufficient to reduce measurement errors below those associated with shot-to-shot variations.

B. AXM9Q-Colloidal Graphite-AXM9Q

As an example of measurement of the contact resistance of a reasonably low resistance layer, we chose a sample with a colloidal graphite layer between two high conductivity pieces of AXM9Q graphite. The sample was fabricated by spreading a thin layer of commercial colloidal graphite liquid on a disk of AXM9Q graphite and then placing a similar disk on top of the liquid layer to form a sandwich structure. The sample was then loaded with a weight, dried with a heat lamp and mounted in the flash diffusivity apparatus. A least squares fit to some typical data is shown in Fig. 5(a) and the sensitivity coefficients for R, α_1, α_2, and T_∞ are shown in Fig. 5(b). It is evident that this is a low loss case, as would be expected from the low temperature and high conductivity of the sample. The sensitivity coefficient curve for R in Fig. 5(b) indicates the high sensitivity at times slightly longer than $t_{1/2}$. A fit to the Lee no-loss theory at $t_{1/2}$ = 44 ms yields a result, R = 0.332 cm^2-K/W which can be compared to the least-squares fit result, R_{LS} = 0.317 cm^2-K/W, a shift of $\approx 5\%$. The sensitivy coefficients in Fig 5(b) show that for a resistance measurement of this type, a single fit between theory and experiment at or near the half-time is as good as a least squares fit because the R sensitivity coefficient peaks near $t_{1/2}$, while the magnitude of the α_2

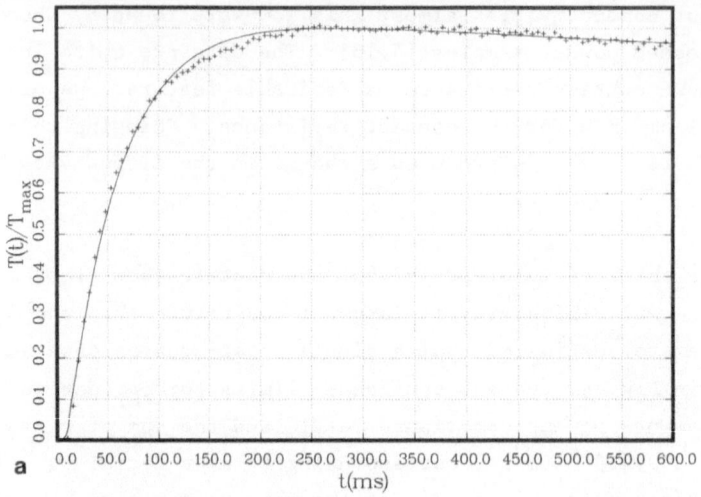

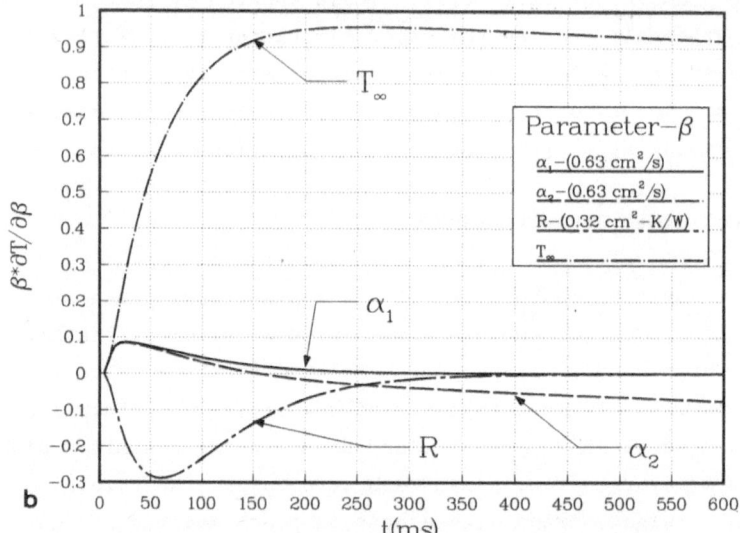

Fig. 5(a) Data(+ signs) for a composite AXM9Q graphite-colloidal graphite-AXM9Q graphite sample at 209°C and least squares curve-fit yielding R = 0.32 cm²-K/W and b_2 = 0.003.

(b) Sensitivity coefficients for the sample in Fig. 5(a). The R sensitivity coefficient has a large magnitude only at short times, t ≤ t_{max}, where t_{max} ≈ 250 ms.

coefficient is small near $t_{1/2}$. In this case, the no-loss theory yields a good result because of the insensitivity of the temperature to R at long times.

To examine the ability to measure the resistance in a system with greater overall thermal resistance, a sandwich of Pyroceram 9606-Colloidal graphite-Pyroceram 9606 was constructed in a manner similar to that used in the case with AXM9Q graphite. Since the Pyroceram sandwich had a much larger thermal resistance, the measurement of the resistance of the colloidal graphite layer would be expected to be more difficult. Examination of the sensitivity coefficients bore out this conclusion; the measurement was much more sensitive to the Pyroceram 9606 diffusivity than to the resistance of the layer, assuming that it was still in the range, $R < 1$ cm^2-K/W. The measurements supported this reasoning, although the data did reveal the presence of the resistive layer. The resistance increased as the furnace temperature was increased from 23°C until about 175°C and then decreased again until it could not be resolved at 300°C. We hypothesize that thermal expansion first opened up the joint, but subsequent high temperature sintering reduced the resistance. When the sample was returned to room temperature, the joint opened and a very high resistance was obtained.

Data reduction for the sample at low temperature showed that the resistance was approximately the same as the previous case with AXM9Q graphite. However, the sensitivity to the Pyroceram diffusivity was so high that, within the limits to which that quantity is known, the resistance could only be bounded, $R < 0.5$ cm^2-K/W.

CONCLUSIONS

In the first part of this paper we have presented arguments which indicate that laser beam non-uniformities should not affect the accuracy of flash diffusivity measurements if an averaging detector is used. The question of whether or not an infrared detector which is viewing the entire sample back-surface truly averages over this surface is an open one which needs to be tested using known asymmetric pulses together with absolute temperature measurements. Studies of this type are planned for the future. The insensitivity to radial losses first predicted by Cape and Lehman[9] appears to be a general result and is consistent with the near linear dependence of these parameters over the entire transient time for all situations which we have examined. Thus, uncertainties in radial losses caused by unknown amounts of heat conduction to sample supports should not affect the accuracy of a thermal diffusivity or contact resistance measurement.

Data for typical multilayer samples have been presented to illustrate the current accuracy of the flash diffusivity technique. These data show that variations in parameters derived from transient data determined under supposedly identical conditions limit the accuracy of the method. Typically, the least-squares fit of a single data set to theory predicts a smaller parameter uncertainty than is actually observed in the shot-to-shot variations of least-squares-values. Future improvements in the measurement accuracy of the technique will require better knowledge of the sources of this variation.

Appendix A

Single Layer Temperature vs. Time Formulas

In this appendix, the various expressions in the single layer temperature vs. time expression, Eq.(2), are defined. The dimensional symbols are defined in Fig. 1. These expressions follow the notation of Ozisik[18]. The prefactor, T_∞, is the long-time, $t \to \infty$, temperature rise, $T_\infty = \Phi_0 \bar{\tau}/\rho c_p$, where ρ = sample density and c_p = sample specific heat.

The radial eigenfunction in Eq.(1) is defined by,

$$R_k(y_m r/a) = J_k(y_m r/a) \qquad (A-1)$$

where J_k is the ordinary Bessel function of the first kind of order k and the y_m are solutions of the radial eigenvalue equation,

$$y_m J_k'(y_m) + b_r J_k(y_m) = 0 \qquad (A-2)$$

In Eq.(A-2), J_k' is the derivative of J_k with respect to its argument. The radial loss factor is defined by $b_r = h_r a/k$, where k=thermal conductivity, $k = \alpha \rho c_p$. The axial eigenfunction, Z, is defined by,

$$Z(x_n z/\ell) = \left(\frac{x_n}{\ell}\right)\cos(x_n z/\ell) + \left(\frac{b_1}{\ell}\right)\sin(x_n z/\ell) \qquad (A-3)$$

The axial eigenvalue equation is,

$$\tan(x_n) = \frac{x_n(b_1+b_2)}{x_n^2 - b_1 b_2} \qquad (A-4)$$

The loss factors in Eqs.(A-3,A-4) are given by, $b_1 = h_1 \ell/k$ and $b_2 = h_2 \ell/k$, where h_1 and h_2 are the heat transfer coefficients for the front and back surfaces, respectively. Usually the assumption is made that these heat transfer coefficients are equal, yielding $b_1 = b_2 = b_z$, the axial Biot number.

The azimuthal eigenfunction is given by,

$$\Theta(k\theta) = \cos(k\theta) \qquad (A-5)$$

The normalization functions in the denominator of Eq.(2) are defined by,

$$N_R(y_m,k) = \frac{J_k(y_m)}{2} \left[\frac{b_r^2 + y_m^2 - k^2}{y_m^2/a^2} \right] \tag{A-6}$$

$$N_Z(x_n) = \frac{1}{2\ell} \left[(x_n^2 + b_1^2)\left(1 + \frac{b_2}{x_n^2 + b_2^2} + b_1 \right) \right]$$

$$N_\theta(k) = \begin{cases} 2\pi, & k=0 \\ \pi, & k\neq0 \end{cases}$$

The time constants, τ_{mn}, are defined by the relation,

$$\tau_{mn} = \frac{\ell^2}{\alpha} \left[y_m^2 (\ell/a)^2 + x_n^2 \right] \tag{A-7}$$

Appendix B

Two Layer Temperature vs. Time Solution, with Contact Resistance and Thermal Losses

In this review of two layer solutions for the axial problem in the presence of thermal losses at the boundaries, we follow the notation of Lee[12] as much as possible. Lee defines the parameters, $H_j = A\rho_j c_j \ell_j$ and $\eta_j = (\ell_j^2/\alpha_j)^{1/2}$, where A=sample area, and ρ_j, c_j, ℓ_j, and α_j are the density, specific heat, thickness, and diffusivity, respectively, of layer j. In our context, j = 1 is the front layer onto which the pulse is incident and j = 2 is the back layer. From the fundamental quantities H_j and η_j, Lee defines the dimensionless ratios,

$$\eta_{i/j} = \eta_i/\eta_j \text{ and,}$$

$$H_{i/j} = H_i/H_j \tag{B-1}$$

Lee then defines the auxiliary quantities,

$$X_1 = H_{1/2}\eta_{2/1} + 1$$

$$X_2 = X_1 - 2, \tag{B-2}$$

and,

$$\omega_1 = \eta_{1/2} + 1$$

$$\omega_2 = \omega_1 - 2, \tag{B-3}$$

together with the dimensionless contact resistance parameter, R_c, which is related to the actual contact resistance, R, by,

$$R_c = R\rho_1 c_1 (\alpha_1 \alpha_2)^{1/2}/\ell_2 \qquad \text{(B-4)}$$

The Lee solution for the no-loss back-surface temperature, relative to the maximum adiabatic temperature rise, T_∞, can be written,

$$V(t)=T(t)/T_\infty = 1 + 2 \sum_{k=1}^{\infty} \left[\frac{(1+PU)Q(x_k,\eta_2,t)}{d_0(\vec{\omega},\vec{X},x_k)+d_R(\vec{\omega},\vec{X},x_k)} \right] \qquad \text{(B-5)}$$

where $P = X_2/X_1$, $U = \omega_2/\omega_1$, $\vec{\omega} = (\omega_1,\omega_2)$, and $\vec{X} = (X_1,X_2)$. The quantity $Q(x_k,\eta_2,t)$ is the convolution integral, as defined by,

$$Q(x,\eta,t) = \int_0^t f(t') \exp\left[-\frac{x^2(t-t')}{\omega_1^2 \eta^2} \right] dt' \qquad \text{(B-6)}$$

In Eq.(B-6), $f(t)$ is the laser pulse shape function. The terms in the denominator of Eq.(B-5) are defined by,

$$d_0 = \cos(x_k) + PU\cos(Ux_k),$$

$$d_R = q\left[\cos(x_k) - \cos(Ux_k) - x_k[\sin(x_k) - U\sin(Ux_k)] \right],$$

$$\qquad \text{(B-7)}$$

where the quantity x_k is the kth root of the eigenvalue equation,

$$\sin(x_k) + P\sin(Ux_k) + qx_k\left[\cos(x_k) - \cos(Ux_k) \right] = 0, \qquad \text{(B-8)}$$

and $q = R_c/\omega_1 X_1$. d_0 in Eq.(B-5) represents the no-loss, no-resistance contribution, while d_R represents the change produced by contact resistance between the two layers. This method of classifying denominator terms and terms in the eigenvalue equation remains in multilayer problems with interface resistance and thermal losses and provides an easy method of keeping track of the contributions of the various processes.

In the case where thermal losses occur at the sample boundaries, the layer loss parameters, $b_1 = h_1\ell_1/k_1$ and $b_2 = h_2\ell_2/k_2$ are introduced. The eigenvalue equation becomes,

$$f_0(x_k) + f_R(x_k) + f_L(x_k) + f_{LR}(x_k) = 0, \qquad \text{(B-9)}$$

where f_0 and f_R are the previous no-loss and contact resistance terms, respectively from Eq. (B-8),

$$f_0(x_k) = \sin(x_k) + P\sin(Ux_k),$$

$$f_R(x_k) = qx_k\left[\cos(x_k) - \cos(ux_k)\right] \qquad \text{(B-10)}$$

The loss term in the eigenvalue equation can be written,

$$f_L(x_k) = -\frac{\omega_1}{x_k}\left\{Y_1\cos(x_k) + PY_2\cos(Ux_k) + \right.$$

$$\left. \frac{Z\omega_1}{x_k}\left[\sin(x_k) - P\sin(Ux_k)\right]\right\}, \qquad \text{(B-11)}$$

where $Y_{1,2} = b_2\left[(b_{1/2}/\eta_{1/2})\pm 1\right]$ and $Z = b_2^2 b_{1/2}/\eta_{1/2}$, with $b_{1/2} = b_1/b_2$. The loss-resistance coupling term is given by,

$$f_{LR}(x_k) = q\omega_1\left\{[Y_1\sin(x_k) - Y_2\sin(ux_k)]\right.$$

$$\left. - \frac{Z\omega_1}{x_k}[\cos(x_k) + \cos(ux_k)]\right\} \qquad \text{(B-12)}$$

The temperature solution no longer has a first term of unity because the Laplace transform does not have a pole at s=0 when b_1 and/or b_2 are non-zero, representing the fact that there is no non-zero steady state component of V(t). V(t) can be written in the form,

$$V(\ell,t) = (1+PU)\sum_{k=1}^{\infty}\left[\frac{2Q(x_k,\eta_2,t)}{d_0(x_k)+d_R(x_k)+d_L(x_k)+d_{RL}(x_k)}\right] \qquad \text{(B-13)}$$

d_0 and d_R are given by Eq.(B-6), while d_L is given by,

$$d_L = \frac{\omega_1}{x_k}\left[Y_1\sin(x_k) + PUY_2\sin(Ux_k)\right]$$

$$+ \frac{\omega_1}{x_k^2}\left[Y_1\cos(x_k) + PY_2\cos(Ux_k)\right]$$

$$+ \frac{Z\omega_1^2}{x_k^3}\left[2[\sin(x_k)-P\sin(Ux_k)]-x_k[\cos(x_k)-PU\cos(Ux_k)]\right]$$

and,

$$d_{LR} = \frac{Zq\omega_1^2}{x_k^2}\left[\cos(x_k)+\cos(Ux_k)+x_k\sin(x_k)+U\sin(Ux_k)\right]$$

$$+ q\omega_1\left[Y_1\cos(x_k)-UY_2\cos(Ux_k)\right]$$

It should be remarked that the solution of the eigenvalue equation, Eq.(B-9), by numerical techniques is not completely straightforward because of the presence of near double-roots for some values of the parameters.

REFERENCES

1. R. E. Taylor, "Heat Pulse Thermal Diffusivity Measurements", High Temperatures-High Pressures 11 43 (1979).
2. R. E. Taylor, K. D. Maglic, "Pulse Method for Thermal Diffusivity Measurement", in Compendium of Thermophysical Property Measurement Methods, Vol. I, Ed. by K. D. Maglic, A. Cezairliyan, V. E. Peletsky, Plenum Press, NY, 1984, pp 305-336.
3. R. E. Taylor, "Critical Evaluation of Flash Method for Measuring Thermal Diffusivity", Rev. Int. Hautes Temp. Refract. 12 141 (1975).
4. J. A. Koski, "Improved Data Reduction Methods for Laser Pulse Diffusivity Determination with the Use of Minicomputers", in Proc. of 8th Symposium on Thermophysical Properties, Vol. II, Ed. by J. V. Sengers, ASME, NY, 1982, pp 94-103.
5. D. A. Watt, "Theory of Thermal Diffusivity by the Pulse Technique", Brit. J. Appl. Phys 17 231 (1966).
6. A. Degiovanni, G. Sinicki, M. Laurent, "Heat Pulse Thermal Diffusivity Measurements-Thermal Properties, Temperature Dependence, and Non-Uniformity of Pulse Heating", in Thermal Conductivity 18, Ed. by T. Ashworth and D. R. Smith, Plenum NY, 1985, pp 537-551.
7. K. Beedham, I. P. Dalrymple, "The Measurement of Thermal Diffusivity by the Flash Method, an Investigation into Errors Arising from the Boundary Conditions", Rev. Int. Hautes Temp. Refract. 7 278 (1970).
8. J. A. McKay, J. T. Schriempf, "Corrections for Non-uniform Surface-Heating Errors in Flash-method Thermal Diffusivity Measurements", J. Appl. Phys. 47 1668 (1976).
9. J. A. Cape, G. W. Lehman, "Temperature and Finite Pulse-Time Effects in the Flash Method for Measuring Thermal Diffusivity", J. Appl. Phys. 34 1909 (1963).
10. J. V. Beck, J. Arnold, Parameter Estimation in Science and Engineering, John Wiley, NY, 1977.
11. J. A. Koski, "Sensitivity and Accuracy Analysis of Pulse Diffusivity Measurements on Layered Samples", in Thermal Conductivity 18, (Ref. 6), pp. 525-536
12. H. J. Lee, "Thermal Diffusivity in Layered and Dispersed Composites", Doctoral Thesis in Mechanical Engineering, Purdue Univ., 1975.
13. A. W. Pratt, Heat Transmission in Buildings, John Wiley, NY, 1981, pp. 149-155.
14. R. L. Shoemaker, J. A. Stark, R. E. Taylor, "Thermophysical Properties of Propellants", High Temp.-High Press. 17 429 (1985).
15. R. L. Shoemaker, J. A. Stark, L. G. Kosigoe, R. E. Taylor, "Thermophysical Properties of Propellants", in Thermal Conductivity 18(Ref. 6), pp. 199-211.
16. L. M. Clark III, R. E. Taylor, "Radiation Loss in the Flash Method for Thermal Diffusivity", J. Appl. Phys. 46 714 (1975).
17. R. D. Cowan, "Pulse Method of Measuring Thermal Diffusivity at High Temperatures", J. Appl. Phys. 34 926 (1963).
18. M. N. Ozisik, Heat Conduction, John Wiley, NY, 1980.

M. Raynaud

Department of Mechanical Engineering
New Mexico State University
Las Cruces, NM 88003

J.V. Beck

Heat Transfer Group
Department of Mechanical Engineering
Michigan State University
East Lansing, MI 48824

R. Shoemaker and R. Taylor

Thermophysical Properties Research Laboratory
Purdue University
West Lafayette, IN 47906

ABSTRACT

The estimation of thermal diffusivity using a flash experiment is
studied. The sequential estimation method is presented and compared to
the original flash method as well as two other methods that account for
heat loss. The potential accuracy of the four methods is tested with a
numerical simulation. The thermal diffusivity of a stainless steel
sample is estimated from experimental data. The estimates given by the
different methods are compared and analyzed. The sequential estimation
method is shown to be less sensitive to random measurement errors and to
provide more information regarding the accuracy of the estimated
parameters.

NOMENCLATURE

b	maximum temperature rise - parameter to be estimated
Bi	Biot number
c	specific heat per unit mass
G	denotes Green's function

h	heat transfer coefficient
k	thermal conductivity
L	thickness of the sample
n	total number of measurement times
Q_o	impulse heat flux
S	surface area of the sample
$t_{1/2}$	half-time
T_j	calculated temperature at time t_j
T_m	maximum temperature rise
T_∞	environment temperature
X_{js}	sensitivity coefficient at time t_j for the parameter β_s
Y_j	measured temperature at time t_j

Greek Letters

α	thermal diffusivity - parameter to be estimated
β_s	parameter s to be estimated (Section II)
β_m	m'th root of the transcendental equation Eq. (16)
ρ	density
τ	instant of the impulse

Subscripts

j	index for time t_j
m	index for time t_m

Superscripts

(k)	iteration index
m	index for value calculated with m measurement times ($1 \leq m \leq n$)
n	index for value with n measurement times

I. INTRODUCTION

The flash method for measuring thermal diffusivity was developed by Parker et al. [1] and is the most widely accepted method. The front face of a small disk-shaped sample (often about the size of a small coin) is subjected to a very short burst of radiant energy. The energy pulse is usually provided by a laser or a xenon flash lamp and irradiation times are of the order of one millisecond or less. The resulting temperature rise of the rear surface (1 or 2°C typically) is measured and the sample thermal diffusivity values are computed from the temperature rise versus time data. The ambient temperature being easily controlled by a small furnace or a chiller, this method has been used to measure thermal diffusivity in the very large temperature range of 80-2500 K. The flash method is shown schematically in Fig. 1 using a laser as the energy source.

In the original equation for estimating thermal diffusivity from the flash method data, heat losses are neglected and the thermal diffusivity,

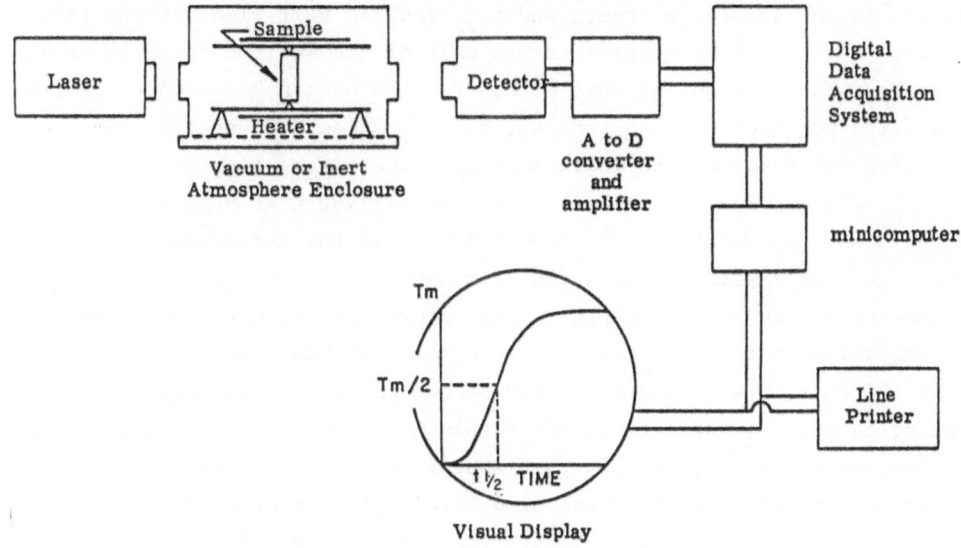

Fig. 1. Experimental set-up for the flash experiment

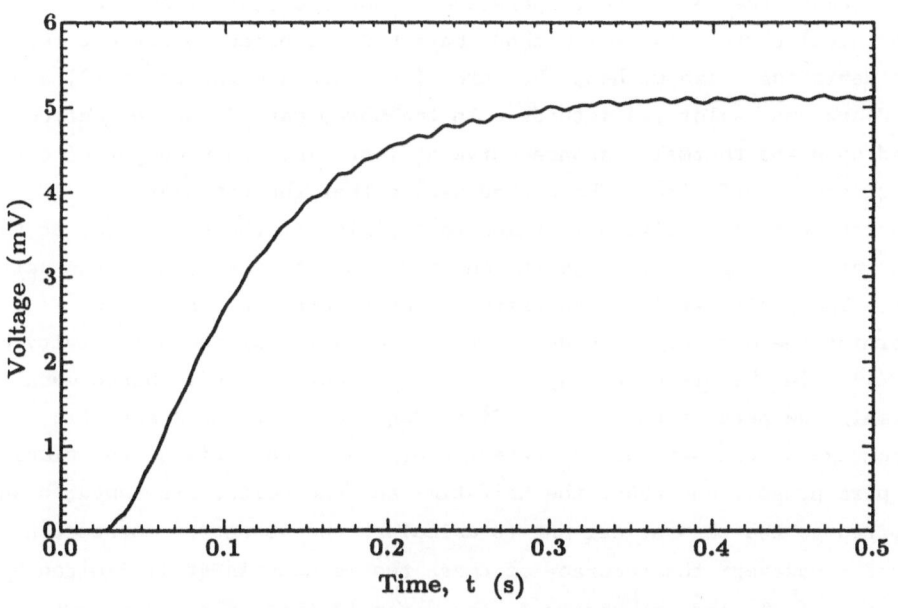

Fig. 2. Voltage rise during the flash experiment given by a
thermocouple placed at the back face of the sample

α, is given by the relation

$$\alpha = 1.37 \frac{L^2}{\pi^2 t_{1/2}} \tag{1}$$

307

where L is the sample thickness and $t_{1/2}$ (called half-time) is the time
at which the rear face temperature is half of the maximum temperature
rise, T_m. The accuracy of this method (called herein uncorrected method)
is limited by three factors. First, it is difficult to accurately
determine the value of the maximum temperature rise. Fig. 2 represents a
typical voltage rise (given by a thermocouple) for this experiment. The
curve is not smooth due to the random errors in the measurements and
approaches the maximum in an asymptotic manner. It is apparent that a
precise estimation of the maximum temperature rise is difficult. The
determination of $t_{1/2}$ will also be in error. It has been proposed by
Cowan [2] that this second error may be reduced by using a relation
similar to Eq. (1) where times other than $t_{1/2}$ are also used to calculate
α. The thermal diffusivity is then approximated by the arithmetic
average of the calculated values of α. The third error is due to
convective and radiative heat losses from the sample. Radiative heat
loss is the major source of error in high temperature thermal diffusivity
measurements using the flash method.

When heat transfer from the rear and front surfaces becomes
significant, the rear face temperature response differs from the
theoretical curve. Several methods have been proposed to correct for
this deviation. Two methods that are often used are the Cowan [2] and
the Clark and Taylor [3] methods. In the Cowan method the correction is
based upon the thermal response curve at long times when compared to the
experimental half-time. The method uses either the ratio of the
temperature rise at five half-times to the rise at the half-time, or the
ratio of the temperature rise at ten half-times to the rise at the half-
time. The ratios are then compared to the theoretical ratios to
determine the correction needed in the calculation of the diffusivity by
Eq. (1). In the Clark and Taylor method the correction is based upon the
thermal rise part of the curve. Times required to reach a specific
percentage of the maximum are determined. Then the ratio of the times at
two percentages, one above the half-time and one below, are computed and
compared to the theoretical one to calculate the amount of correction
needed. However, the accuracy of these two methods is still limited by
the error in T_m and the errors in the determination of the required
times.

The objective of this work is to show how a powerful technique
called sequential estimation can be used to estimate the thermal
diffusivity of the sample based on the experimental temperature rise of
the rear face. The method incorporates the estimation of T_m and all the
measurement times are used, rather than at specific times such as $t_{1/2}$.
Consequently, the effects of the first and second errors are reduced.
The sequential estimation method is also used to estimate the heat

transfer coefficient so that heat losses from the sample can be considered. It also has a statistical basis so that statistical statements regarding the errors can be made.

The sequential estimation technique is presented in Section II. In Section III, a numerical simulation is done to determine the potential accuracy of the a) uncorrected, b) Cowan, c) Clark and Taylor, and d) sequential estimation methods. An experimental data case is studied in Section IV and some conclusions are given in Section V.

II. SEQUENTIAL PROCEDURE

There are several ways to develop the sequential procedure, two of which are discussed in Beck and Arnold [4]. See also Beck [5]. One of these is called the direct method and the other is called the matrix lemma method. The simpler of these is the former and is the one now described. Both methods give exactly the same estimates. Actually the estimates were obtained using program NLINA (Beck and Arnold, [4]) which uses the more general matrix lemma method. Since only a few parameters were simultaneously estimated (1, 2 or 3) for this paper, the generality of the matrix lemma to treat many parameters was not fully exploited.

The method as employed is an extension of nonlinear least squares. The procedure starts with the sum of squares function

$$S = \sum_{j=1}^{n} (Y_j - T_j)^2 \tag{2}$$

where Y_j represents the measured temperature at the rear surface (x=L) at time t_j and T_j is the corresponding calculated temperature at the same location and time. In general, T_j could be obtained by any numerical (finite difference or element) method or from the exact solution. In this paper T_j is obtained from the exact solution. The upper limit n in Eq. (2) is the total number of measurement times. For simplicity assume that there are two parameters of interest, β_1 and β_2; there can be more (or less) parameters but the procedure is the same.

Two aspects have to be simultaneously considered. The first is that the problem is nonlinear and must be treated in an iterative manner. The second is that there is to be a sequential procedure; that is, the effect of adding one measurement after another is considered. At first one might be confused between the concepts of iteration and sequential analysis but the ideas are quite different.

The method of least squares states that the parameter estimates are where S is a minimum with respect to the parameters. The minimum occurs

to zero. Let s denote the index for a parameter and then

$$\frac{\partial S}{\partial \beta_s} = 2 \sum_{j=1}^{n} \left(Y_j - \hat{T}_j \right) \left(-\frac{\partial \hat{T}_j}{\partial \beta_s} \right) = 0 \tag{3}$$

where $s = 1$ and 2 and symbol $\wedge$ denotes evaluation at $\beta_s = \hat{\beta}_s$. When $\partial T/\partial \beta_s$, which is called the β_s sensitivity coefficient, is a function of β_1 or β_2, a set of nonlinear algebraic equations is given by Eq. (3). For such cases an effective way to solve Eq. (3) is to replace $\hat{T}_j$ by the Taylor series

$$T_j^{(k+1)} = T_j^{(k)} + X_{j1}^{(k)}(\beta_1^{(k+1)} - \beta_1^{(k)}) + X_{j2}^{(k)}(\beta_2^{(k+1)} - \beta_2^{(k)}) \tag{4}$$

$$X_{js} = \partial T_j / \partial \beta_s \tag{5}$$

where the superscript k denotes an iteration index and X_{js} is the β_s sensitivity coefficient; β_1 and β_2 values for $k=0$ are assumed and values are calculated for $k=1$ and so on. The superscripts (k) on T_j, X_{j1} and X_{j2} indicate that these quantities have been evaluated at $\beta_1^{(k)}$ and $\beta_2^{(k)}$. Now Eq. (4) is used in Eq. (3) for $p=1$ and 2 and the two equations are solved simultaneously for $\beta_1^{(k+1)}$ and $\beta_2^{(k+1)}$ to obtain

$$\beta_1^{(k+1)} = \beta_1^{(k)} + (H_1 C_{22}^n - H_2 C_{12}^n)/\Delta \tag{6a}$$

$$\beta_2^{(k+1)} = \beta_2^{(k)} + (H_2 C_{11}^n - H_1 C_{12}^n)/\Delta \tag{6b}$$

where

$$\Delta = (C_{12}^n)^2 - C_{11}^n C_{22}^n \tag{7a}$$

$$C_{st}^n = \sum_{j=1}^{n} X_{js}^{(k)} X_{jt}^{(k)} \tag{7b}$$

$$H_s^n = \sum_{j=1}^{n} (Y_j - T_j^{(k)}) X_{js}^{(k)} \tag{7c}$$

The iterations on k are continued until negligible variations occur between $\beta_s^{(k+1)}$ and $\beta_s^{(k)}$ for $s=1$ and 2. The converged $\hat{\beta}_1$ and $\hat{\beta}_2$ values are the estimates of β_1 and β_2, respectively. Using Eqs. (6 and 7) is called nonlinear estimation. Then one last calculation is performed in a sequential manner. In the sequential procedure, the upper index n is started at 1 and is increased by one until it again equals n. The sequential values of $\hat{\beta}_1^{\,m}$ and $\hat{\beta}_2^{\,m}$ (estimates of β_1 and β_2 using m measurements) are found using Eq. (6) with C_{st}^n and H_{st}^n replaced by C_{st}^m and H_{st}^m, respectively. In this calculation all the terms on the right hand side are evaluated at the converged $\hat{\beta}_1$ and $\hat{\beta}_2$ values. Note that $\hat{\beta}_1^n = \hat{\beta}_1$ and $\hat{\beta}_2^n = \hat{\beta}_2$. The insights provided by the sequential procedure are discussed in Section IV.

In order to minimize the computation time, Eq. (7b) can be written as

$$
\begin{aligned}
C_{st}^m &= \sum_{j=1}^{m} \hat{X}_{js} \; \hat{X}_{jt} - \sum_{j=1}^{m-1} \hat{X}_{js} \; \hat{X}_{jt} + \hat{X}_{ms} \; \hat{X}_{mt} \\
&= C_{st}^{m-1} + \hat{X}_{ms} \; \hat{X}_{mt}
\end{aligned}
\tag{8}
$$

where C_{st}^m is for the summation with m terms. Notice that Eq. (8) gives a recursion relation that can be used for $m=1,2,\ldots,n$. For $m=1$, C_{st}^o is needed which is

$$
C_{11}^o = \delta_1, \qquad C_{22}^o = \delta_2, \qquad C_{12}^o = 0
\tag{9}
$$

where δ_1 and δ_2 are small values compared to C_{11} and C_{22}. A recursion relation for H_s is

$$
H_s^{(m)} = H_s^{(m-1)} + (Y_m - \hat{T}_m) \; \hat{X}_{ms}
\tag{10}
$$

In Eqs. (8) and (10) the symbol $\hat{\ }$ indicates that the sensitivity coefficient X and the temperature T are evaluated using the converged $\hat{\beta}_1$ and $\hat{\beta}_2$ values.

Statistical statements can be made regarding the estimates of β_1 and β_2 obtained by using least squares in the above manner. The confidence regions depend upon the character of the measurement errors, some of which can be determined by examining the residuals, $Y_j - \hat{T}_j$, for the converged parameter values. See Beck and Arnold [4] for further discussion.

III. SIMULATION

The objective of the simulation is to determine the heat conduction model that should be used to describe the temperature rise in the sample during the flash experiment. Two one-dimensional models are tested. The thermal diffusivity estimated by the sequential estimation method with these two models is also compared to the thermal diffusivity determined by the ratio, Cowan, and Clark and Taylor methods. In the first model, called the X22 model, it is assumed that there is no heat loss from the sample. (The number system, XIJ, is for the x-coordinate system and boundary condition of the Ith kind at x = 0 and of the Jth kind at x = L; the values of I, J = 1 denote a temperature condition, I, J = 2 for a $\partial T/\partial x$ condition and I, J = 3 for a convective condition. See Beck [6].) The X22 model is the one used in the flash method. The temperature rise at x=L due to an impulse heat flux at time t=0 and x=0 in term of the Green's function G is (Beck, [7])

$$T(L,t) - T_\infty = \frac{\alpha}{k} \int_{\tau=0}^{t} q(\tau) \; G_{X22}(L,t|0,\tau) d\tau \tag{11}$$

where T_∞ is the initial uniform temperature. For the case of q(t) being

$$q(t) = Q_0 \delta(\tau) \tag{12}$$

where δ is the Dirac delta function, Eq. (11) becomes

$$T(L,t) - T_\infty = \frac{\alpha Q_0}{k} \; G_{X22}(L,t|0,0) \tag{13}$$

with

$$G_{X22}(L,t|0,0) = \frac{1}{L} \left[1 + 2 \sum_{m=1}^{\infty} \cos(m\pi) \; \exp(-m^2\pi^2 \frac{\alpha t}{L^2}) \right] \tag{14}$$

In the second model, called X33, there are convective heat losses both at x=0 and x=L with the same heat transfer coefficient h at x=0 and L. Radial heat losses are neglected. The temperature rise due to the impulse heat flux is given by Eq. (13) with the appropriate Green's function now being

$$G_{X33}(L,t|0,0) = \frac{2}{L} \sum_{m=1}^{\infty} \frac{(\beta_m \cos\beta_m + Bi \sin \beta_m)\beta_m}{(\beta_m^2 + Bi^2)(1 + \frac{Bi}{\beta_m^2 + Bi^2}) + B} \exp\left(-\beta_m^2 \frac{\alpha t}{L^2}\right) \tag{15}$$

where the β_m satisfy the transcendental equation,

$$\tan \beta_m = \frac{2\beta_m Bi}{\beta_m^2 - Bi^2} \quad \text{with} \quad Bi = \frac{hL}{k} = \frac{h}{\rho c} \frac{L}{\alpha} \tag{16}$$

For small Biot number values, Bi < 0.05, the β_m's can be approximated by

$$\beta_1 = \left[\frac{3(2Bi + Bi^2)}{3 + 2 Bi}\right]^{1/2}; \; \beta_m = (m - 1)\pi + \frac{2Bi}{(m - 1)\pi}, \; m = 2,3,\ldots \tag{17a,b}$$

The sample considered in the simulation is a .0006m thick cylinder with the thermal conductivity value of k=400 W/m K and thermal diffusivity of α=1.1775 10^{-4} m^2/s. The input data are generated using

Eqs. (18) and (19) given in Cape and Lehman [8]. Radial heat losses were not neglected to obtain these equations, thus the input data results from a two-dimensional temperature field in the sample. Three large values of the heat transfer coefficient are studied: h=1000, 10000 and 30000 W/m^{2}K, which respectively correspond to Biot numbers of .0015, .015 and .045. (Large values have the greatest deviation from the simple X22 model.)

In both models there are three unknowns: the magnitude of the input heat flux Q_o, the thermal diffusivity α and the thermal conductivity k. It can be shown that k and α cannot be simultaneously determined unless Q_o is known. Since the goal is the estimation of α alone, the number of unknowns in the models can be reduced by estimating only the two following parameters:

α, the thermal diffusivity

and

$b = \dfrac{\alpha Q_o}{kL}$ maximum temperature rise

If Q_o and L are known, then the value of k can be obtained from the estimated values of b.

Table 1 shows the results of a simulation using nonlinear estimation based on all 200 data points (time step of .012 s) for the three values of the Biot number. As the Biot number increases, the heat losses become greater. The estimation of α becomes less accurate as the Biot number increases, particularly when the X22 model is used. The errors are reduced by a factor of almost four when the X33 model is used instead of the X22 model. In Table 2 are the results of the thermal diffusivity determined by the uncorrected, Cowan, and Clark and Taylor methods. The uncorrected method gives results similar to the one obtained with the sequential estimation based on the X22 model. This is logical since the heat losses are totally neglected in both procedures. Using the Cowan method to account for the heat losses gives little improvement. On the other hand, the Clark and Taylor correction procedure gives good results.

Two conclusions are drawn from this simulation. First, heat losses cannot be neglected except for extremely small Biot numbers. The X33 model should be used which requires the estimation of the heat transfer coefficient. Second, the Clark and Taylor method accounting for heat losses is more accurate than the Cowan method. Third, the use of a one-dimensional model is valid for this experiment.

Table 1. Nonlinear estimation for simulation

Bi	X22 model		X33 model	
	$\alpha 10^4$	% error	$\alpha 10^4$	% error
.0015	1.179	+ 0.16	1.178	+ 0.04
.015	1.196	+ 1.58	1.182	+ 0.42
.045	1.234	+ 4.76	1.190	+ 1.06

Table 2. Flash methods for simulation

Bi	uncorrected		Cowan [2]		Clark & Taylor [3]	
	$\alpha 10^4$	% error	$\alpha 10^4$	% error	$\alpha 10^4$	% error
.015	1.197	+ 1.66	1.195	+ 1.49	1.174	- 0.30
.045	1.226	+ 4.12	1.207	+ 2.50	1.174	- 0.30

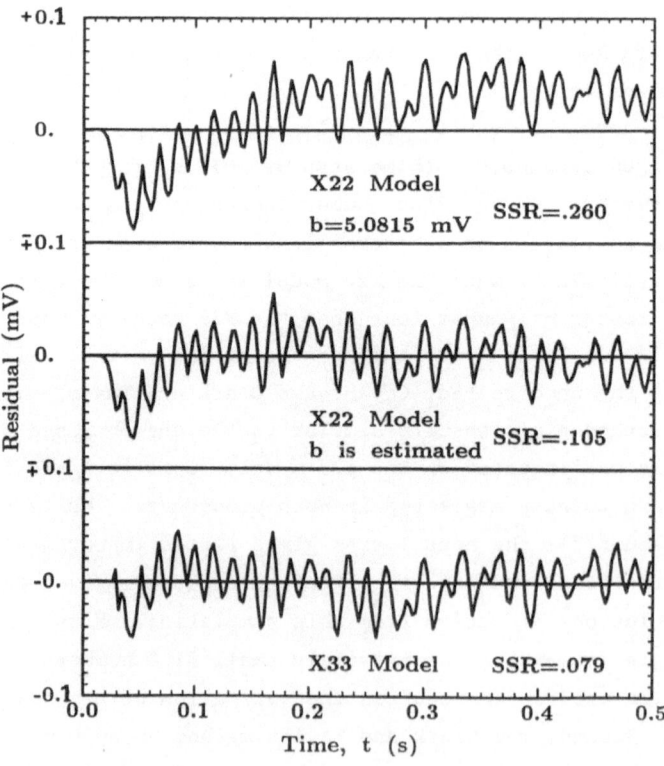

Fig. 3. Residuals given by the sequential estimation method for the different models

IV. EXPERIMENTAL DATA CASE

The objective of this section is to estimate the thermal diffusivity of a stainless steel sample from data collected during a flash experiment. The sample is $1.542 \cdot 10^{-3}$ m thick. Two sets of temperatures measured at the back face are available for the estimation. The first set represents the 0.52 second that follows the impulse heat flux and contains 200 data points, Fig. 2. The second set has 100 data points measured every 0.05 second while the sample cools down. Actually, as shown in Figure 2, the voltage given by the thermocouple through the data acquisition system is used directly and not converted to temperature. This is possible since in the sequential estimation the parameter b is estimated and therefore can include a constant to account for the sensitivity of the thermocouple. Similarly, the uncorrected, Cowan, and Clark and Taylor methods are all based on temperature ratios which are identical to voltage ratio. The only assumption made is of a constant sensitivity of the thermocouple in the temperature rise considered (around 2°C) which is reasonable.

There is no information concerning heat losses so in the first stage the X22 model is used in estimating the thermal diffusivity. The sequential estimation using the 200 data points gives $\alpha = .03357 \ 10^{-4} \ m^2/s$ and b=5.1221 mV. The residuals are shown in the center of Fig. 3. They oscillate around zero which indicates that the model is appropriate. The magnitude of the residuals should be compared with the magnitude of the data points shown in Fig. 2. Except for the beginning of the time period, the residuals do not exceed 1% of the maximum rise.

The uncorrected and Clark and Taylor methods require the determination of the maximum voltage b. Due to oscillations in the experimental data, Fig. 2, the maximum is determined by smoothing all of the data with a weighted averaging algorithm then scanning the data for the maximum point. With this procedure the magnitude of the maximum is b = 5.0815 mV which is .8% smaller than the one given by the sequential estimation. In order to appreciate the effect of an error in b on the evaluation of α, the sequential algorithm is run with the X22 model, but instead of estimating both α and b, b is made equal to 5.0815 and only α is estimated; the thermal diffusivity is found to be $\alpha = .03397 \ 10^{-4} \ m^2/s$. The residuals shown at the top of Fig. 3 are not centered around zero, suggesting a less accurate model. The Sum of Squared Residuals (SSR) is also a convenient way of comparing models. As indicated in Fig. 3 the SSR is doubled when b is not estimated with the sequential algorithm.

The simulation has indicated that higher accuracy could be obtained if the X33 model were used. This requires the knowledge of the global heat transfer coefficient h. An approximate value of h can be determined from the data taken while the sample cools down. The sample can be modeled by a lumped system which is governed by the equation

$$\rho c V \frac{dT}{dt} = - hS (T - T_\infty) \tag{18}$$

The solution is

$$T - T_\infty = T_o \exp\left[- \frac{h}{\rho c} \frac{t}{L}\right] \tag{19}$$

There are two unknowns in this model : h and ρc. It is not possible to estimate simultaneously h and ρc with the information available. However it is not necessary to determine both h and ρc but only the ratio $h/\rho c$ which is involved in Eq. (15). The sequential estimation based on Eq. (19) and the 100 data points gives $h/\rho c = 30$ W/m^{2}K. The order of magnitude of the Biot number can now be calculated,

$$Bi = \frac{hL}{k} = \frac{h}{\rho c} \frac{L}{\alpha} = 0.014$$

Note that an approximate value of α is needed to calculate the order of magnitude of the Biot number. For a Biot number of this magnitude, the simulation has shown that it is preferable to use the X33 model. The sequential algorithm is run using the above value for $h/\rho c$ and Eq. (15) as the model. The estimates of the two parameters are $\alpha = .03322 \ 10^{-4}$ m^2/s and b = 5.24035 mV. The residuals are shown in the bottom of Fig. 3. They oscillate around zero and the SSR is slightly lower than the one given by the residuals of the X22 model. There are other parameters that can be studied to gain more insight in sequential estimation.

A good deal of information is given by the sensitivity coefficients shown as a function of time in Fig. 4. First of all, the sensitivity coefficients of the two parameters α and b are linearly independent, thus a unique determination of both parameters is possible. The sensitivity coefficients of the parameter b for the models X22 and X33 are not distinguishable on the plot. They start from zero and increase up to the end of the time period indicating that all the data should be used for an accurate estimation of b. The sensitivity coefficients for α have the same shape for models X22 and X33 with the X33 sensitivity coefficient being just a little smaller. Hence, the X33 model as shown by the simulation is adequate for this experiment. Another interesting fact is that the sensitivity coefficients of α tend toward zero for t > 0.3s

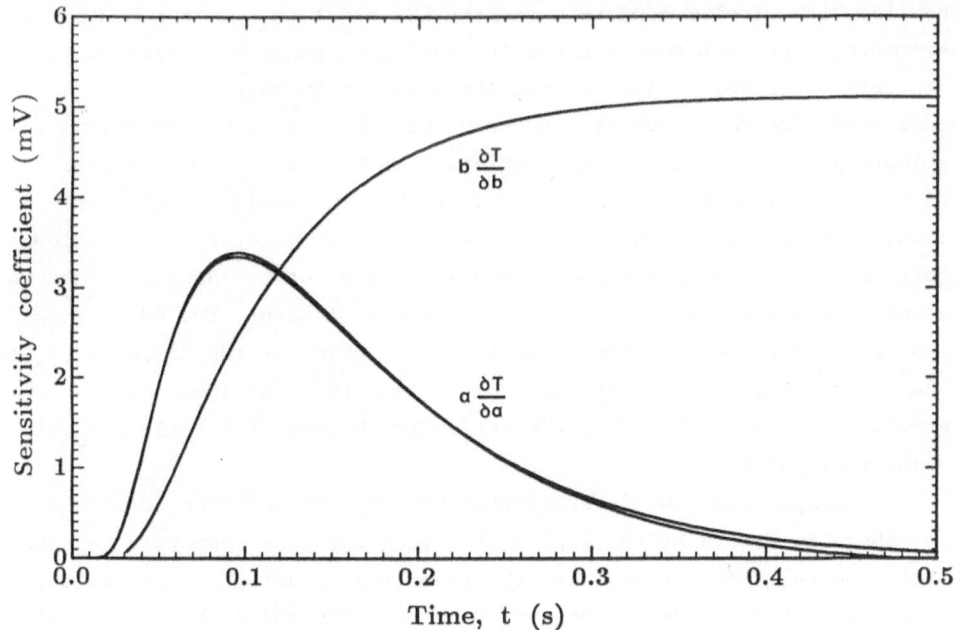

Fig. 4. Sensitivity coefficients

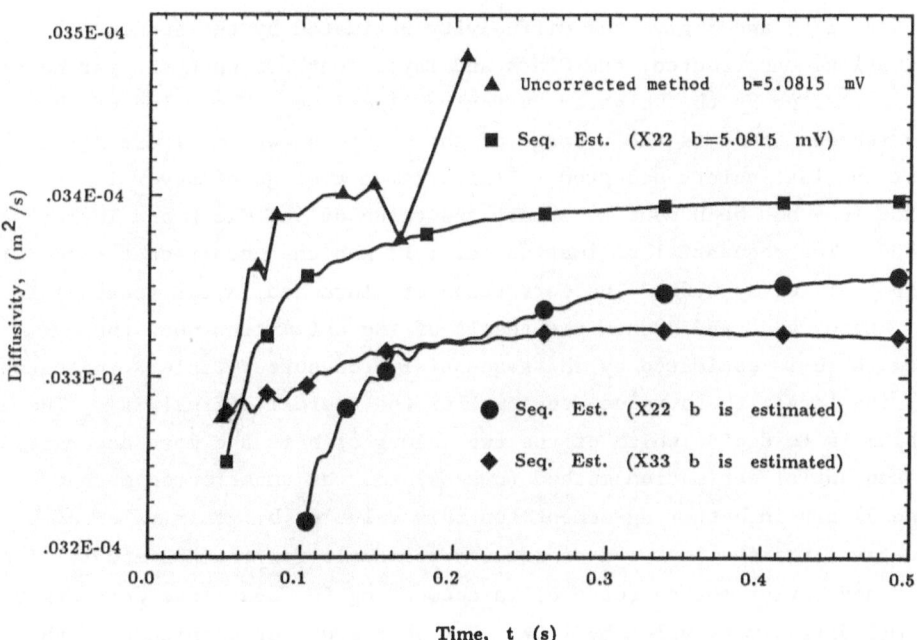

Fig. 5. Sequential estimates of the thermal diffusivity of the
uncorrected method and of the sequential method for
different models.

indicating that these data are not useful in estimating α after that
time.

The sequential results of the estimation are also of great interest.
Fig. 5 shows the estimation of α versus time for the different models.

The line with circles depicts α for the X22 model when both α and b are
estimated. The fact that α keeps increasing slightly with time (more and
more data are used) indicates that the model is inadequate. The X22
model used with the predetermined value of b (b=5.0815 mV and only α is
estimated) gives the line with squares. For t > .3s α barely varies
which is appealing but one must remember that the residuals (top of Fig.
3) were not centered around zero. These two contradictory observations
indicates that this is not the appropriate value of b. However, the two
curves show how sensitive is the estimation of α to b. The solid line
with diamonds represents the estimation of α given by the X33 model. For
0.2s > t > 0.5s, α varies by less than .15%. This constancy of the
parameter estimate along with the residuals (bottom of Fig. 3) suggest
adequacy of model X33.

For comparison, the determination of α by the uncorrected method is
also shown in Fig. 5 by the line with triangles. The dispersion of the
points due to random errors, and the inability of using all of the data
are the two striking disadvantages of the uncorrected method compared
with the sequential estimation.

A summary of the results of the different methods is given in Table
3. Case 1, 2 and 3 give the diffusivity estimated by the sequential
estimation, uncorrected, and Clark and Taylor methods when the parameter
b is estimated by the weighted averaging algorithm. The value of the
uncorrected method is the average of the points shown in Figure 5, the
first and last points accepted. Similarly an average of several
corrections has been done to obtain the value of the Clark and Taylor
method. The sequential estimation (case 1) and the uncorrected (case 2)
methods differ by 0.6%. The correction of Clark and Taylor (case 3) is
.7%. Cases 4, 5 and 6 show the result of the estimation when the
parameter b is estimated by the sequential procedure. Table 3 indicates
that the larger the b value, the smaller the thermal diffusivity. The
problem is to decide which of the two values of b is the most accurate.
The sequential estimation method (case 4) and the uncorrected method
(case 5) are in better agreement for this value of b. This value of b
also gives the smallest sum of squared residuals for the X22 model. The
Clark and Taylor method (case 6) in accounting for heat loss corrects the
thermal diffusivity value by 1.4%. For this order of magnitude of the
Biot number, the simulation has given a correction of 2%. Thus this
value of b gives results that are in better agreement with the results of
the simulation. Case 7 is for the sequential technique when used with
the X33 model to account for heat loss. The estimated diffusivity is .3%
larger than the value of Clark and Taylor when b = 5.1221. This again is
in better agreement with the results of the simulation of this Biot
number than when b = 5.0815 mV, see Tables 1 and 2.

Table 3. Comparison of estimation methods - experimental data

Case	Method	b	α
1	Sequential estimation with X22 model b is fixed	5.0815	.03397
2	Uncorrected	5.0815	.03378
3	Clark & Taylor (1975)	5.0815	.03353
4	Sequential estimation with X22 model b is estimated	5.1221	.03357
5	Uncorrected	5.1221	.03362
6	Clark & Taylor (1975)	5.1221	.03315
7	Sequential estimation with X33 model b is estimated	5.2395	.03324

V. CONCLUSIONS

Four methods for estimating thermal diffusivity based on the temperature rise of a sample during a flash experiment have been compared. First, simulation using exact data has been done to study the potential accuracy of the four methods. The simulation has shown that except for extremely small Biot number, heat loss should be taken into account. Of the two methods studied to account for heat loss and to correct the results given by the uncorrected method, the Clark and Taylor method worked well and proved to be the most efficient for the particular case chosen. Similarly, the use of a model that accounts for heat loss in the sequential estimation significantly improved the accuracy. Comparable accuracies have been obtained with the Clark and Taylor correction method and with the sequential estimation method.

Next, the thermal diffusivity of a stainless steel sample has been estimated from experimental data. The results given by the uncorrected method and the sequential estimation when used with the X22 model are very similar. The heat transfer coefficient has also been determined with the sequential method. With this value, the sequential estimation procedure has been used with a model that accounts for heat loss. The result of this estimation is in excellent agreement with the one given by the Clark and Taylor method.

However, the agreement between the methods is good only when the same value of the parameter b, which represents the maximum temperature rise, is taken. Because of the random errors in the data, the value of the parameter b estimated by the weighted averaging algorithm differs by 0.8% from the value given by the sequential estimation method. The latter value gives results that are in better agreement with the simulation than the former. A more accurate value for b can be obtained by the sequential method because it uses all the data points to estimate b, whereas the weighted averaging algorithm uses only a few. This is one major advantage of the sequential procedure over the other methods.

Another advantage of the sequential method is shown in Fig. 5. The sequential procedure uses an increasing number of data points as the estimation is carried out. The more data points involved in the estimation, the smaller the effect of random measurement errors. When the model is correct, the estimated parameters as more data points are used eventually stabilize at a given value. This is shown in Fig. 5 for the X33 model. On the other hand the uncorrected method uses only one measurement at a time and as shown in Fig. 5 is very sensitive to random measurement errors.

The sensitivity coefficients are also of great interest especially for the design of optimal experiments. The sensitivity coefficients should be as large as possible and must be linearly independent for a unique minimum point to exist. As shown in Fig. 4, the sensitivity coefficients are sufficiently large compared to the measured data (Fig. 2) and are linearly independent. It has been found from experience that difficulty encountered in convergence is frequently due to approximate linear dependence. Most of the presented results have converged within 10 iterations with initial guesses for the parameters off by as much as 100%. Fig. 4 shows that the data for $0.05 < t < 0.2s$ are the most useful for the estimation of the thermal diffusivity. On the other hand the data for $t > 0.2s$ are the most useful for the estimation of b. Consequently all the data should be used to obtain the most accurate estimates of the two parameters.

The sequential estimation method is quite straightforward to use. The method has a statistical basis and thus can provide more information regarding the accuracy of the estimates than the other cited methods.

REFERENCES

1. W.J. Parker, R.J. Jenkins, C.P. Butler and G.L. Abbott, "Flash method of determining thermal diffusivity, heat capacity and thermal conductivity," J. App. Physics 32: 1679-1684 (1961).

2. R.D. Cowan, "Pulse method of measuring thermal diffusivity at high temperatures," J. App. Physics 34: 926-927 (1963).

3. L.M. Clark III and R.E. Taylor, "Radiation loss in the flash method for thermal diffusivity," *J. App. Physics* 46: 714-718 (1975).

4. J.V. Beck and K.J. Arnold, "Parameter Estimation in Engineering and Science," John Wiley & Sons, NY (1977).

5. J.V. Beck, "Sequential estimation of thermal parameters," *J. Heat Transfer* 99C: 314-321 (1977).

6. J.V. Beck, "Green's functions and numbering system for transient heat conduction," *AIAA Journal* 23: 1609-1614 (1986).

7. J.V. Beck, "Green's function solution for transient heat condition problems," *Int. J. Heat Mass Transfer* 27: 1235-1244 (1984).

8. J.A. Cape and G.W. Lehman, "Temperature and finite pulse-time effects in the flash method for measuring thermal diffusivity," *J. App. Physics* 34: 1909-1913 (1963).

IMPROVEMENT OF LASER BEAM PROFILE USING OPTICAL FIBER

FOR LASER PULSE THERMAL DIFFUSIVITY MEASUREMENTS

Tetsuya Baba, Teruo Arai and Akira Ono

Thermophysical Metrology Department
National Research Laboratory of Metrology
Tsukuba, Ibaraki 305, Japan

ABSTRACT

Beam profiles of a YAG-laser, which was used as a pulse heat source of a laser pulse apparatus were measured with two methods. One is a two-dimensional profile measurement using a TV camera. Another is a new application of a thermographic technique to measure one-dimensional energy distributions of pulsed laser beams. Observed profiles of direct beams from the YAG-laser were too irregular to apply to a precise thermal diffusivity measurement by the laser pulse method. It was demonstrated that such irregular beams were converted to uniform beams after transmission through a step-index optical fiber.

INTRODUCTION

Nonuniform heating of a specimen surface is one of the most serious error sources in thermal diffusivity measurements by the laser pulse method[1],[2]. Because high-power pulsed lasers used in laser pulse apparatus are usually operated in multi-mode, it is very unlikely that laser beams are spatially uniform enough to satisfy the ideal initial and boundary conditions where heat flows one-dimensionally in the specimen. Although there have been quite a few theoretical studies of nonuniform heating error[3],[4], experimental approach has been very limited. K. Beedham and I. P. Dalrymple measured laser beam profiles using a photographic plate and a densitometer. They also made numerical calculations to estimate nonuniform heating errors induced by the measured nonuniformity of the laser beam[5]. Similar study was also performed by R. E. Taylor[6].

However, there has been only one systematic measurement where nonuniform heating error is quantitatively estimated and corrected, and where the beam with guaranteed uniformity is used to heat the specimen surface[7]. A TV camera method was applied to monitor two-dimensional profiles of pulsed laser beams[8]. A thermographic method was newly developed to measure one-dimensional energy distribution of the beams qualitatively[9]. Simultaneously, it was tried to improve the spatial uniformity of laser beams. In the present paper, improvement of laser beam profiles using a step-index optical fiber is described.

EXPERIMENTAL SETUP

Fig. 1 shows a functional diagram of a beam profile measurement system. When a two-dimensional profile of the pulsed laser beam was measured, the beam emitted from the laser head was reflected by mirrors, attenuated by neutral density filters and incident on the detective area of a Super-CHALNICON E5476 tube. The video signal was recorded by a VCR and monitored on a TV display.

When a one-dimensional energy distribution of the pulsed laser beam was measured quantitatively, first, the laser beam was absorbed by a thin receiving film in a vacuum chamber. Then, the transient temperature increment of the receiving film was radiometrically detected by a thermographic instrument before the temperature increment faded out. More details about the thermographic method are described in reference 9.

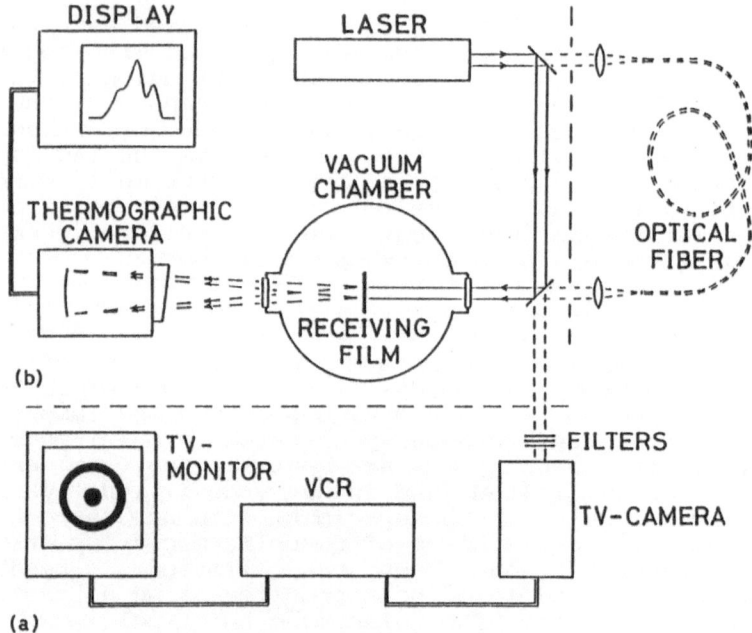

Fig. 1. Function diagram of the beam profile measurement system.

When an optical fiber was used to improve the laser beam profile, the direct beam was focused on one end of the fiber by a concave lens of 20 mm focal length and the ejected beam from the other end was collimated by another concave lens of the same focal length.

RESULTS AND DISCUSSION

Profile measurements were made for the pulsed YAG — laser, whose specifications are listed as follows:

wavelength: 1.064 μm,
YAG rod: 100 mm in length and 8 mm in diameter,
excitation: xenon flash lamp,
condenser: single ellipsoid mirror,
pulse duration: 1 ms (max),
oscillation: multimode,
energy per pulse: 10 J (max).

Photo 1 shows two-dimensional beam profiles measured with the TV camera method. Photo 1(a) is one example of the direct beam profiles, 55 cm from the YAG laser head, where the pulse duration was set at 100 μs and the pulse energy before attenuation was about 0.2 J. Such an irregular profile of the direct laser beam is attributed to the multimode oscillation of high power lasers. Photo 1(b) and (c) are two-dimensional profiles of the transmitted beams through a step-index optical fiber whose length was 1.5 m, core radius was 400 μm and numerical aperture was 0.2. Photo 1(b) is the result when the incident beam was focused on the end of the optical fiber with 2° inclination to the central axis of the fiber. When the inclination angle was increased to 6°, the transmitted beam profile changes into Photo 1(c).

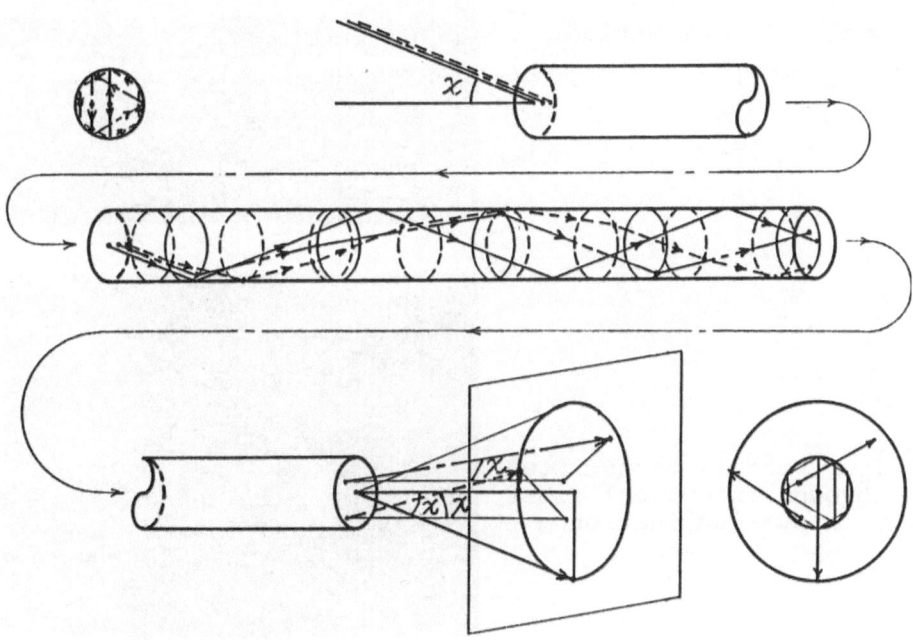

Fig. 2.   Trajectories of lights travelling through a step-index optical fiber.

These results show that beams transmitted through an optical fiber are axially symmetric. The process in which an irregular beam is converted into an axially symmetric beam is illustrated

(a) Direct beam profile
    from a YAG laser.

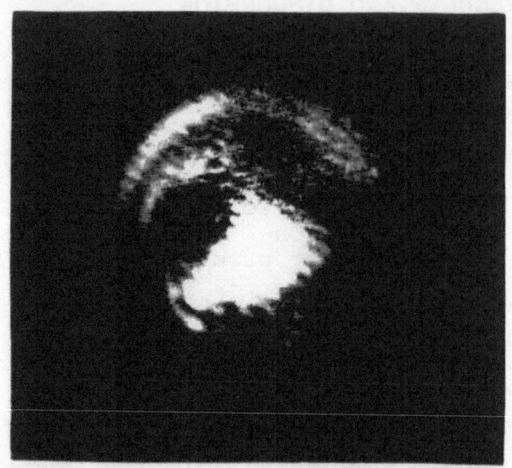

(b) Beam profile transmitted
    through an optical fiber
    with ~2° inclination.

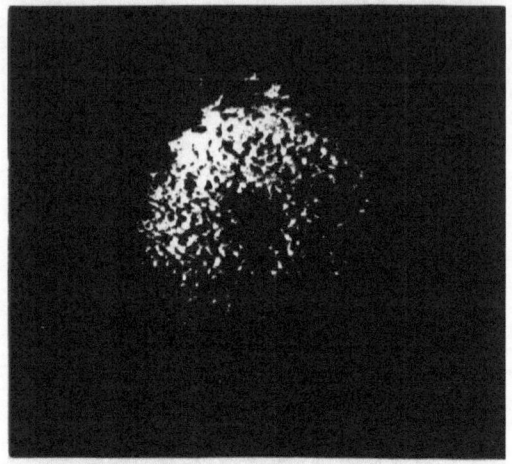

(c) Beam profile transmitted
    through an optical fiber
    with ~6° inclination.

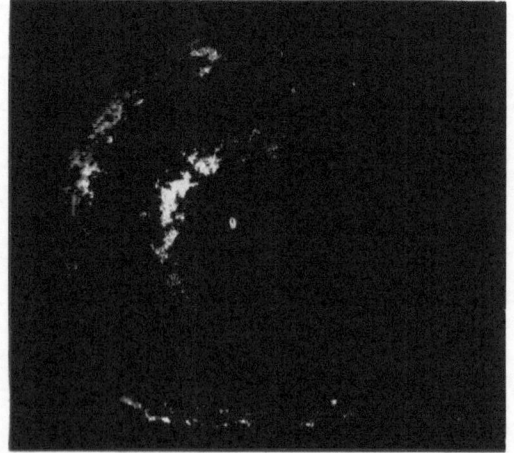

Photo 1. Two-dimensional laser beam profiles
          measured with the TV camera method.

in Fig. 2 and explained as follows. When a light is incident on
one end of an optical fiber with an incident angle, χ, it travels
with multiple reflection in the optical fiber. Most lights travel
rotating like corkscrews, which are called scew rays. Even if
incident angles, χ's, are the same, a small difference of the
entering point to the fiber end results in large difference of
the ejected direction of the transmitted light as shown in Fig. 2.
Whereas, the angle between the ejected direction of the
transmitted light and the central axis of the optical fiber is
conserved at the original incident angle, χ. Consequently, lights
are ejected like a cone and a circular profile is obtained with
projection on a perpendicular plane.

Photo 2 shows one-dimensional energy distributions of laser
beams along a diameter measured with the thermographic method.
Photo 2(a) is one example of the direct beam profile 83 cm from
the YAG-laser head, where the pulse duration was set at 1 ms and
the pulse energy was about 1 J. If the central portion marked in
the photo 2(a) is expanded and incident on the surface of the
specimen of 15 mm in diameter and 2 mm in thickness, the
nonuniform heating error is estimated to be about 8 % according
to the calculation described in reference 8.

Photo 2(b) is the profile of the transmitted beam through
the optical fiber when the incident beam was normally focused on
the fiber end without inclination. The profile is close to the
gaussian distribution. Photo 2(c) is the profile when the
incident beam was focused with 6° inclination to the central axis
of the fiber. Flat area in Photo 2(c) is much larger than Photo
2(a) and (b).

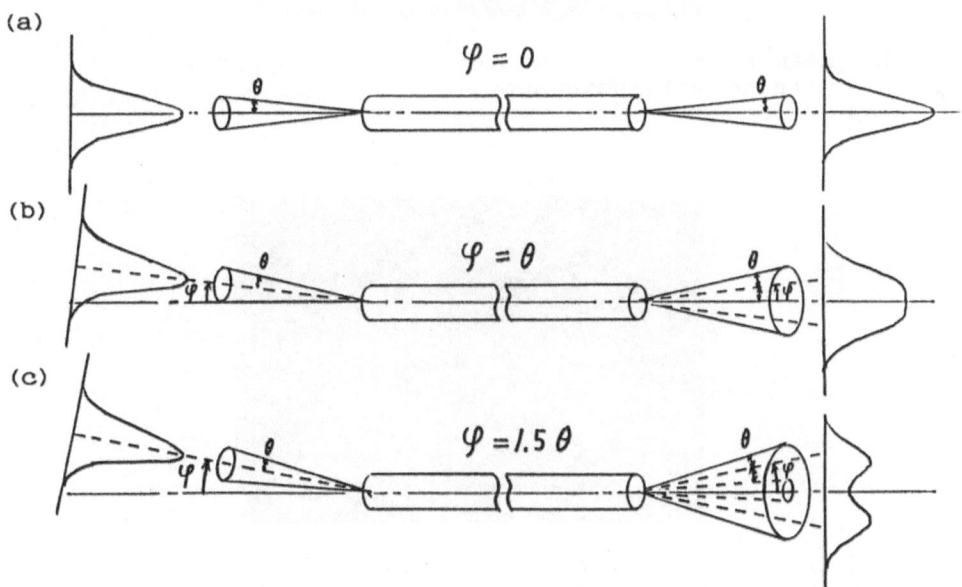

(a)

$\varphi = 0$

(b)

$\varphi = \theta$

(c)

$\varphi = 1.5\,\theta$

Fig. 3.   Profiles of transmitted beams through an optical fiber
dependent on the incident angle, φ, to the fiber end,
where θ is the characteristic angle of the gaussian
distribution.

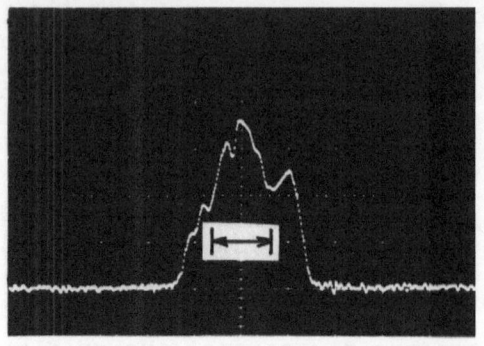

(a) Direct beam profile from a YAG laser.

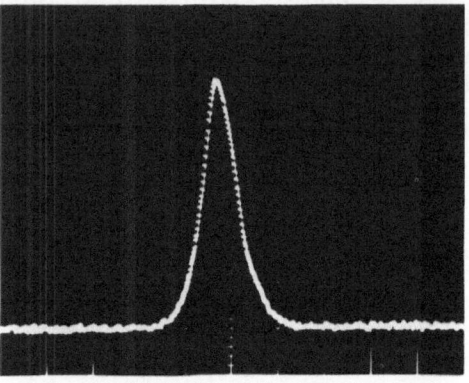

(b) Beam profile transmitted through an optical fiber
with normal incidence.

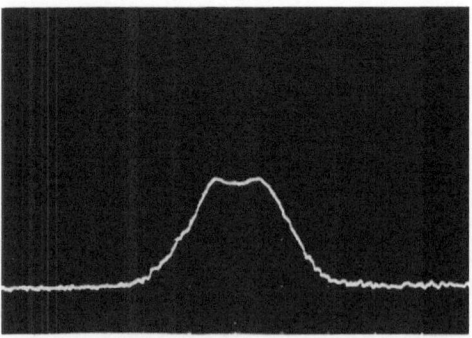

(c) Beam profile transmitted through an optical fiber
with ~6° inclination.

Photo 2. One-dimensional laser beam profiles
measured with the thermographic method.

The dependence of the radial distribution of the laser beam energy on the incident angle, φ, is illustrated in Fig. 3 and explained by the following calculation. For the convenience of calculation, it is assumed that the angular intensity distribution $F_0(\chi)$ of the focused beam to the fiber end is the gaussian with characteristic angle, θ.

$$F_0(\chi) = \exp\left(-\frac{\chi^2}{\theta^2}\right) \qquad (1).$$

When the incident beam is normally focused on the fiber end as shown in Fig. 3 (a), the transmitted beam has the same distribution as the incident beam formulated by Eq. (1). When the incident beam is inclined with the angle, φ, the angular distribution of the ejected direction of the transmitted beam is expressed by the following equation:

$$F_\varphi(\chi) = \exp\left(-\frac{\chi^2 + \varphi^2}{\theta^2}\right) I\left(\frac{2\chi\varphi}{\theta^2}\right) \qquad (2),$$

where $I(\chi)$ is the modified Bessel function of the first kind of order 0. Fig. 3 (b) and (c) illustrate the process where the distribution expressed by Eq. (2) is obtained when $\varphi = \theta$ and $\varphi = 1.5\,\theta$, respectively. If $\varphi = \theta$, beam is flat within 1 % over the area where 9 % of the total beam energy is contained. If $\varphi = 1.1\,\theta$, beam is flat within 3 % over the area where 27 % of the total beam energy is contained.

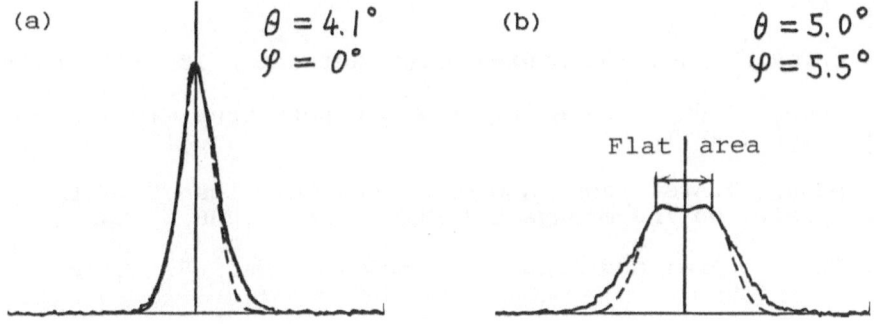

(a) $\theta = 4.1°$ (b) $\theta = 5.0°$
       $\varphi = 0°$ $\varphi = 5.5°$

Flat area

Fig. 4. Measured beam profiles fitted with calculated curves based on Eq. (1) and Eq. (2).

Measured distributions (solid lines) and calculated distributions (broken lines) are compared in Fig. 4. In the case of normal incidence, the measured distribution of the transmitted beam is fitted with the gaussian where the characteristic angle $\theta = 4.1°$ as shown in Fig. 4 (a). When the incident beam is focused with 6° inclination to the central axis of the fiber, the measured distribution of the transmitted beam is fitted with Eq. (2) substituting θ by 5.0° and φ by 5.5° as shown in Fig. 4 (b).

If the marked flat area of the beam in Fig. 4 (b) is expanded and incident on the surface of the specimen of 15 mm in diameter and 2 mm in thickness, the nonuniform heating error is estimated to be less than 1 % according to the calculation described in reference 8. It is substantial improvement compared with the error of 8 % when the direct beam of Photo. 2 (a) is used.

SUMMARY

1. Beam profiles of a pulsed YAG-laser were measured with a TV camera method and a thermographic method.

2. Observed profiles of direct beams from the YAG-laser were too irregular to apply to a precise thermal diffusivity measurement by the laser pulse method.

3. Irregular direct beams can be converted to more uniform beams after travelling through a step index optical fiber.

REFERENCES

1. F. Righini and A. Cezairliyan, High Temperatures- High Pressures, **5**, 481 (1973).

2. R. E. Taylor and K. D. Maglic, Compendium of Thermophysical Property Measurement Methods, Vol. 1. Survey of Measurement Techniques, edited by K. D. Maglic, A. Cezairliyan and V. E. Peletsky (Plenum, New York, 1984), Chap. 8, p. 305.

3. D. A. Watt, Brit. J. Appl. Phys., 17, 231 (1966).

4. J. A. McKay and J. T. Schriemph, J. Appl. Phys., **47**, 1668 (1976).

5. K. Beedham and I. P. Dalrymple, Rev. Int. Hautes Temp. Refract., **7**, 278 (1970).

6. R. E. Taylor, Rev. Int. Hautes Temr. Refract., **12**, 141 (1975).

7. T. Arai, T. Baba and A. Ono, High Temperatures-High Pressures, **19**, 269 (1987).

8. T. Baba, T. Arai and A. Ono, Proc. of the Seventh Japan Symposium on Thermophysical Properties, 1986, p. 235.

9. T. Baba, T. Arai and A. Ono, Rev. Sci. Instrum., **57**, 2739 (1986).

THE AUTOMATION OF A LASER FLASH DIFFUSIVITY APPARATUS

Robert L. Shoemaker

Thermophysical Properties Research Laboratory
School of Mechanical Engineering
Purdue University
West Lafayette, IN

ABSTRACT

The method used to automate a laser flash diffusivity apparatus is presented. The diffusivity apparatus is fully automated by using a PC/AT equivalent and a PDP 11 computer. The thermal diffusivity over a temperature range of room temperature to 2000°C can be determined. The sample under test is heated at a constant rate of up to 500°C/minute and the diffusivity data is collected while the sample is heating. The data are checked for accuracy and noise and the diffusivity determined while the test is in progress. If the data errors are excessive the experimental parameters are automatically adjusted and the data collected again. Radiation heat loss corrections are also applied to the data during the test. The raw data is saved for further analysis.

An infrared detector is used as a temperature sensor, and is autobiased during the test. The sample is heated by a Ta tube furnace under the control of a programmable controller which is interfaced to the PC/AT for automatic shut down in case of a fatal experimental error. Various equipment status sensors are constantly monitored by the PC/AT and warnings are issued and corrective action taken if an unacceptable condition occurs. A PDP 11 system collects and analyzes the diffusivity data and monitors the PC/AT's activity.

INTRODUCTION

Presently, in the defense community there is much interest in the thermophysical properties of materials that are undergoing transient heating. The interest is primarily in engineering applications where the effective diffusivity at various heating rates is desired for use in modeling surface ablation. Conventional thermophysical property measurement techniques cannot determine a material's properties while a sample is undergoing transient heating. However, the flash diffusivity technique is well-suited for this application. Much work has been done in refining this technique and in understanding its limitations (1,2,3,4). The effective thermal diffusivity of a material that is being heated at a constant rate can be obtained by automating the flash diffusivity apparatus. The automated device can be operated at a low heating rate, simulating the conventional operation, thus gaining the benefits of unattended operation, improved speed of measurement, and hopefully improved accuracy of the diffusivity. How one determines the actual diffusivity from data taken while a sample is undergoing a high heating rate is left for another paper. The design criteria, the automation of each of the apparatus components, the system integration and the data handling are presented here.

DESIGN CRITERIA

The standard flash diffusivity test involves several steps that are done manually by the experimenter. The automation of some of these steps is essential to achieving high heating rate results. The primary objectives for the automation is: 1) to determine the thermal diffusivity of a sample undergoing a heating rate of up to 500°C/minute; 2) to ramp heat the sample up to a maximum temperature of 2300°C with heating rates from 1°C/minute; 3) to collect and analyze the diffusivity data in real time and at predetermined temperatures and 4) to automatically correct or repeat bad data points. A schematic of the complete experimental setup is shown in Figure 1.

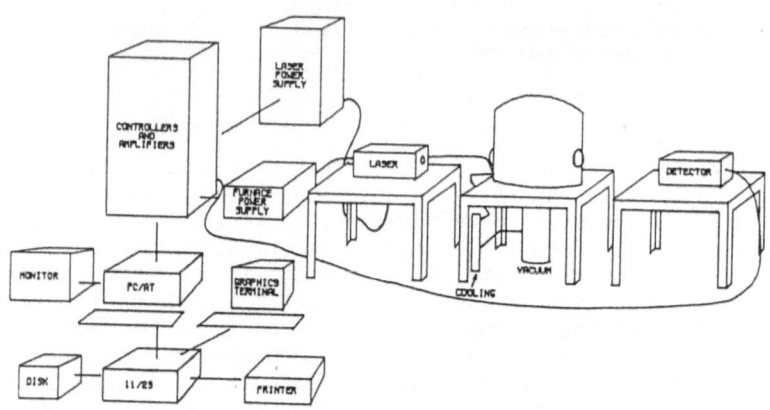

Fig. 1. Automated Diffusivity System

EXPERIMENTAL EQUIPMENT AUTOMATION

The experimental steps considered for automation were; the sample temperature setting, the transient temperature sensor biasing, the amplifier gain and filter settings, laser power setting, data collection duration for the baseline and main data, sample position and the data analysis. The automation design proceeded from that which was desirable to that which was essential. The decision to automate a particular component was reached by minimizing the amount of automation necessary to accomplish the primary objectives and by maximizing the system's reliability. The failure history for each component of the system was examined, looking particularly for the degree of failure (from non-critical to catastrophic) and for the failure rate. The failures were ranked according to the degree of effect they would have on the experimental results. A fatal failure was assigned to those failures that were unrecoverable and would cause further testing to be in error. The results of this ranking along with the essential nature of some of the components determined the level of automation and monitoring required of each component. The ranking are presented in Table 1.

The components 1, 2 and 6 of Table 1 were considered the most critical to the automation and have been automated. The laser power controller has also been completed because it can be used to improve the data's signal to noise ratio. All of the components will be discussed even though the decision to automate some has not yet been made.

DETECTOR BIAS

The automation of the detector's bias is crucial. The detector signal is amplified to a maximum of 250,000 times so that the small transient temperature, which is a result of the laser pulse, can be read by an analogue to digital converter. The detector sees both the radiant temperature of the sample as it is heated and the transient radiant temperature as a result of the laser pulse. Because the transient temperature is so small in comparison to the sample temperature, the signal from the sample temperature must be nulled out before it is amplified. The biasing for the conventional tests is accomplished by manually adjusting the detector's bias to zero once the sample has reached thermal equilibrium. In the high heating rate experiment the sample never comes to equilibrium so that the detector signal is constantly rising. This requires an autobias circuit to keep the signal near zero as the sample's temperature rises until the diffusivity data is to be collected. At this point the autobias is set so that the temperature change as a result of the laser pulse can be read. This signal is composed of the transient temperature change from the laser superimposed upon the heating ramp. Once the data are collected the autobias is again activated. The autobias circuit logic is presented in Figure 2.

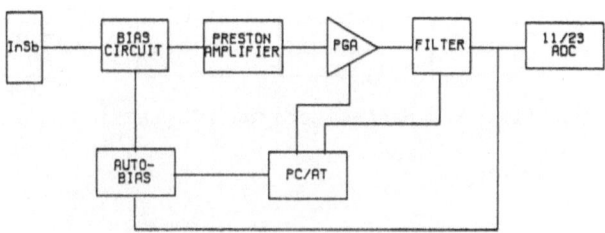

Fig. 2.   Detector Auto-Bias Logic

FURNACE

The furnace requires a programmable temperature controller that can be set for heating rates over the entire range from 1°C/minute to 500°C/minute. Since the furnace radiantly heats the sample, a tungsten/3% rhenium vs. tungsten/25% rhenium thermocouple is attached to the sample holder which is likewise radiantly heated. The furnace controller is fully programmable over the desired heating rate range and is capable of detecting an open thermocouple as well as generating an error if the heating rate is out of control. The furnace control was accomplished by using an Omega temperature controller with two Sorensen power supplies and an interface. The interface was required to condition the Omega output signal for the Sorensen power supplies. The furnace control circuit is shown in Figure 3.

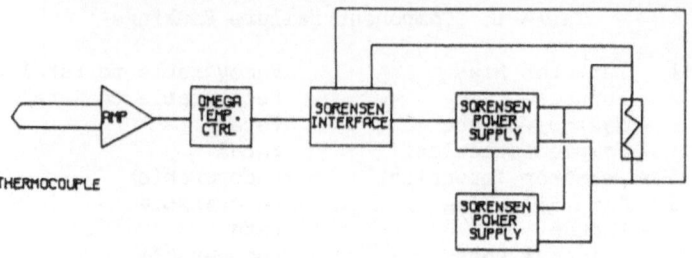

Fig. 3.   Furnace Control Logic

SYSTEM COOLANT

The parts of the apparatus that must be cooled with running water are: the bell jar, the power electrodes, the baseplate, the diffusion pump and the windows. A schematic of the cooling system is shown in Figure 4. If there is a coolant loss then there could be catastrophic failure resulting in a loss of vacuum, a loss of diffusion pump and a loss of furnace by oxidation, just to name a few of the possibilities. So monitoring the water flow in the cooling lines is important to minimize failures. Flow rate switches have been selected for installation on the exit water lines.

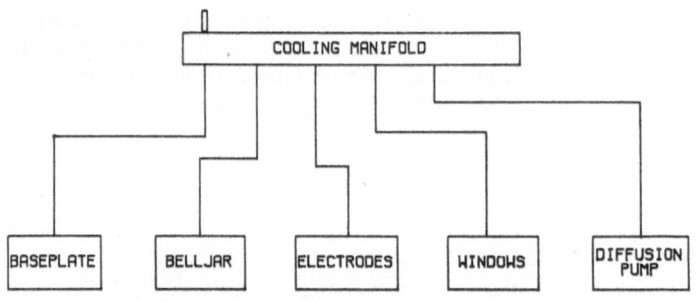

Fig. 4.   System Cooling Diagram

SAMPLE DETECTION

Occasionally a sample will soften, crack, or melt which results in the sample or pieces of the sample falling on the furnace. This allows the laser beam to strike the temperature sensor directly which will destroy an i.r. detector. A HeNe laser sample detection system has been designed but not tested or implemented.

DETECTOR INTENSITY

As a sample is heated to high temperatures the light emitted that is out of the sensitive band for the InSb i.r. detector increases as T^4. It is useful to minimize this light since it results in the detector being heavily biased. This is accomplished by an iris that is closed accordingly. This effectively extends the temperature range of the detector. A commercially available motor controlled iris has been selected for automating the radiant intensity to the detector.

GAIN/FILTERING

The signal from the detector requires amplification and filtering. The amplification optimizes the signal to be read by an analogue to digital converter. The filtering helps eliminate unwanted noise that degrades the signal to noise ratio. The signal amplitude changes as the sample is heated because of changes in the sample's diffusivity, emissivity and the detector's responsibility. As the sample's temperature rises the noise increases, primarily due to the increase in power going to the furnace. The gain and filtering are controlled by using programmable gain amplifiers which are set by the CS. The MS determines the amplifier gain, the signal filtering and the laser power settings to maximize the signal to noise ratio. A 35 db notch filter set at 60 Hz is also part of the system.

VACUUM

The system vacuum is important primarily for protecting the furnace. The furnace used is a low mass tantalum tube that is directly heated. Tantalum oxidizes readily and therefore must be protected when going to temperatures above a few hundred degrees C. A loss of vacuum could also oxidize the sample under test.

NITROGEN LOSS

The detector used requires cooling to liquid nitrogen temperatures to obtain the most sensitivity. As the nitrogen boils off, the detector's temperature gradually rises, decreasing the sensitivity. This process is fairly predictable. However, for long unattended tests it is desirable to monitor this process.

LASER POWER

The amount of energy, Q, deposited upon the sample surface is dependent upon the laser power and the sample's emittance. As a sample is heated the energy lost by reradiation increases as T^4, and the detector's response increases as the blackbody curve shifts into the detector's most sensitive region. This often requires a reduction in the laser power to optimize the signal to noise ratio.

SYSTEM INTEGRATION

Two computers were used to automate the experiment. One is a master system for the high level experiment control and a control system for the low level experimental component control. The master system, MS, is a Digital Equipment Corporation PDP 11/23 computer using the RSX11M multiuser real time operating system. The control system, CS, is a IBM PC/AT equivalent computer using DS-DOS. A schematic of the system interaction is shown in Figure 5.

The MS is the system that controls the experiment and collects and analyzes the data. The user has direct access to the MS but not to the CS. The operator sets up the experiment and has override control by interacting with the MS. The MS notifies the operator of errors in the system and displays the experimental results on a graphics terminal in real time. Table 2 contains the initial parameters that the operator enters into the MS.

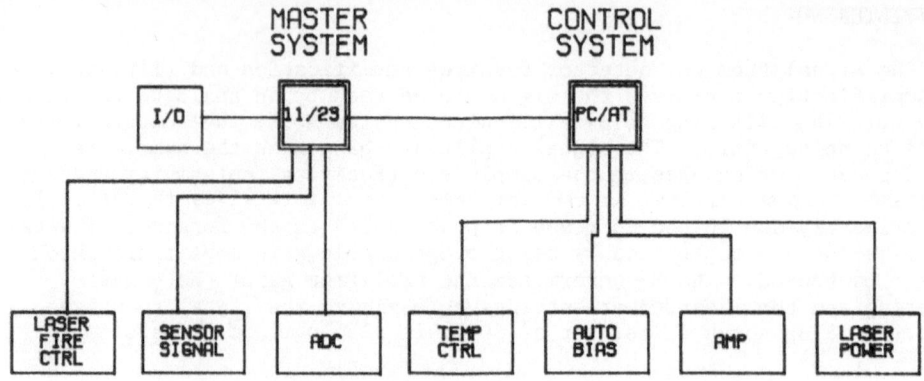

Fig. 5.   System Logic Diagram

Once the experimental parameters are entered the MS takes over experimental control. The MS will then transfer the initial parameters to the CS which in turn loads the appropriate controllers. The MS takes a test laser shot to determine if the initial settings are correct. The data is analyzed and plotted on the terminal. The MS then waits for the user's response before continuing. Once the user has given his consent, the MS takes over control and starts the experiment. The furnace controller's heating rate is set, the MS disk is checked for sufficient storage room, the starting temperature is read, the signal to noise ratio is determined, the amplifier gain and filtering and the laser power are set and another laser shot is taken. The MS then starts the furnace and analyzed and plots the data while monitoring the CS for the command to take the next data point.

Table 2.   Initial Parameters

1 - Sample Description
2 - Data Collection Duration
3 - Starting Temperature
4 - Final Temperature
5 - Temperature Rate
6 - Temperature Interval
7 - Amplifier Gain and Filtering

Each experimental component has a controller that was designed to interface to the CS. The CS loads each controller with the parameters it received from the MS for the experiment and monitors the controllers for error conditions. The experimental parameters are displayed on a color monitor and are updated every few seconds. The CS has a special mode of operation for diagnostics. In this mode the user can enter commands at the keyboard and the CS will display pertinent data on the monitor.

The CS reads the error status from each controller several times a second. The controllers to be monitored are; the autobias, the furnace, the laser power, the coolant, the vacuum, the sample, the detector intensity, and the liquid nitrogen in the detector. Presently only the autobias and furnace controllers have been implemented. These are critical controllers, in that if the furnace loses control the experiment must be aborted. The loss of autobias will invalidate the current data collection but if it can be re-established, it will not abort the experiment.

DATA HANDLING

The CS monitors the temperature of a tungsten – 3% rhenium vs. tungsten – 25% rhenium thermocouple attached to the sample holder to determine when the laser is to be fired for each laser shot. This value is stored with each data point. The MS collects the data from the i.r. detector after the laser has been fired. The signal to noise ratio is determined and the signal size is evaluated. If the signal to noise ratio is too low or the signal is out of range, the MS will determine the corrective action and transmit the appropriate commands to the CS which in turn will adjust the amplifier's gain, filtering or the laser's power. If the current data is acceptable, the MS will determine if any adjustments are necessary for the next shot and will take the necessary action.

The data is stored on the MS disk for further analysis after the experimental run. However, as much real time analysis will take place as time permits. If the heating rate is too high, then the MS does not have enough time to do complete analysis. If the heating rate is low enough, then the MS smooths the data to find the maximum and minimum points, to determine the amount of heat losses, and if laser flash-by is present. The diffusivity is then determined, corrected for heat losses if present, the uncertainty is computed and the diffusivity versus temperature is plotted on the graphics terminal. While the MS is analyzing the data, the smoothed data is plotted against the theoretical on the terminal, which remains displayed until the diffusivity computations are done. Once the diffusivity is plotted, the MS writes the results to an ongoing file which will be printed at the end of the test.

SUMMARY

The automated diffusivity system design, construction and integration have been completed. Some of the non-critical components of the system have not yet been built. However, the MS and CS systems are fully operational and the first data collection has been accomplished. The measurement of the diffusivity of a well-characterized material at various heating rates is underway.

ACKNOWLEDGEMENT

This work is funded by the U.S. Department of Defense. The author appreciates the assistance of Thomas Goerz, David Wells, Alan Brisco and Mark Bigler in getting the apparatus operational.

REFERENCES

1. Parker, W.J., Jenkins, R.J., Butler, C.P. and Abbott, G.L., J. App. Physics, Vol. 32, 9, pp. 1679-84, Sept. 1961.
2. Cowan, R.D., J. App. Physics, Vol. 34, 4, pp. 926-27, April 1963.
3. Cape, J.A. and Lehman, G.W., J. App. Physics, Vol. 34, 7, pp. 1909-13, July 1963.
4. Taylor, R.E. and Maglic, K.D., "Pulse Method of Thermal Diffusivity Measurement", Compendium of Thermophysical Property Measurement Methods, Vol. 1, Chpt. 8, Kosta D. Maglic, Editor, Plenum Publishing Corp., 1984.

ELECTRONIC FLASH: A RAPID METHOD FOR MEASURING THE THERMAL CONDUCTIVITY AND SPECIFIC HEAT OF DIELECTRIC MATERIALS

R. E. Giedd and David G. Onn

Applied Thermal Physics Laboratory[†]
Department of Physics and Astronomy
University of Delaware
Newark, DE 19716

ABSTRACT

"Electronic Flash", a new technique for the simultaneous determination of thermal conductivity and specific heat between about 20 K and 450 K, is described. The experimental apparatus is outlined and the limitations and resulting accuracy discussed. Data obtained on several materials is presented and compared with reference values. The technique is particularly valuable for rapid scanning of new advanced materials.

INTRODUCTION

Renewed interest in the thermal conductivity (κ) of dielectric materials, motivated in part by applications in electronic packaging, has stimulated a need for a fast, accurate method of measurement. Historically there have been two approaches to measurement of κ: (i) slow steady state direct measurement of κ and (ii) faster time-temperature transient techniques which measure the thermal diffusivity (α). Thermal conductivity (κ) is then inferred from separate specific heat (C_p) and density (ρ) measurements since

$$\kappa = \rho C_p \alpha. \tag{1}$$

The new technique described here, "Electronic Flash", belongs to category (ii). It has the added advantage that it provides simultaneous values of C_p and α.

The idea of substituting an electrical energy pulse for the laser pulse as a method of energy deposition has been suggested previously[1] but due to experimental limitations its use was confined to temperatures below about 4 K. Capacitor discharge techniques[2] which do not produce square electrical pulses have also been used but deconvolution of the $\Delta T(t)$ is much more difficult than for square pulse sources. Furthermore both techniques were only used for measurements of α although in principle C_p could have been determined.

---

*Fellow, Center for Advanced Study, University of Delaware.
[†]Supported by Delaware Research Partnership (DRP) and E. I. Du Pont de Nemours and Co.

Our present unique implementation of the concept of using electronically generated square pulses with accurately known energies allows us to make κ and C_p measurements between about 20 K and 450 K. Extensions to higher and lower temperatures are possible in principle.

THEORY

The theoretical basis of "electronic flash" is similar to that of the well known "laser flash" method[3]. In both cases one dimensional heat flow is produced in a planar sample by pulse heating of one face. The time-temperature transient response of the opposite face $\Delta T(t)$ is monitored and analyzed to determine a half rise-time, $t_{\frac{1}{2}}$. Provided that radiation losses are small and the heating pulse length, τ, is less than $0.1 t_{\frac{1}{2}}$ the thermal diffusivity is obtained from

$$\alpha = \frac{1.37 \ell^2}{t_{\frac{1}{2}}} \tag{2}$$

where ℓ is the sample thickness. Further analysis of the transient response permits corrections for finite pulse lengths to be applied even if $\tau \sim t_{\frac{1}{2}}$[4].

If two further conditions are met namely that:

a) heat losses from all causes are small and can be corrected for and,

b) the total energy ΔQ deposited by the heat pulse is accurately known

it is possible to measure C_p from the overall temperature rise of the sample $\Delta T(\infty)$ since:

$$C_p = \frac{1}{m} \left(\frac{\Delta Q}{\Delta T(\infty)} \right) \tag{3}$$

where m is the sample mass.

It is very difficult in the "laser flash" experiment to determine the energy input accurately due to differing reflection coefficients of specimen surfaces and variations in pulse energy.

With our "electronic flash" technique heat losses from all causes are reduced by careful design of the specimen holder. In addition ΔQ is accurately measured for every electronically generated energy pulse by using micro-computer data acquisition.

EXPERIMENTAL

For accurate use of our "electronic flash" method to measure α and C_p simultaneously four basic criteria must be met:

a) a high energy, short ($\tau < 50$ msec) electrical pulse must be generated,

b) the energy of the pulse must be measured quickly to within 2%,

c) a uniform heater must be applied to the sample face so that during the energy pulse a spatially uniform current density exists in the heater and,

d) heat losses from all sources must be minimized.

If α alone is to be measured then b) and d) above are less critical. If C_p alone is to be measured then a) and c) are less critical. We discuss each of these criteria in turn.

Energy Pulse

To determine the maximum energy requirement of the electronic pulse we set extreme upper limits on values for C_p ($\sim$ 1.5 J/g K), temperature rise $\Delta T(\infty) \sim 5$ K and sample mass (3g). From equation (2) $\Delta Q \sim 23$ J. In order to measure α accurately we must also require τ to be less than $0.1t_{\frac{1}{2}}$, although, as in the "laser flash" method finite pulse effects can be compensated for to some extent[4]. Data acquisition limitations in measuring $\Delta T(t)$ with an intrinsic thermocouple lead to a minimum measurable $t_{\frac{1}{2}}$ of 0.5 sec. This in turn sets a maximum pulse length of 50 msec. In addition the pulse must be very square with maximum rise and fall times three orders of magnitude less than the pulse length (i.e. 50 μsec for the 50 msec pulse). Figure 1 shows a typical energy pulse for the "electronic flash" experiment.

The above conditions require that we use a maximum 150V, 3A electronic pulse. Normal transistors cannot switch such high energies. However, with the advent of SCR's the management of a.c. energies comparable to this was accomplished. Details of the circuit we use to generate our pulses will be published elsewhere[5].

Voltage, Current, and Temperature Measurement

At a given temperature, the measurement of one transient is designed to last a maximum of 30 sec. This minimizes radiation loss corrections. During this 30 sec the energy input of the pulse to the heater and the temperature rise of the sample must be measured. The energy of the pulse is determined by 50 separate voltage and current measurements made during the pulse. We used a Metrabyte Dash 16F high speed A/D converter combined with an IBM XT micro-computer. The Dash 16F A/D is capable of a maximum

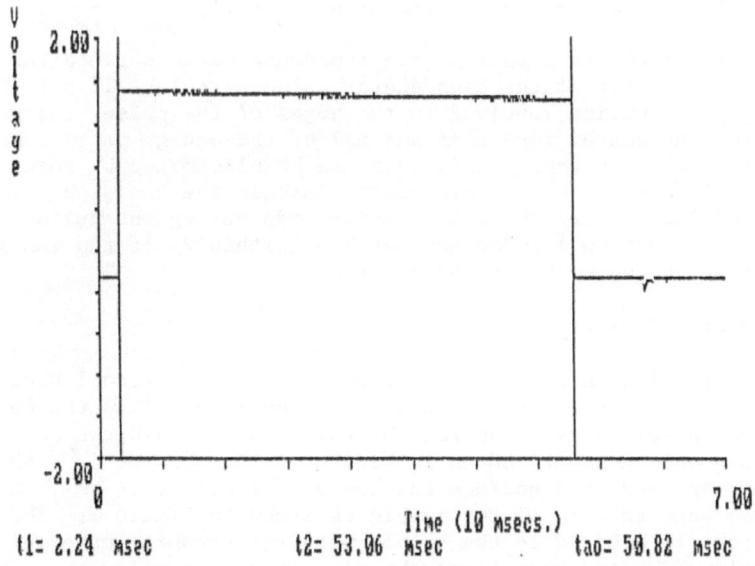

Figure 1. Shape of electronic flash heat pulse. Vertical scale is proportional to the electrical current flowing through the graphite heater.

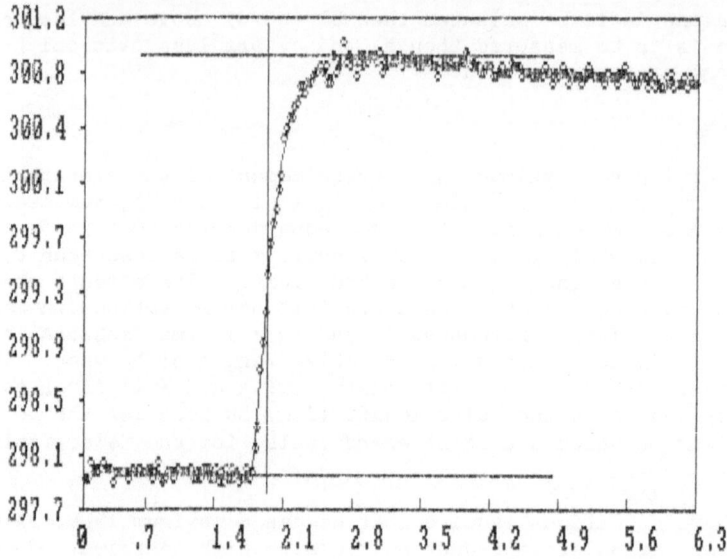

Figure 2. Time-temperature transient. Smooth curve is the filtered transform of the data to cut off high frequency noise.

sampling rate of 100,000 samples per second using direct memory access. It also has two D/A outputs that can be used to trigger the high energy pulse. Initially, the computer reads the temperature of the back face of the sample before energy is delivered to the heater. The computer next triggers the high energy pulse and measures the voltage and current in the sample heater. The computer then switches back to reading the temperature of the back face of the sample until the end of the transient. Figure 2 shows a typical time-temperature transient $\Delta T(t)$.

It is important to insure a good impedance match between the sample heater and the output of the high energy pulse circuit. If not, because of the high frequencies involved in the edges of the pulse, the pulse may reflect from the heater impedance and all of the energy may not be dissipated in the sample heater. This loss can be eliminated by insuring that the d.c. resistance of the sample heater matches the resistance calculated from the voltage and current measurements made during the pulse. We found that the system was well impedance matched (within 2% if the sample heater resistance was between 50 and 150 ohms).

Specimen Heater Geometry

The expression in equation (2) requires one dimensional heat flow in the sample. In order to achieve one dimensional heat flow the heat must be supplied uniformly over one face of the sample. This can be achieved by using a sample geometry shown in Figure 3. The heater is a thin layer of graphite sprayed to a uniform thickness of 1 mil (22.4 µm). Current is passed from edge to edge of the sample as shown in Figure 3. The 5 mil copper wires are adhered to the sample with the graphite material itself. The variation in resistance along the wire is much smaller than the variation in resistance across the heater. This system allows for nearly parallel current flow between the two wires, resulting in a constant current density in the graphite heater.

The sample is suspended on the two heater wires as shown in Figure 4. A Type K (chromel-alumel) thermocouple is placed on the top surface of the specimen usually in an intrinsic mode. Thus only six 5 mil wires attach the sample to the background. This insures that the specimen is in good thermal isolation for times shorter than the transient time (about 30 sec). The wires that attach the sample to the background do provide a weak thermal link to allow the specimen to be heated and cooled slowly. Another advantage of this system is that the time of measurement (about 30 sec) is short enough not to require temperature control. The data can be obtained while the sample is ramped in temperature through the desired range at ~ 50 K/hr.

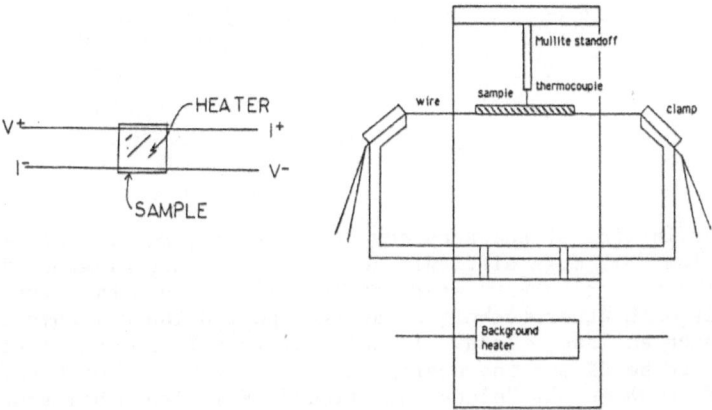

Figure 3.   Sample geometry and mounting arrangement.

Additional thermal isolation is achieved by evacuating the sample chamber to better than 10^{-5} torr and including two passive shields around the sample.

RESULTS AND DISCUSSION

The accuracy of the C_p measurement was determined by using a sapphire specimen tested with different heater resistances between 50 and 150 ohms. Over six measurements at 300 K, the sapphire was determined to have a C_p of 0.78 with a precision of 3%. Sapphire is quoted elsewhere[6] to have a C_p of 0.779 J/g.K at 300 K with an accuracy of better than 1%. The "electronic flash" values are within the 5% accuracy calculated for the C_p determination. The sapphire standard was measured from 400 K to 90 K to test the accuracy of the C_p determination at low and high temperatures. The results are shown in Figure 4. The measured C_p was within 5% of the values obtained for the C_p of sapphire cited by NBS[6]. α for this sapphire sample could not be determined because the small thickness of the sample (0.09 cm) resulted in a $t_{\frac{1}{2}}$ too fast to measure.

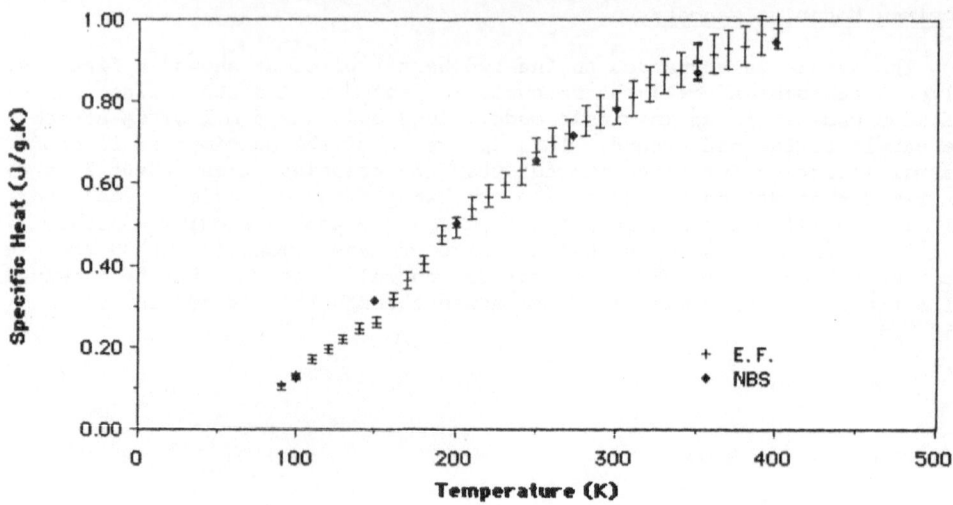

Figure 4: Specific heat of sapphire. Solid points are from NBS[6].

The determination of the accuracy of the κ or α measurement from "electronic flash" is more difficult due to the absence of good dielectric standards. The best accuracy determination was achieved by cross calibrating materials with other techniques at the Applied Thermal Physics Laboratory (ATPL) such as laser flash. In this way we calculated the accuracy of α measurement to be 5% and the overall accuracy of the κ measurement to be 10%. We also compared our "electronic flash" (EF) with other tabulated values for the material wherever possible. Table 1 is a summary of a comparison between laser flash (LF) and differential scanning calorimetry (DSC), and "electronic flash" (EF) for the Du Pont Company series of Green Tape[TM] substrate materials[7,8].

The κ of N.B.S. 710 soda-lime glass was determined at 300 K to be .91 $\pm$ .09 W/m.K which is in the range of 0.84 to 1.0 W/m.K given in the reference tables for this material.

Another silica based glass, Corning 7070, which is often cited as a reference material, was determined to have a room temperature κ of 1.0 $\pm$ 0.1 W/m.k and C_p of 0.78 $\pm$ .04 J/g.K. C_p and κ of the 7070 glass from 90 K to 450 K is shown in Figure 5. κ for 7070 glass is nearly proportional to the C_p over this temperature range which implies a small constant mean free path since

$$\kappa = \frac{1}{3}C_p\lambda V \qquad\qquad (4)$$

where λ is the mean free path and V is the sound velocity. A constant small mean free path is typical of an amorphous material.

C_p and κ for an alumino-silicate ceramic (Aremco 502-1100[9]) was measured from 300 K to 80 K as shown in Figure 6. The results indicate a room temperature κ of 1.5 $\pm$ 0.15 and C_p of 0.67 $\pm$ 0.04 J/g.K. Aremco lists the material as having a room temperature κ of 1.61 W/m.K (no error cited).

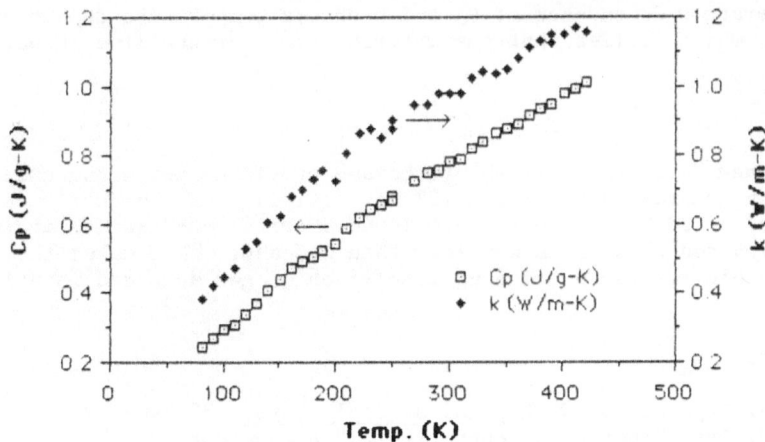

Figure 5. Thermal conductivity and specific heat of Corning 7070 glass.

Table 1. Thermal Properties of Green Tape™ at 300 K

Technique	κ(W/m.K)	C_p(J/g.K)	$\alpha \times 10^{-3} cm^2 sec$
EF	2.2 ± .2	.81 ± .04	9.50 ± .05
LF (ATPL)			9.55 ± .05
LF and DSC[7]	2.2 ± .25	.85 ± .04	9.45 ± .05

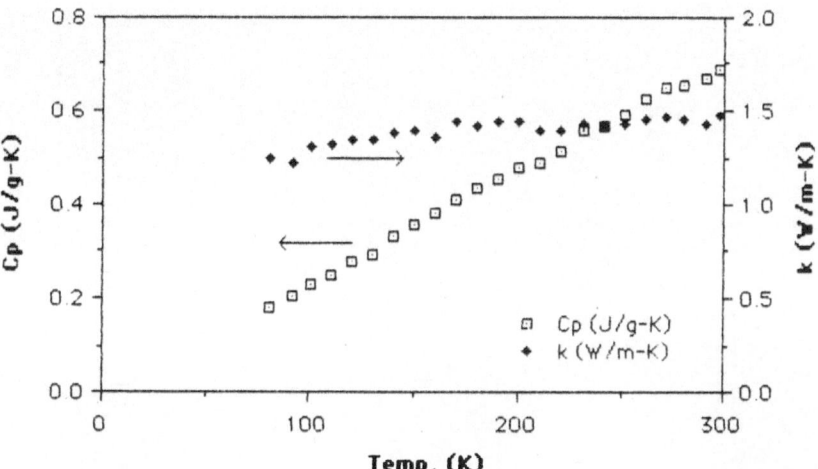

Figure 6. Thermal conductivity and specific heat of an alumino silicate ceramic.

The temperature dependence of C_p and κ shows a temperature dependent mean free path which implies that the material has a crystalline phase.

SUMMARY

Our new "electronic flash" technique provides fast measurements of C_p and κ with accuracies of 5% and 10% respectively for dielectric materials between about 20 K and 450 K. Sample geometrical constraints are not severe provided that $t_{\frac{1}{2}}$ is not less than 0.5 sec. This makes the technique very valuable for rapid scanning comparison of new advanced materials.

REFERENCES

1.   J. A. Cape and G. W. Lehman, J. Appl. Phys., 32:1909 (1963).
2.   R. P. Tye, "Thermal Conductivity 2," Academic Press, New York (1969).
3.   W. J. Parker, R. J. Jenkins, C. P. Butler, and G. L. Abbott, J. Appl. Phys., 34:1909 (1963).
4.   R. E. Taylor and J. A. Cape, Appl. Phys. Lett., 5:212 (1964).
5.   R. E. Giedd and D. G. Onn, to be published.
6.   L. Furukawa, J. Res. Nat. Bur. Stand., 57:67 (1956).
7.   E. L. Rich, S. K. Suko, A. J. Martin, B. H. Smith, R. E. Giedd, A. J. Whittaker, D. G. Onn, and F. K. Patterson, "Thermal Management Considerations for a Low-Temperature, Co-Fireable Ceramic System," ISHM Proceedings 1987 (1987).
8.   E.I. Du Pont de Nemours and Co., Inc., Electronics Materials Division, Wilmington, DE 19898.
9.   Aremco Products, Inc. Ossining, N.Y. 10562.

A COMPUTERIZED THERMAL DIFFUSIVITY APPARATUS

Vladimir V. Mirkovich
Mineral Sciences Laboratories, CANMET
Energy, Mines and Resources Canada
Ottawa, Canada

Peter S. Gaal and James H. Kareis
Anter Laboratories, Inc.
Pittsburgh, PA
U.S.A.

ABSTRACT

A previously developed apparatus, based on the concept of an infinite cylinder and radial symmetry, was used for measurement of thermal diffusivity of 25-mm diameter cylindrical specimens. For the heat flow model in which the mathematical boundary conditions can be satisfied best, the surface temperature of the cylinder is a harmonic function of time. The temperatures are measured at a point near the surface and in the centre of the cylinder and are simultaneously plotted against time. Thermal diffusivity is related to the time interval needed for the heat wave to penetrate from one to the other measuring point, the time period of the heat wave and the geometry of the specimen.

The slowness of measurement was the principal disadvantage of the original version. It would take one to two hours for a determination, whereas in the computerized model only 3 to 5 minutes are required.

The computerized system performs the following functions automatically:

1) Controls the rate of the temperature increase of the specimen.
2) Bucks the base emf from thermocouples in the specimen.
3) Amplifies and stores the (two) harmonic portions of emf.
4) Determines the time interval from the two harmonic emfs and calculates the thermal diffusivity.
5) Calculates the temperature for the average of a desired number of determinations.
6) Stores all information and, if requested, plots the harmonic portion of emf.

The entire test sequence, comprising measurements at any number of temperature levels over a desired temperature range, can be programmed to perform without attendance. Operation of the system's components is discussed and the results of measurements on Pyroceram 9606 are presented.

INTRODUCTION

A large number of non-steady-state, heat-flow boundary conditions have been discussed in the literature. In all cases, one of the three basic heat-flow modes are identified: converging, parallel or diverging. Conditions favoring minimum attenuation of the amplitude of the heat pulse during its progress through the specimen improves the precision of the thermal diffusivity measurement. This work deals with the converging heat flow case.

In the converging heat flow in a sphere, the attenuation of a heat pulse is the lowest. However, the mathematical boundary condition in this case cannot be readily satisfied experimentally. Also, the preparation of spherical specimens and temperature measurement would be difficult.

In the case of a cylinder (with radial symmetry) the attenuation of a converging heat pulse is somewhat greater than that in a sphere. Due to the relative ease of satisfying the theoretical boundary conditions experimentally, the simplicity of specimen preparation, and of the temperature measurement, a system based on the concept of an infinite cylinder with radial symmetry was chosen.

The periodic measuring method is rather slow and tedious due to extensive calculations and graphing needed. In the non-automated configuration (1,2) it required more than an hour to obtain a data point. However, by computerizing the operation, not only can the duration of the measurement be reduced to just a few minutes, but the measurement can be performed without attendance and the computations are performed on-line. It therefore becomes an attractive method for measurement of thermal diffusivity.

Theoretical Background

In the periodic mode the surface temperature of the cylinder of radius "a" varies according to:

$$T = T_0 \sin (\omega t), \tag{1}$$

where T_0 is the steady average temperature level of the cylinder, $\omega = 2\pi/t_c$, and t_c is the time of one full heat wave cycle. The temperature distribution through the cylinder, given by Carslaw and Jaeger (3), is

$$T(r,t) = T_0\{M_0(\omega'r)/M_0(\omega'a)\}\sin\{\omega t + \theta_0(\omega'r) - \theta_0(\omega'a)\} \tag{2}$$

where $\omega' = (1/\alpha)^{0.5}(2\pi/t_c)^{0.5}$, M_0 and θ_0 are forms of Bessel functions (4), and α is thermal diffusivity.

When temperature is observed at the center of the cylinder, i.e., at $r = 0$, Eq. (2) reduces to

$$V(0,t) = \{1/M_0(\omega'a)\}\sin\{\omega t - \theta_0(\omega'a)\}, \tag{3}$$

where $V = T/T_0$.

The solid curve in Fig. 1 is the plot of Eq. (1) and the dotted curve represents the plot of Eq. (3). The time differential between the two temperature waves, Δt_c, is a function of thermal diffusivity. Conversely, thermal diffusivity can be calculated from experimentally obtained Δt_c. In order to equate the dimensionless forms of Eqs. (1) and (3), both equations have to be equal to the same V. If at time t Eq. (1) equals V, then, as can be deduced from Fig. 1, Eq. (3) will equal V at time $t + \Delta t_c$. Thus one obtains

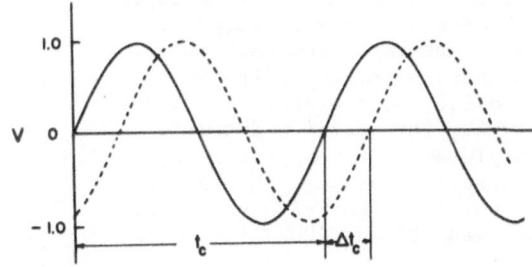

Fig. 1 - Plot of the surface and centre relative temperatures, V, against time for an infinite cylinder subjected to a periodically varying temperature.

$$\sin(\omega t) = V = \{1/M_0(\omega'a)\sin\{\omega(t + \Delta t_c) - \theta_0(\omega'a)\}$$

Experimentally, the time differential Δt_c is determined from the graph most conveniently at V = 0. In such a case t = 0, t_c, $2t_c$,.... For t = 0, the above equation becomes

$$\sin(\omega 0) = 0 = \{1/M_0(\omega'a)\}\sin\{\omega(0 + \Delta t_c) - \theta_0(\omega'a)\}.$$

Because $1/M_0(\omega'a) \neq 0$, then

$$\sin\{\omega(\Delta t_c) - \theta_0(\omega'a)\} = 0$$

and also

$$\omega(\Delta t_c) - \theta_0(\omega'a) = 0 \tag{4}$$

The radius a in Eq. (4) is to be taken as one reaching to the measuring point near the surface (not the actual radius of the cylinder).

The value of $(\omega'a)$ is determined from the product $\omega(\Delta t_c)$, which is in turn obtained by using the measured quantity Δt_c. Because a and t_c are known, α can be readily calculated.

EXPERIMENTAL

The Measuring System

The specimen column and heat pulse generating methods. The specimen is a cylindrical column, 25 mm in diameter and approximately 75 mm long. It may be composed of several discs. Two 0.75 mm diameter thermocouple holes, one in the centre and the other near the surface extend axially from one end halfway into the column. The thermocouple holes are about 11 mm apart.

The surface heat pulses can be generated in three ways: In one of the configurations (Figure 2) cylindrical specimens forming the column with ceramic insulators at the end (to minimize axial heat flow) are held together by two pairs of stainless-steel wires (not shown). A Nichrome heating wire is wound (non-inductively) in a helix directly around the sample column. A thin layer of refractory cement keeps the wires separate from each other. To measure thermal diffusivity at different temperature levels, the specimen assembly is placed in a temperature controlled furnace. Measurements in this configuration were made between 30° and 900°C.

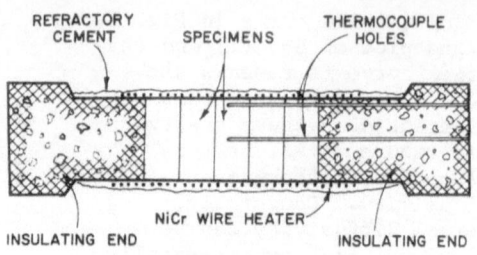

Fig. 2 – Heat-pulse generating Nichrome wire heater wound directly onto the specimen column.

In the second configuration (Figure 3) the specimen column is placed in a snugly fitting stainless-steel jacket. The jacket is approximately 8.5 cm long, with the heat-pulse-generating Nichrome wire wound on the outside. A thin layer of refractory cement separates the wires from the jacket and from each other. The stainless-steel jacket containing the specimen column is placed concentrically within a 5-cm diameter and 8.5-cm high fused-silica tube. Two Nichrome heating wires are wound from top to bottom and around the circumference of the silica tube. One of these is used to maintain the required temperature level. The assembly is insulated from the ambient. Measurements with this system were performed in the range of 30° to 450°.

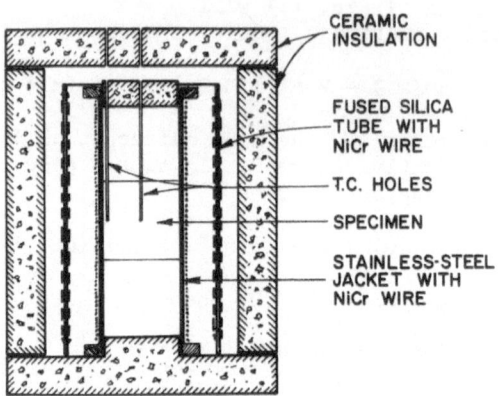

Fig. 3 – Specimen column inside the stainless-steel pulse-generating jacket.

In the third configuration, the stack of specimens forming the specimen column is placed concentrically within the fused-silica tube. One of the heating wires on the fused silica tube serves as a heat-pulse generator, while the other maintains the desired temperature level. The assembly is insulated from the ambient. Measurements in this mode were made in the 350° to 900° temperature range.

Method of Operation

In response to the advance of a sinusoidal type heat wave into the specimen, emf signals are generated by the near surface and centre thermocouples. Each emf signal consists of a base emf and a small harmonic emf. The magnitude of the base emf varies between 0 and, say 40 mV in the case of chromel-alumel thermocouples, depending on the steady average temperature of the specimen. The harmonic emfs, which reflect temperature oscillations at the measuring points, are usually 0.150 mV (caused by about 2°C variation) and 0.075 mV (caused by up to 1°C variation) for the surface and centre thermocouples, respectively.

The input to the computer must be in volts and therefore the surface emf is amplified 2,000 times while the centre emf (because of attenuation) is amplified 4,000 times. Furthermore, because the computer can accept signals only up to 2.5 Volts, the base emf must be cancelled out first so that only the amplified harmonic emf reaches the computer.

The harmonic portions, usually 2.5 to 3.5 wavelengths, are temporarily stored for each determination, then analyzed to determine the exact time of one wave length (t_c) and the time differential (Δt_c) for the wave to travel from the surface to centre measuring points. The thermal diffusivity is calculated automatically and reported together with the temperature of the specimen. The digitized wave forms can be plotted immediately and/or stored for later examination.

As stated above, the computer accepts signals from 0 to 2.5 Volts (i.e. the equivalent of non-amplified thermocouple output of up to 1.25 mV). At the onset of measurements the bucking system brings the amplified harmonic emf to a near zero position of the operating range. As the temperature of the specimen increases, the amplified harmonic emf moves toward the 2.5 Volt limit. However, once 75 percent of the range is reached, the computer increments the bucking emf, bringing the amplified harmonic emf back to the near zero position.

The average temperature of the sample is obtained from the absolute value of the bucking potential and the unbucked portion (overflow) of the base emf.

The measurements are usually made at a constantly increasing temperature. The rate of the temperature increase is regulated by programming for each series of measurements the amount of power increment to the furnace, the time intervals at which the power is to be incremented, and the maximum rate of temperature increase.

Upon reaching the programmed top temperature level, the system either turns itself off, reverses the direction, or, if so requested, repeats the series of measurements in the same temperature range.

Main and auxiliary electronic devices. The heat pulses are generated by means of two function generators and a suitably rated audio power amplifier. One of the function generators provides a constant signal with the frequency of 1000 Hz. This signal is then modulated by the second function generator producing a high-frequency sinusoidal pulse of the required time length, usually between 45 and 200 seconds. The modulated signal is amplified in an audio power amplifier. It was determined that this type of power supply causes the least amount of electric interference in the thermocouple output.

A motorized variable transformer supplies power to the electric furnace containing the specimen (or to the Nichrome heater on the fused-silica tube). The motor increments the transformer output as required to achieve a slowly increasing temperature of the environment.

A schematic representation of the analog measuring circuitry is shown by the block diagram in Figure 4.

As stated above the signals from the outside and inside thermocouples are amplified 2,000 and 4,000 times, respectively. Because it is the minute variation of each signal that is needed as data, the base emf must be cancelled. To extract the small variation riding on top of a several hundred times larger signal is a very difficult task. In this case it is accomplished as follows: the computer controlled high speed Analog-to-Digital (A/D) Converter samples the output of each thermocouple first and, if it proves to be above the 75% full scale value, it instructs the Digital-to-Analog (D/A) Converter to inject a bucking voltage of a preset increment with opposing polarity to each thermocouple circuit. The sample/buck process is repeated, until the signal remains within the useful range of the A/D. Once the bucking has stabilized, the time interval for approximately 3 waves are computed and 300 evenly spaced data points are taken for each thermocouple in an interweaved fashion. Thus a data set consists of 300 data pairs with their associated time values.

<u>The operating program</u>. The operating program, prepared to operate with the Apple IIe computer, is made up of four segments. The first part controls the experiment, data taking and storage. The second portion analyzes the data immediately after it is taken and performs all calculations detailed below. The third portion contains trouble shooting and maintenance routines that are not generally executed as part of an experiment. The fourth section contains a plotting routine specifically written for this application, allowing a cross check on the logical decisions made by the computer in evaluating the wave patterns.

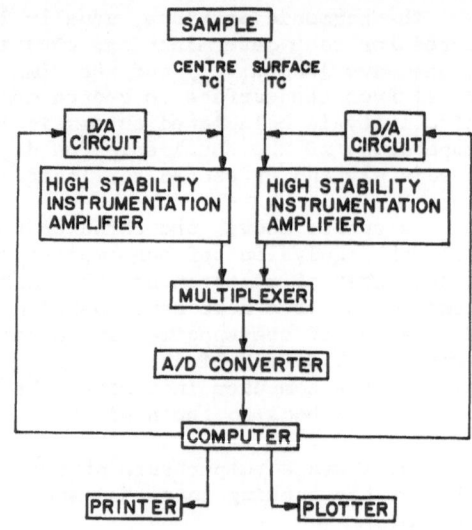

Fig. 4 – Block diagram of the computerized thermal diffusivity measuring system.

The most complex task is the data analysis. First, each set of data points is searched and evaluated to determine the maxima and minima present for each signal pattern representing individual thermocouples. From the maxima and minima the midpoints are computed to define the X-axis for the sine wave. Using the rationalization developed in earlier work and used with manual computational methods, straight lines are fitted to the points clustered close to the midpoint, and these simultaneous linear equations are solved for the intercepts. The 300 data points are spaced to cover 2 1/2 to 3 waves, providing five to six lines with four to five intercepts. In this manner it is assured that under any circumstances, the four intercepts necessary for defining a full wave are present. The time coordinates of the intercepts are correlated between the two sets and the average phase shift determined. Through a look–up table of the Bessel functions, the diffusivity is then computed.

Using the plotting functions, one can duplicate graphically the above process and cross check the appropriateness of the fits. It also allows computation of the phase lag by other methods, such as straight differencing of the zero crossing points of the waves. In general, this latter method also works well, except when the measurement is taken on the rise. A steady heating range always leads to a systematic distortion of the sine wave, having one lobe shorter than the other, plus effectively shifting the zero crossing point as well. However, the distortion is present for both signals and tends to cancel out. The tangent method is found to be more general and therefore it is used most often.

RESULTS AND DISCUSSION

Two Pyroceram 9606 specimens from the same batch were used in this investigation. The results of the measurements with the automated system are shown in Figures 5, 6, and 7. For purposes of comparison, thermal diffusivities of Pyroceram 9606 (but from different lots), measured by Plummer (4), Flieger (5), and Rudkin (6) are also included in each figure. All measurements were made using a 105-second long wave.

The results in Figure 5 were obtained with specimen P-3. The pulse-generating Nichrome wire was found directly on the specimen column. Each circle represents an average of three measurements. At lower temperature, the present data are bout 5% higher, whereas at 900°C they are some 4% lower than Flieger's and Rudkin's results. In general, the agreement is good.

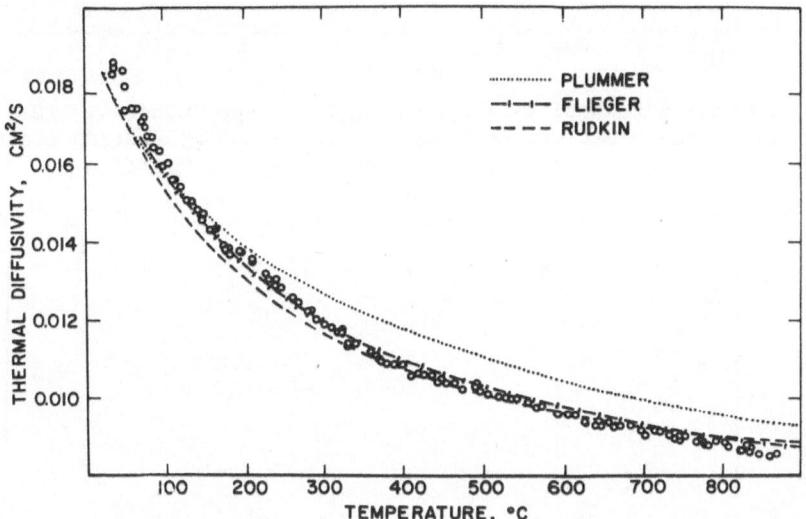

Fig. 5 – Thermal diffusivity of Pyroceram 9606, specimen P-3, obtained with pulse-generating Nichrome wire wound directly onto the specimen column. Each circle represents three determinations.

Figure 6 represents the results obtained with specimen P-4. In this case, the specimen column was contained in the stainless-steel heat-pulse generating jacket. Two hundred and twenty-five measurements were made in the temperature range of approximately 40° to 560°C. All the points are plotted in the diagram. The average deviation is ±1.7%. The absolute value is a little higher than that of specimen P-3 in Figure 5. The difference has been attributed to the variation in the time response and positioning of the thermocouples. For accurate measurements of thermal diffusivity by this method it is essential that the time response of the thermocouples to temperature pulses be the same. Because of small differences in the heat capacities between individual thermocouples, the changes that occur

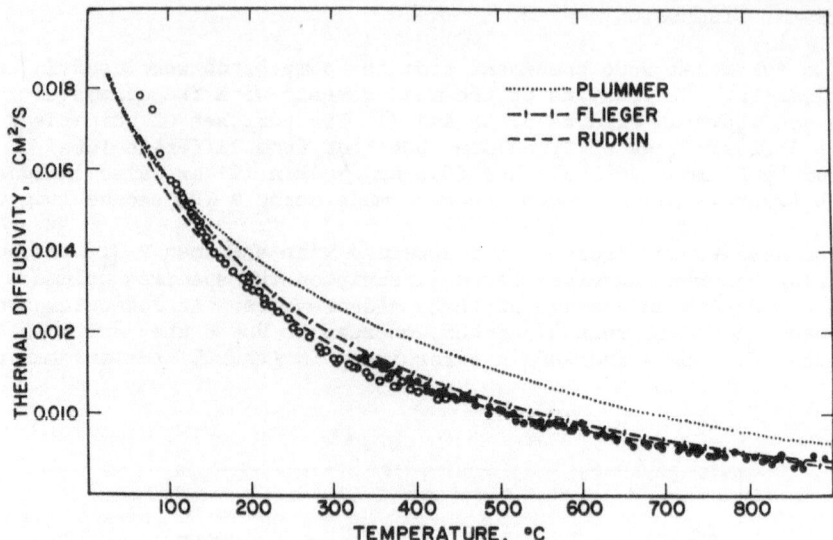

Fig. 6 – Thermal diffusivity of Pyroceram 9606, specimen P-4. The
specimen column was contained in the pulse-generating stainless-
steel jacket. Each dot represents one determination.

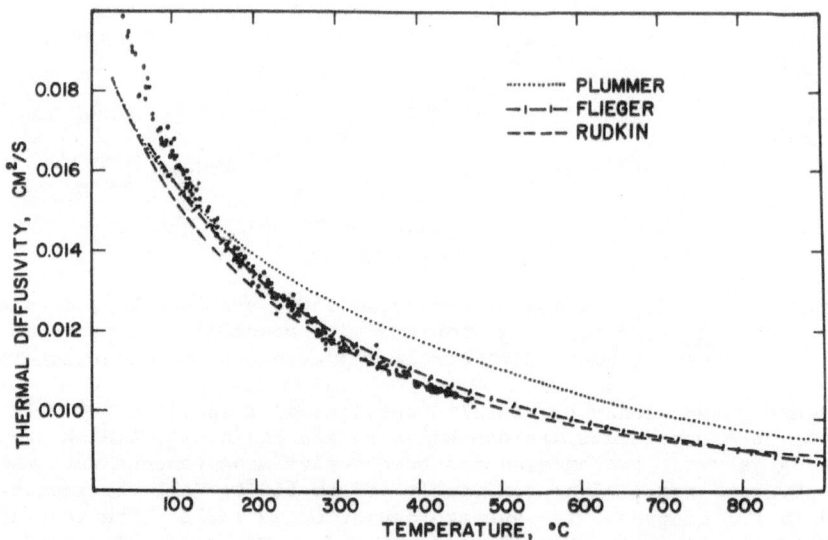

Fig. 7 – Thermal diffusivity of Pyroceram 9606, specimen P-4. Circles
represent data obtained with pulse-generating jacket, but with
a different set of thermocouples from those in Fig. 6. Solid
dots are data obtained with heat-pulse radiated onto the
surface of the specimen. Each circle and dot represent three
determinations.

in the course of measurements and the slightly different positioning of thermocouples within the thermocouple holes each time the measurement assembly is set-up, the variations of the absolute value are difficult to eliminate.

The results in Figure 7 were obtained with specimen P-4 placed in the heat-pulse-generating jacket. The specimen column, however, was first removed and ater inspection placed back into the jacket. Furthermore, another set of thermocouples was used. The measurements were made in the 60° to 450°C range. Each circle is the average of three determinations. The results are between 3% and 4% lower than the ones shown in Figure 5, although the measurements were performed with the same specimen and under apparently identical conditions.

The dots in Figure 7 represent thermal diffusivities of specimen P-4 obtained in the experimental configuration in which the heat pulses were radiated on the surface of the specimen. Each dot is the average of three determinations. In the overlapping region, in the 350°C to 400°C temperature range, these results appear to be a little higher than those obtained with the pulse-generating jacket. This difference, however, is considered to be within the measurement variations. The data are in excellent agreement with Flieger's and Rudkin's results.

CONCLUSIONS

The automated thermal diffusivity system performed quite satisfactorily in these measurements. Although it performs without attendance, the design allows monitoring at all stages of measurement. Furthermore, stored data could be readily recalled for subsequent analysis. The results obtained on Pyroceram 9606 conform well with the data obtained by other authors.

REFERENCES

1. Mirkovich, V.V., Thermal Diffusivity Measurement of Armco Iron by a Novel Method, Rev. Scient. Instr.; 48 (5) 560-5 (1977).

2. Mirkovich, V.V., An Apparatus for Measuring Thermal Diffusivity in Air, CANMET, Energy, Mines and Resources Canada, CANMET Report 77-21 (1976).

3. Carslaw, H.S. and Jaeger, J.C., Conduction of Heat in Solids, (Oxford U.P., New York, 1959), 2nd ed., p. 201.

4. Plummer, W.A., Campbell, D.E. and Constock, A.A., Method of Measurement of Thermal Diffusivity to 1000°C, J. Am. Ceram. Soc.; 45 (7) 310-6 (1962).

5. Flieger, H.S., Jr., The Thermal Diffusivity of Pyroceram at High Temperatures, Proceedings of the 3rd Conference on Thermal Conductivity (Oak Ridge National Lab., Oak Ridge, Tennessee). 2 769-83 (1963).

6. Rudkin, R.L., Thermal Diffusivity Measurements on Metals and Ceramics at High Temperature, Technical Documentary Report No. ASD-TDR-62-24, Part II (available from the Office of Technical Services, U.S. Dept. of Commerce, Washington, D.C.).

EXPERIENCES GAINED WHILE EXAMINING THERMOPHYSICAL

PROPERTIES OF STATE-OF-THE-ART MATERIALS

Hans Groot

Thermophysical Properties Research Laboratory
School of Mechanical Engineering
Purdue University
West Lafayette, IN 47906

ABSTRACT

TPRL has investigated the thermophysical properties of a wide range
of materials from various organizations, including industrial and govern-
mental. Because of the diverse nature of these materials, it has been
necessary to modify techniques and apparatuses. In many cases, it is
desirable that the results be cross-checked either by using several different techniques or by varying the experimental conditions substantially.
While the use of standard reference materials is valuable for checking
apparatuses, it has been found that the use of these materials is no
guarantee that accurate results are being obtained on other samples. The
determination of several different properties, combined with theory and
experience, permits cross-checking and helps establish the reliability of
the results. A number of examples are given.

INTRODUCTION

The Thermophysical Properties Research Laboratory (TPRL) is part of
the School of Mechanical Engineering at Purdue University. As such it
conducts graduate level research programs on thermophysical properties.
In addition, TPRL maintains a permanent technical staff which responds to
unsolicited requests from industrial and government organizations. As a
result of such requests, more than 700 reports containing original data
have been generated. While much of this data concerns proprietary or
sensitive materials and were never published, some of the experiences and
techniques resulting from these efforts may be of interest to the thermophysics community.

TPRL possesses a highly computerized state-of-the-art facility fea-
turing two laser flash diffusivity apparatuses, a high temperature multi-
property apparatus, Kohlrausch device, heated probe, thermal comparator,
push-rod dilatometer, differential scanning calorimeter, pulse heating
devices, etc., and is capable of testing materials by a variety of techniques over wide temperature ranges.

357

INTERNAL STANDARDS

TPRL maintains a number of standards for conductivity, diffusivity, specific heat, and expansion. In addition to the SRM's available at one time or another from the Department of Commerce, a number of materials became de facto internal standards by virtue of long running quality control programs or concerted efforts involving widely different techniques. Examples include a proprietory graphite and fine-weave carbon/carbon. Thermal diffusivity data on a fine-weave carbon/carbon measured by the laser flash technique are compared in Fig. 1 to diffusivity values calculated from measured thermal conductivity, specific heat, and bulk density values. The agreement is excellent over the temperature range 300 to 2500 K, indicating that the various techniques are capable of yielding reliable results on fine weave composite samples [1].

Similar intercomparisons have been made internally for SRM materials. Also there have been a number of external intercomparisons made during both national and international round-robin programs covering a variety of materials.

FIBER-REINFORCED MATERIALS

The concept of effective diffusivity obviously applies to fine-weave fiber-reinforced composites where the normal size sample thickness exceeds four unit cell spacings. However, in the case of fiber-reinforced composites where the weave structure is much coarser (about 0.22 cm), the situation is considerably different. For example, Fig. 2 shows the normalized rear-face temperature rises for various locations in respect to a fiber bundle in a proprietary fiber-reinforced material. Note that these rise

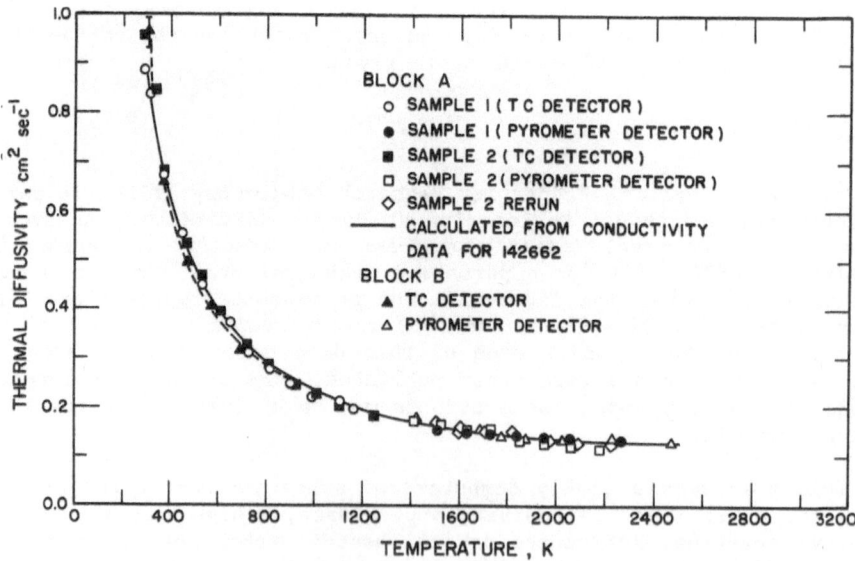

Fig. 1.   Intercomparison of Thermal Diffusivity Results for a Fine Weave Carbon/Carbon.

curves do not follow the theoretical model for homogeneous materials, in contrast to that observed for a fine-weave carbon/carbon composite. Concurrently, the diffusivity values for the experimental data represented in Fig. 2 decrease markedly as the percent rise time increases in marked contrast to the results for a fine-weave composite.

The degree of mismatch between the normalized experimental rise time curve and the theoretical model depends upon three parameters, namely (1) the relative magnitudes of the diffusivity values of the fiber reinforcement and the matrix, (2) the thickness of the sample, and (3) the rear face area of sample sensed by the temperature monitor. When the sample is reasonably thick or the ratio of diffusivity values for fiber and matrix approach 1, then the response curve approximates the theoretical model for a homogeneous material. In this case the diffusivity values calculated at increasing percent rise times asymptotically approach limiting values, and these values correspond to the diffusivity values which would be calculated from measured conductivity values.

If we use highly localized temperature sensors, it may be possible to measure the diffusivity of the fiber-reinforcement in situ [2]. Figure 3 (curve A) shows the voltage rise values from an amplified thermocouple output where the thermocouple is located on a fiber bundle in a fiber-reinforced composite. The results for a duplicate experiment taken over a shorter time duration is also shown in Fig. 3 (curve B). Curve A of Fig. 3 can be described as the addition of two transients, one from the fiber bundle (curve B) and a second from the matrix. The transit half-time through the matrix is about 10 times longer than that along the fiber bundle. Consequently, the heat flow through the matrix, measured at the detector, does not affect curve B until approximately 0.07 s, hence, its increasing slope beyond this time. When normalized, the data of Fig. 3 (curve B) approximates the theoretical model, and diffusivity values calculated at various rise times were generally within a few percent of each other. These diffusivity values, which apparently apply to the fiber bundle, were about 10 times greater than the average values for the composite.

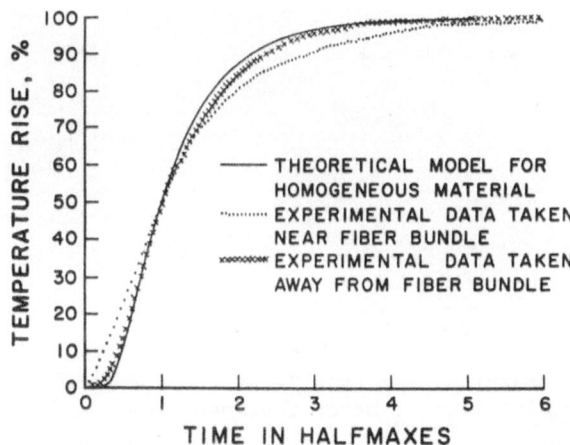

Fig. 2. Normalized Rear Face Temperature Rise Measured by Point Source Detector at Various Locations in Relation to a Fiber Bundle in a Coarse Weave Carbon/Carbon.

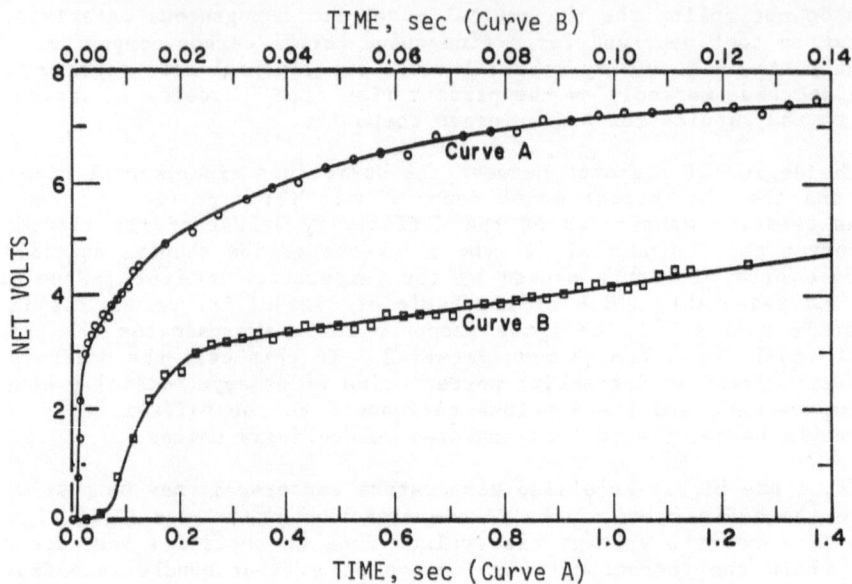

Fig. 3. Temperature Detector Response for a Point Source Detector
Located in a Fiber Bundle in a Coarse Weave Carbon/Carbon.

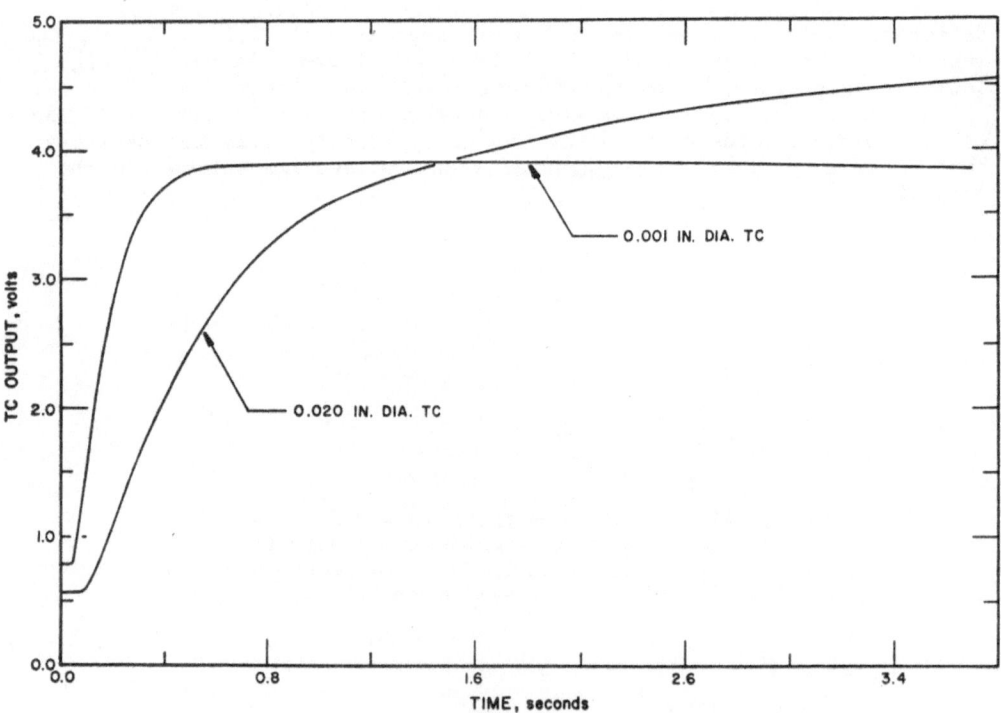

Fig. 4. Measured Rear Face Temperature Response Using Two
Different Diameter Thermocouples.

TRANSIENT TEMPERATURE MEASUREMENTS

Good techniques for making transient temperature measurements, particularly using thermocouples, are not generally understood. A simple equation describing response time (to 95% of final reading) is

$$\tau_{0.95} = \frac{25}{\pi} \ \frac{D_T^2}{\alpha_S} \ \frac{\lambda_T}{\lambda_S}$$

where α is the thermal diffusivity of the sample, λ is the thermal conductivity, D is the wire diameter, and the subscripts T and S refer to thermocouple and sample. The importance of using small wire diameter is shown in Fig. 4 where Curve 1 was obtained with a 1 mil chromel-alumel thermocouple while Curve 2 was obtained with a 20 mil chromel-alumel thermocouple on the rear surface of a SS sample. The diffusivity value obtained using the 20 mil thermocouple is a factor of three off.

Comparison of diffusivity values obtained using an i.r. detector, spot-welded thermocouple and spring-loaded thermocouples is given in Table 1. It can be seen that it is possible, using good techniques, to get the same results using thermocouples and i.r. detectors.

TEMPERATURE RISE EFFECTS IN THE LASER FLASH DIFFUSIVITY TECHNIQUE

The laser flash diffusivity technique assumes that the temperature rises are small. Then the temperature dependency of the properties are negligible and it can be assumed that the rear face detector output is a linear function of temperature. If the rear face temperature rise is

Table 1. Comparison on Diffusivity Values Obtained Using Different Temperature Detectors

Sample (Type)	Thickness (cm)	Diffusivity (I.R. Detector)	Diffusivity (WTC)[a]	Diffusivity (SLTC)[b]
S Steel	0.254	0.0364	0.0123[c]	----
S Steel	0.254	0.0364	0.0366	0.0345
S Steel	0.635	0.0383	0.0382	0.0365
Iron	0.127	0.205	0.202	0.183
Iron	0.635	0.226	0.225	0.220
Mo	0.152	0.495	0.465	0.423
Mo	0.330	0.538	0.530	0.503
Graphite	0.318	1.00	----	0.097
Ceramic	0.508	0.0253	----	0.0263
Ceramic	0.508	0.0675	----	0.0642
Asbestos-X	0.2521	0.00815	----	0.00824
Asbestos-Z	0.2521	0.00482	----	0.00488

[a] WTC = welded thermocouple (0.002 inch diameter).
[b] SLTC = spring-loaded thermocouple.
[c] 0.020 inch diameter.

large, experience has shown that the non-linearity of the i.r. detector causes the largest errors. Figure 5 shows the normalized i.r. voltage detector rise for a thin carbon bonded carbon fiber sample. Normally we consider the voltage output to be linear related to temperature and use this in the analysis. However, when the actual temperature rise was computed from the i.r. output (a laborious process), the results are quite different (curve marked "Temperature" in Fig. 5. The use of the detector voltage signals directly leads to substantial errors in this case. By attenuating the laser output, using CuSO$_4$ solutions of various strengths, it was possible to decrease the temperature rise to the point where the i.r. output was a linear function of temperature. This is indicated in Table 2. The results at 23°C show a substantial dependence on laser power (ΔT rise), since rises up to 41°C were observed and the detector is not very sensitive at room temperature. At higher temperatures, where the rise is less due to the increased heat capacity of the sample and the i.r. detector is more sensitive, the dependency is less.

EFFECT OF SAMPLE THICKNESS (LASER BEAM NON-UNIFORMITY)

It was noted that for thin samples of high diffusivity materials, measured values of diffusivity depended upon sample thickness. This is mainly due to laser beam inhomogeneity. A study using two SRM's was undertaken. The results for SRM 1461 (austenitic stainless steel) are given in Table 3. Samples of 0.25 inch thickness yielded results in good agreement with those calculated from the SRM's conductivity values. An important part of the differences between the measured values for the 0.25 inch thick sample and the SRM's quoted values between 200 and 400°C may be due to the use of NBS supplied specific heat values to calculate diffusivity. These values are 1.5% larger than our measured specific heat values in this range. The fact that the differences between the measured

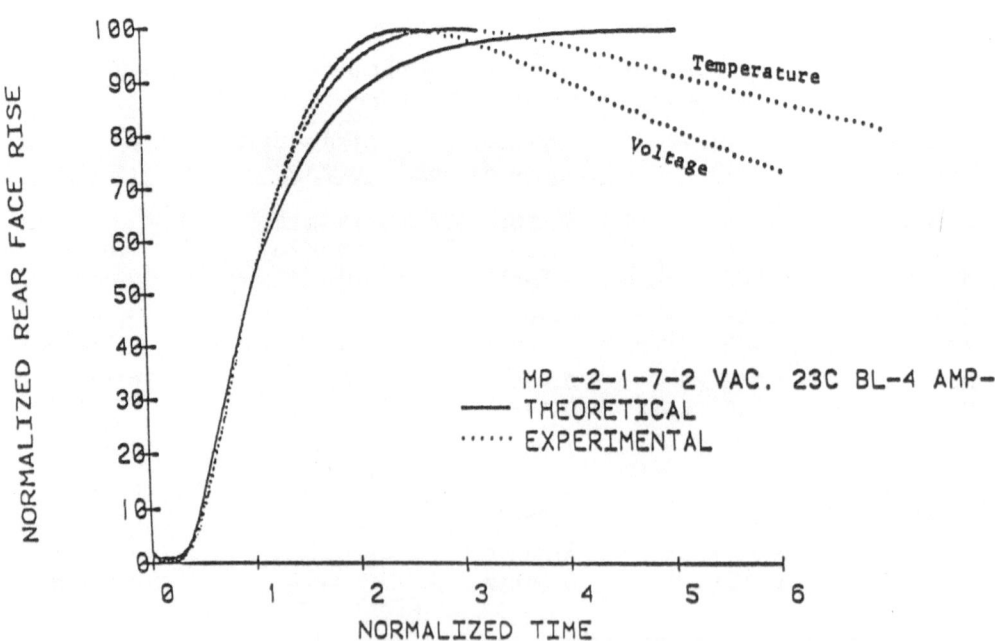

Fig. 5. Normalized Rear Face Temperature Rise Compared to Normalized Rear Face Voltage Rise (MP-2-1-7-2 at 23°C, no absorber).

values persist over an extended temperature range indicating that i.r. linearity is not responsible. It can be argued that thin samples would be affected more by changes caused by machining since the relative sample volume near the surface is larger. However, we have observed this thickness dependence for several different metals and the dependency remains at higher temperatures where annealing could occur. Non-uniformity of the laser beam is a serious difficulty with the flash technique. Attempts to homogenize the beam using fiber optics are now underway.

Table 2. Thermal Diffusivity Results for Sample MP-2-1-7-2 as a Function of Temperature Using Various Liquid Absorber Concentrations

Temp. (^{o}C)	Liquid No.	Max ΔT Rise (^{o}C)	Diffusivity (cm^2 sec^{-1})
23	0	41	0.00315
23	2	17	0.00368
23	4	9.5	0.00393
23	5	6.7	0.00402
23	6	1.2	0.00422
100	0	30	0.00342
100	2	12.4	0.00370
100	4	8.4	0.00380
100	5	4.9	0.00385
200	2	9	0.00356
200	5	3.4	0.00361
200	6	0.6	0.00368
300	2	6.9	0.00340
300	4	3.8	0.00340
300	5	2.7	0.00342
300	6	0.5	0.00342

Table 3. Diffusivity* Results for SRM 1461 (Austenitic Stainless Steel)

Temp. (^{o}C)	0.050 in. Thick	0.100 in. Thick	0.250 in. Thick	Calculated from NBS
23	0.0338	0.0364	0.0383	0.0381
100	0.0371	0.0376	0.0400	0.0409
200	0.0386	0.0403	0.0424	0.0438
300	0.0410	0.0428	0.0446	0.0460
400	0.0435	0.0452	0.0472	0.0488
500	0.0452	0.0481	0.0499	0.0506
600	0.0466	0.0501	0.0522	0.0531
700	0.0497	0.0519	0.0560	0.0551
800	0.0509	0.0536	0.0579	0.0578

* cm^2 sec^{-1}.

In order to measure the thermal diffusivity of samples of transparent or translucent materials by the flash technique, it is necessary to apply an opaque layer to the front surface. In some cases applying a layer so thin that its effect is negligible as far as response time measurements are concerned is possible. However, a thin layer applied to an insulating surface may become very hot during laser firing and either act as a radiant energy source itself or burn off, and thus a thicker layer is preferred. Attaching an opaque layer presents several problems including contact conductance between layers, affect of bond coat, penetration of adhesive into the sample (especially if it is porous), increased heat losses during the experiment, etc. The relative thicknesses and diffusivities of the coating and sample are important and one must use care in optimizing the choice of coating.

Results obtained using three different thicknesses of graphite front layers attached to samples of different diffusivities are shown in Fig. 6. The data points on the far left are for one-layer samples, and the other data points are obtained using two-layer samples. From left to right these data were calculated using no heat loss correction, ratio heat-loss correction method and Cowan heat loss correction method. The results show that errors occur with higher diffusivity samples and the use of thick front layers.

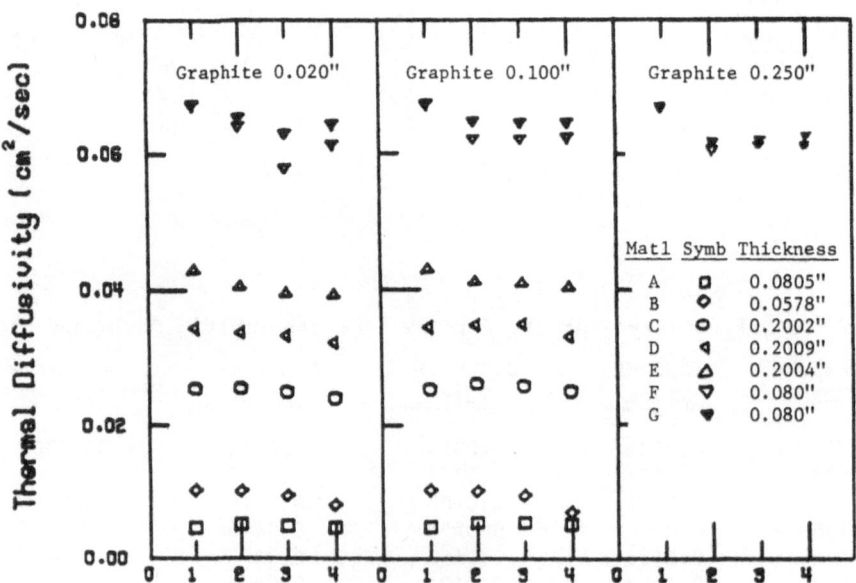

Fig. 6. Results for Diffusivity for Samples of Various Materials Using Different Thicknesses of Graphite Layers.

CARBON BONDED CARBON FIBER (CBCF)

TPRL has performed a number of studies of CBCF material in conjunction with ORNL. Thermal diffusivity values were measured in various atmospheres and temperatures. Originally the laser flash data near room temperature were in error resulting from excessive temperature rises in the small opaque, low density (0.2 gm cm^{-3}), low heat capacity samples. However, where the absorbed power was reduced, the results became reliable. A typical set of data are shown in Figure 7. In this figure, diffusivity values for a sample measured in vacuum (10^{-5} torr), 0.05 atm and 1 atmosphere of nitrogen at various temperatures are shown. Taylor [3] has obtained good agreement between experimental and calculated conductivity values at 1 atmosphere measured in various gases and temperatures using the vacuum values for the solid and literature values for the conductivity of the gases. In addition, ORNL measured the thermal conductivity of CBCF in their radial heat flow apparatus [4]. They concluded that "A series of direct comparisons of several samples between the laser flash technique and radial heat flow technique using identical conditions of temperature and gas atmosphere are in reasonable agreement and shows similar trends between sample-to-sample" (Table 4).

DIFFUSIVITY OF INSULATIONS

TPRL is engaged in extending diffusivity techniques to insulations. This is an area where diffusivity and even conductivity results must be viewed with great skepticism. This is particularly true for fibrous insulations and foams where radiation and gas conduction effects can be significant. In a number of such cases, it is obvious that the laser flash technique is not suitable and we have developed step-heating methods [3,4]. While such methods are much more tedious and time-consuming than the flash techniques, they are still much quicker than steady-state methods. When we use the flash technique to measure low diffusivity/conductivity materials such as teflon and plexiglass, we obtain excellent

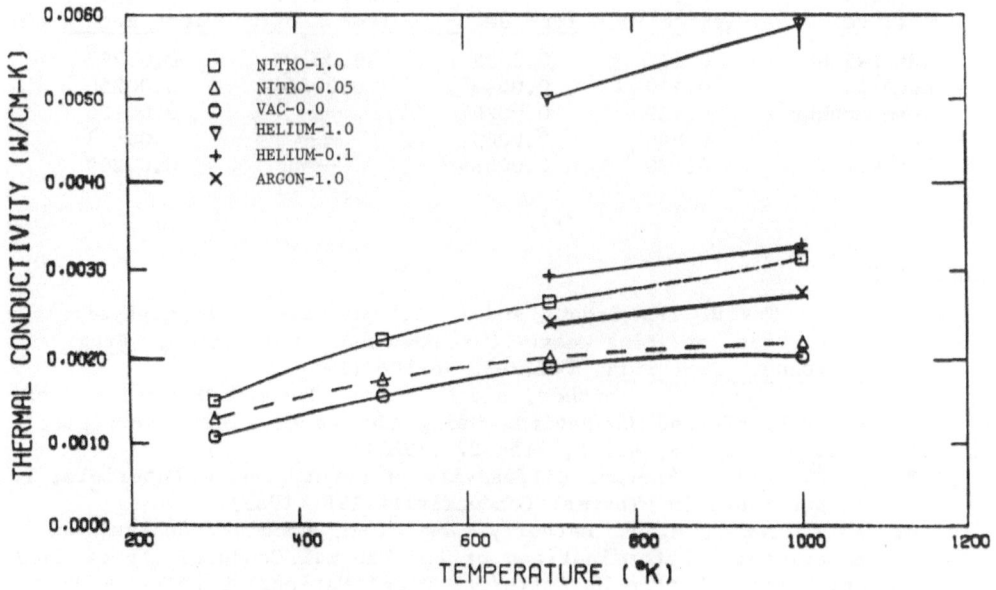

Fig. 7. Thermal Conductivity of CBCF.

results. Even when we measure CBCF we obtain good results. However, our attempts to measure certain insulation materials with the flash technique have not been very satisfactory to date. The laser flash diffusivity results at one atmosphere are usually about a factor of two larger than values calculated from other methods (Table 5). The reasons for this are not clear and we are attempting to ascertain the cause. In particular we wish to make measurements in vacuum and various gases in order to delineate effects due to gas conduction, and we wish to examine the effects of radiation by making measurements at various temperatures and by controlling front face temperature rises.

ACKNOWLEDGMENTS

The author wishes to acknowledge the support of the National Science Foundation, Air Force Office of Scientific Research and Oak Ridge National Laboratory for support of portions of the work described here. In addition, the author wishes to acknowledge the help of the TPRL staff, including Dr. Taylor for his contributions.

Table 4. CBCF Comparison

Temp. (^{o}C)	TPRL Diffusivity (cm^2 sec^{-1})	TPRL Conductivity, Calculated (W m^{-1} K^{-1})	ORNL Conductivity (W m^{-1} K^{-1})
27	0.00845	0.152	----
227	0.00748	0.222	0.223
427	0.00699	0.263	0.259
727	0.00684	0.313	0.297

Table 5. Comparison of Diffusivity Values for Various Insulations

Sample (Type)	Density (gm cm^{-3})	Flash (cm^2 sec^{-1})	Literature (cm^2 sec^{-1})	Probe (cm^2 sec^{-1})
SRM 1450b	0.137	0.00523	0.00294	0.00292
Manville	0.240	0.00467	0.00244	0.00256
Foam Rubber	0.129	0.00261	---	0.00126
Foam	0.046	0.00657	---	0.00423
Blanket	0.129	0.00594	---	0.00209

REFERENCES

1. R. E. Taylor, H. Groot, and R. L. Shoemaker, Thermophysical Properties of Fine-Weave Carbon-Carbon Composites, Prog. Astronaut. Aeronaut., Vol. 83, 96-108 (1982).
2. R. E. Taylor, J. Jortner, and H. Groot, Thermal Diffusivity of Fiber-Reinforced Composites Using the Laser Flash Technique, Carbon, Vol. 23, No. 2, 215-222 (1985).
3. R. E. Taylor, Thermal Diffusivity of Heterogeneous Materials, to be published in "Thermal Conductivity 19" (1985).
4. R. E. Pawel, W. P. Eatherly, and J. M. Robbins, Analysis of Experimental Determinations of the Thermal Conductivity of CBCF Insulation, Martin Marietta/Oak Ridge National Laboratory Rept. ORNL-6302, 65 pp. (1986).

ON THE APPLICATION OF THE LASER FLASH METHOD FOR DIFFERENT MATERIALS

T. W. Wojtatowicz and K. Rozniakowski

Institute of Physics
Technical Univ. of Lodz
Lodz, Poland

ABSTRACT

The work presents theoretical foundations for the Axial and Radial Heat Flow Methods for measuring thermal diffusivity and their applications in experimental investigations. The thermal diffusivity of St3SX steel sample was determined with the aid of the Radial Heat Flow Method. Temperature distribution in the surface layer was monitored with liquid crystal indicators. The Axial Heat Flow Method was used for the evaluation of thermal parameters of porous materials utilized in the building industry, especially gypsum slurry of various porosity. The influence of porosity on thermal diffusivity was observed. The results obtained with both methods are in good agreement with those obtained by other methods. The assumption that the surface heat source model in the case of porous materials is valid is discussed.

INTRODUCTION

The Laser Flash Method (also called Pulse Technique) is a comparatively simple method which enables simultaneous determination of three thermophysical properties of materials: thermal diffusivity α, heat capacity (specific heat) C_p and thermal conductivity k.

Considering the mode of heat transfer and the experimental methods we can distinguish the standard Axial Flash Method (with one-dimensional flow of the heat) proposed by Parker et al.[1], the Radial Flash Method (with two-dimensional flow of the heat) developed and used by Donaldson and Taylor[2] and the new Converging-Thermal-Wave Technique (where the heated surface temperature is monitored) presented by Cielo[3].

The rapid development of the Laser Flash Method in the last decades involved not only theoretical but also experimental investigations of its possible applications. Over the last years this method was implemented for liquids (Tada et al.[4]), molten semiconductors (Taylor et al.[5]), ceramics (Elchinger et al.[6]), composites (laminated and undirectional reinforced composites — Luc and Balageas[7], carbon/carbon fibre composities — Whittaker et al.[8]), cermets (Roth and Smith[9]), and for building materials (hardboard, corkboard, brick, asbestos powder and gypsum — Rozniakowski and Wojtatowicz[10]). The history of the rear surface temperature in the Axial Method was monitored with the use of thermocouple[1,4-10] or photodetector[11] and in the Radial Method with the use of thermocouples[2].

This work shows the possibility of measuring the rear surface temperature with the aid of liquid crystal temperature indicators[12] in the Radial Heat Flow Method's measurements. The authors of this paper propose also the application of the Axial Laser Flash Method for the evaluation of the α coefficient for porous thermoinsulating materials. The sample from gypsum slurry of different porosity (internal structure) was used as a testing material. Methods used up-to-now, eg. Transient Hot Strip Method[13], required powdered samples, which can cause a discrepancy between the values obtained in such measurements and the realistic thermoinsulating properties of "ready-for-use" building elements. The Laser Flash Method allowed the determination of the thermal parameters of the samples from ready-to-use elements.

THEORY OF LASER FLASH METHOD

Thermal conduction under non-equilibrium thermal condition in absence of heat sources in the sample volume is described by:

$$\nabla^2 T(x,r,\varphi,t) = \frac{1}{\alpha} \frac{\partial T}{\partial t} \tag{1}$$

where α is the thermal diffusivity. The sample can be regarded as an infinite plate (with axial symmetry) limited by the thermally insulated planes $x=0$ and $x=L$. Assuming that the heat absorption takes place in the thin film (of thickness g) near the sample surface and there are no heat losses, and that the temperature of this film has Dirac delta function's shape and $Q/(\rho\, C_p\, g)$ value (where Q is the power density of the heat pulse, ρ is the material density and C_p is the specific heat), we can obtain from (1) the temperature of the rear surface of the sample (x=L) described as follows[14]:

$$T(L, t) = Q \cdot (\rho C_p L)^{-1} [1+2\Sigma(-1)^n \exp(-n^2\pi^2 L^{-2}\alpha t)] \tag{2}$$

and $Q/(\rho\, C_p\, L)$ stands for the maximum temperature of the rear surface T_{max}:

In the case when the energy pulse is instantaneous and uniform, the flow of heat is one--dimensional, the surface heat losses are negligible and material properties are temperature independent, the temperature of rear surface is only a function of a Fourier number:

$$Fo = \alpha t/L^2 \tag{3}$$

This yields that dimensionless temperature $\theta(x, Fo)$ has only axial component and is described by equation:

$$\theta(L, t) = T(L, t)/T_{max} = f(Fo) = f(\alpha t/L^2) \tag{4}$$

which gives thermal diffusivity as follows:

$$\alpha = L^2 t^{-1} f^{(-1)}(\theta(t)) \tag{5}$$

where $f^{(-1)}$ means inverse function of a function f defined by a series:

$$f(Fo) = 1 + 2\Sigma(-1)^n \exp(-n^2\pi^2 Fo)$$

Function $f^{(-1)}$ has not an analytical form and may be obtained for example by numerical interpolation of the tabulated values $[z, f(z)]$ or by using polynomial approximations. For example, when $\theta=0.5$ one has $f^{(-1)}(0.5)=0.139$ and from (5)

$$\alpha = L^2 (t_{1/2})^{-1} f^{(-1)}(0.5) \simeq 0.139 L^2 (t_{1/2})^{-1} \tag{6}$$

where $t_{1/2}$ is the time in which the temperature of rear surface increases to half its maximum value, T_{max}. This relation is basic for the standard Axial Laser Flash Method proposed by Parker et al.[1] in which the temperature of the rear surface is monitored in one place, often at the point where the optical axis crosses the sample.

In the case when the diameter of the sample is sufficiently large in comparison to the diameter of the heated region we have two-dimensional flow of the heat. Assuming that there exists an anisotropy of thermal properties of material and that the surface heat losses are linear dependent with temperature the dimensionless temperature θ may be described as a superposition of a radial (φ) and axial (σ) components:

$$\theta(\epsilon, \psi, \text{Fo}) = \varphi(\epsilon, \text{Fo}) \cdot \sigma(\psi, \text{Fo}) \tag{7}$$

where $\epsilon = y/R$ is nondimensional radial position (R is the radius of heated region equal to the radius of laser beam), $\psi = x/L$ is nondimensional axial position, Fo is the Fourier number given by Eq. 3.

The radial component of temperature distribution is given by equation[15]:

$$\varphi(\epsilon, \text{Fo}) = \int_{\epsilon=0}^{R/L} \left(\frac{\epsilon'}{2\text{Fo}}\right) \exp\left(-\frac{\epsilon^2 + \epsilon'^2}{4\text{Fo}}\right) \text{Io}\left(\frac{\epsilon\epsilon'}{2\text{Fo}}\right) d\epsilon' \tag{8}$$

After integration Watt[17] derived the radial distribution of dimensionless temperature described by:

$$\theta(\epsilon, \text{Fo}) = \Sigma \frac{\text{Io}(\omega_n \epsilon) I_1(\omega_n)}{[\text{Io}^2(\omega_n) + I_1^2(\omega_n)]} \exp(-\omega_n R^2 L^{-2} \text{Fo}^{-1}) \tag{9}$$

where ω_n are the solutions of eigenfunction $\omega \text{Io}(\omega) = I_1(\omega) R/L$.
In the axis point ($\epsilon = 0$) function (8) reduces to[15]:

$$\varphi(\theta, \text{Fo}) = 1 - \exp(-0.25 R^2 L^{-2} \text{Fo}^{-1}) \tag{10}$$

The ratio of the temperature at two different radial positions ϵ_1 and ϵ_2 in the rear surface of sample is dependent only on Fo and ϵ (the axial component σ has been eliminated). When ϵ_1 is the point where the optical axis crosses the sample as a reference point and ϵ_2 is the second point outer of laser beam radius we now have:

$$\frac{\theta(\epsilon_2, \text{Fo})}{\theta(\epsilon_1, \text{Fo})} = [1 - \exp(-0.25 R^2 L^{-2} \text{Fo}^{-1})] \cdot \Sigma \frac{\text{Io}(\omega_n \epsilon) I_1(\omega_n)}{[\text{Io}^2(\omega_n) + I_1^2(\omega_n)]} \times$$

$$\times \exp(-\omega_n R^2 L^{-2} \text{Fo}^{-1}) \tag{11}$$

In analogy, as in the Axial Heat Flow Method, the thermal diffusivity can be described:

$$\alpha = R^2 t^{-1} f^{(-1)}(\theta(\epsilon_2, \text{Fo})/(\epsilon_1, \text{Fo})). \tag{12}$$

This is basic equation for the Radial Laser Flash Method, which need the registration of the history of temperature at two points on the rear surface of the sample: the temperature T_r in known radius $r > R$ and the temperature To in the centre of the sample. For the measured ratio $\gamma = \Delta T_r / \Delta \text{To}$ where ΔT_r and ΔTo are the increases of temperature T_r and To, respectively, one can calculate value $f^{(-1)}(\gamma)$ and from Eq. (11) obtain thermal diffusivity of the material.

PULSE–DURATION AND HEAT-LOSS EFFECTS

The problem of radiant and convective heat losses from the sample surfaces, finite pulse-time effects and the procedures for the reduction of experimental results has been discussed by Cowan[16] (the case of different radiation losses at the front and rear surface of the specimen with instantaneous heat pulses and no radial flow). Watt[17] (axially symmetric pulse heating and time independent surface conditions) and Heckman[18] (the coupling of heat losses and finite-time effects) and other authors.
The solution of Eq. 1 in this case (for $t \geqslant \tau$) has the following form:

$$T(L,t) = Q (\rho C_p L)^{-1} \int_0^t F(t) \cdot G(t') dt \tag{13}$$

where F is a pulses shape function and G is the Green's function.

The thermal diffusivity is now a function of the time $t_{1/2}$ and radiation or convective loss parameters. The radiation loss parameter h_x (where x equal 0 or L means the front or rear surface) is given by relationship[18]:

$$h_x = 4L\sigma_B T_{av3} k^{-1} \tag{14}$$

and convective loss parameter by:

$$h_x = L\ h\ k^{-1} \tag{15}$$

where h is the film coefficient, σ_B is the Stefan-Boltzmann constant and T_{av} is the average temperature of point on front or rear surface respectively. Function $f^{(-1)}$ in Eq. 5 has in this case value different from the value obtained for ideal conditions. Cowan[16] proposed a simple technique for the determination of these parameters on the basis of theoretical curves and the experimentally determined ratio $\theta(5t_{1/2})/\theta(t_{1/2})$. Heckman[18] presented some alternative analytical methods and gave results of numerical analysis of the finite pulse and heat loss effects which aids the data correction processes.

CONDUCTION OF HEAT IN POROUS MATERIALS

The conduction of heat by a gas closed in pores is the principal way of heat transfer for ideal porous materials. Convection and radiation are the negligible components in this materials (esp. when the pores have small size). The thermal conductivity of thermoinsulating materials depends on ratio of gas volume V_p and total sample volume V. This ratio is called the total porosity of material:

$$P = V_p/V \tag{13}$$

In practice the pore's walls which are not isothermal surfaces conduct the heat too. Moreover, the humidity (not bounded water vapour included in pores) of the sample must be considered, because the water is vaporized on the "hot" wall and then condensed on the "cold" wall, and the heat is transferred. Thus, the total thermal conductivity of porous material is given by:

$$k = k_f + k_d + k_m \tag{14}$$

where k_f is the thermal conductivity of a gas closed in pores, k_d is the thermal conductivity component of heat exchange caused by water vapourization and k_m is the thermal conductivity of solid phase.

The thermal conductivity of homogenous (non-porous) material may be calculated from the thermal diffusivity α and specific heat C_p measured with the use of Laser Flash Method by:

$$k = \alpha \cdot C_p \cdot \rho \tag{15}$$

where ρ is the density.

In the case of porous materials and heat pulses of short duration the thermal conductivity obtained from (15) would be different from the realistic value for solid phase. Sakai and Hirosaki[19] made correction for the porosity of the compact using Maxwell-Eucken relationship for thermal conductivity of composite material:

$$k_t = k\ (1+P)/(1-P) \tag{16}$$

where k_t is the true value of thermal conductivity (equal to the thermal conductivity of solid phase, t_m) and k is the thermal conductivity obtained from Eq. 15.

Other authors[20] used constant correction factor equal 1.05 for porosity of ceramics and received a good agreement with values given in tables.

EXPERIMENT

The experimental setup for determination of thermal diffusivity of materials (fig. 1) consist of a neodymium glass laser working in free generation regime (laser light length $\lambda=1.06$ μm, pulse duration $\tau \simeq 2$ ms, pulse energy $E \simeq 6$ J) and a unit for measurement and recording the temperature.

In the Radial Heat Flow Method (fig. 1a) the laser beam was focused on the front surface of the sample (focal point was inside the sample) until the melted region had occurred. The temperature distribution in outer layer was monitored with the thin film of liquid crystal indicator (the mixture of propionate chloride and cholesterin oleate) and recorded on negative colour ORWO CHROM film of 15 DIN sensivity with the use of Arriflex film camera. Table 1 includes liquid crystal indicator colours and the corresponding temperatures. Before measurement the sample was held in temperature T_R and the laser beam radius was matched in such a way that the maximum temperature of the rear surface should not exceed the T_v value.

Table 1. The colours of liquid crystal temperature indicator and the corresponding temperatures and wavelengths

Colour	Wavelenght (nm)		Temperature (°C)
Red	640	T_R	30.7
Orange	595		31.3
Yellow-green	575		31.7
Green	535		32.4
Blue	475		33.2
Violet	440	T_v	33.5

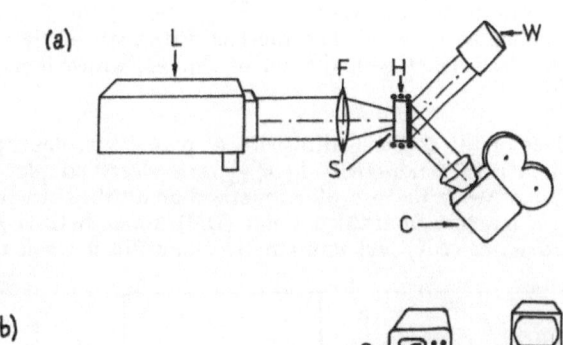

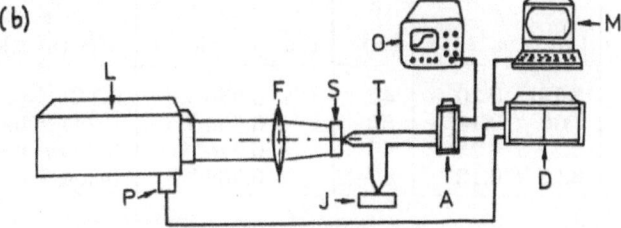

Fig. 1. Experimental setup for: (a) Radial and (b) Axial Laser Flash Method.
L, laser; F, lens; S, sample; P, photodetector; H, heater; T, thermocouple; J, cold junction; A, amplifier; D A/D converter; M, microcomputer; W, light source; C, film camera; O, oscilloscope.

In the Axial Heat Flow Method (fig. 1b) the beam was focused to illuminate the entire sample surface. The temperature of rear surface was determined using the copper-constantan thermocouple, one junction of which was covered with conducting adhesive and pressed close against the sample and the other junction was maintained at $0^{\circ}C$. The electromotive force of the thermocouple connected to the ambient temperature junction was compensated by a constant voltage. The additional component that responded to the temperature increase caused by heat pulse was amplified 1000 times and recorded as a function of time in the memory of the oscilloscope or of the microcomputer. The sample was fixed in a special turbax holder assuring perpendicular projection of light into the sample and a close contact with the thermocouple placed in the sample's center.

The possibility of application of the liquid crystal indicator in the Radial Flash Method was tested with the use of a sample of St3SX carbon steel, a plate 90 mm wide, 80 mm high and 3.7 mm thick. The sample had surfaces made chemically black which improved the laser light absorption and improved the conditions of temperature field visualisation.

The Axial Flash Method was used for determining the thermal diffusivity of gypsum slurry of various porosity. The changes in porosity were caused by using different mass ratios of water and dry gypsum[24] ($CaSO_4$ $2H_2O$) or by using water prepared by distillation and/or magnetisation (with magnetic induction 1 T during 5 minutes). The samples had the shape of disc 15 mm in diameter and about 5 mm thick.

The structural examination of the gypsum samples was carried out with the use of the scanning electron microscope. The structure of the sample fractures was analyzed. The examined surfaces were covered with a thin film of gold by means of the vacuum evaporation technique. The porosity of the sample was measured by using the point method[21] and micrographs.

RESULTS

From the analysis of the distribution of temperature shown on the sequences of movie frames we can obtain the diameters D_p of zones bounded by the isotherms T_R (when the temperature T_v occured in the centre). These diameters are a function of the time starting from the beginning of laser pulse to the time of the exposed frame. The time was determined on the basis of frame number N. The dimensionless radius $\epsilon = r/R = D_p/(2R)$ for different values of time was calculated and the constant $\gamma = \Delta T_r/\Delta To = (To-T_R)/(T_v-T_R) = 0.214$ was determined.

On the basis of curves shown in work Chu et al.[22] (in fig. 2) the Fourier number Fo was determined corresponding to ϵ and γ. The thermal diffusivity of the steel sample equal $0.104 \cdot 10^{-4}$ $m^2 \cdot s^{-1}$ was calculated with the aid of Eq. 11, which is in good agreement with α given in tables[23].

Table 2. Values of the thermal diffusivity α, porosity P, density ρ and thermal conductivity k of gypsum slurry samples obtained with the use of magnetized-un-distilled water (WM), magnetized-distilled water (DM), unmagnetized-distilled water (DN) and unmagnetized-undistilled water (WN).

Sample	α (10^{-7} m^2s^{-1})	P (%)	ρ (10^3 $kg \cdot m^{-3}$)	k (W/(m $\cdot$ K))
WM	3.73 ± 0.20	42	0.950	0.340 ± 0.024
DN	3.06 ± 0.14	50	0.903	0.284 ± 0.020
WN	3.37 ± 0.23	45	0.916	0.314 ± 0.028
DM	3.25 ± 0.12	47	0.909	0.304 ± 0.018

Using the Axial Flash Method we obtaiend (from recorded temperature history curves) time $t_{1/2}$ and calculated the thermal diffusivity for four samples from gypsum slurry applying Eq. 6. Nine measurements were made for each sample. Table 2 gives mean values of thermal diffusivity and values of porosity and density. Table 2 also presents thermal conductivity k of gypsum samples obtained from Eq. 14 with the use of specific heat value for gypsum of $C_p = 1031$ J/(kg K) given in tables[25].

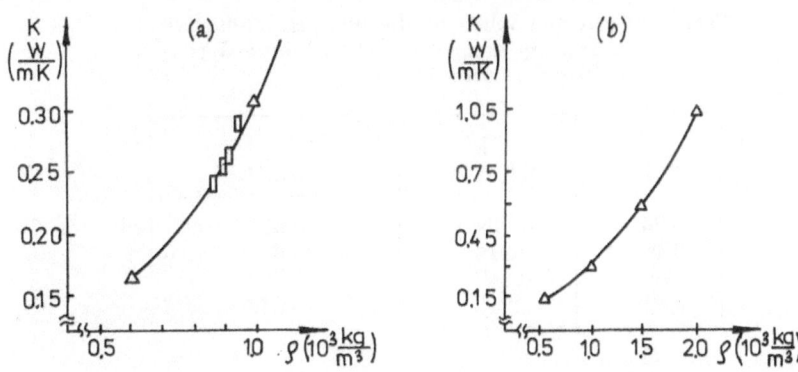

Fig. 2. Thermal conductivity of gypsum slurry (a) obtained experimentally (□) and given in tables[25] (△); and (b) given in tables[25].

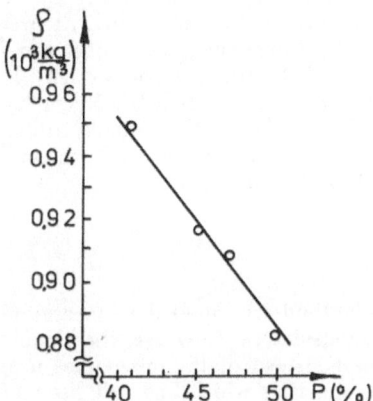

Fig. 3. The plot of gypsum slurry density ρ versus porosity P.

The thermal conductivity of solid phase k_m obtained from relationship (16) is approximately equal 0.8 W/(m · K) (see Table 3) which compares favorably with the value given in tables[25] for gypsum of $1.6 \cdot 10^{-3}$ kg · m^3 density but not for crystalline $CaSO_4 \cdot 2H_2O$.

DISCUSSION

Since the analysis of the gypsum samples structure was made from selected photographs of fracture surfaces the values of porosity are only approximate. Real porosity can be obtained by analysis of many photographs made for a few fractures of the same sample. Thus, we can observe that the sample made with the use of undistilled-magnetized water (WM) has the highest porosity and the sample with undistilled-unmagnetized water (WN) has the lowest porosity, which is confirmed by the changes in density.
From equation (13) one can obtained that the porosity is approximately defined by[26] $P = 1 - \rho_p/\rho_0$ where ρ_p is the density obtained from the total mass and the geometrical volume of sample and ρ_0 is the density of solid phase. The dependence between density and porosity is collinear and as shown in fig. 3 the plot of experimental values has similar character.
The energy pulse duration τ is comparable to the $t_{1/2}$ time in some applications of the Laser Flash Method. In such cases the numerical factor in Eq. 6 increases[18]. In investigations with the use of Axial Method, presented here the ratio $\tau/t_{1/2}$ is lower than 1/6000 (since the time $t_{1/2}$ equals 6 s for gypsum samples) and Parker's criterion $\tau \ll t_{1/2}$ is

Table 3. Corrected values of the thermal conductivity of gypsum samples k and the thermal conductivity of solid phase k_m.

Sample	k (W/(m K))	k_m (W/(m K))
WM	0.289	0.832
DN	0.214	0.852
WN	0.267	0.827
DM	0.258	0.843

satisfied. From this condition it follows that an error of the thermal diffusivity evaluation is low (since Heckman[18] calculated taht this error was not greater than 2% when the ratio $\tau/t_{1/2}$ equaled 0.05).

On the other hand, time $t_{1/2}$ is so long that the heat-loss effects have occurred. Since the temperature of the sample is low the radiation losses are negligible but the convective losses are significant. The corrections for thermal diffusivity were made with the use of α technique proposed by Heckman[18]. Correction factor obtained in this way was equal 1.17. Table 3 presents corrected values of the thermal conductivity k. The values obtained in this way are in good agreement with those (k = 0.314 W/(m K) for $\rho = 1.0 \ 10^3$ kg m^{-3} and k = 0.176 W/(m K) for $\rho = 0.6 \ 10^3$ kg m^{-3}) found by other methods[25] (see fig. 2a). The shape of experimentally obtained curve in figure 2a is similar to those obtained on the basis of values given in tables[25] (fig. 2b).

CONCLUSIONS

The well-known Laser Flash Method was applied in measurements of thermophysical parameters of metals and thermoinsulators. However, the liquid crystal indicator recording temperature distribution was used instead of thermocouples usually applied in investigations of metals. Several porous materials which had not been often examined were also investigated. As an example the results for St3SX steel and a few kinds of gypsum slurry were presented. The values obtained in these measurements were in good agreement with those found by other methods.

The results indicate that the liquid crystal temperature indicators applied in the experiment enable to record temperature fields generated in metals by pulse laser radiation. Such a way of temperature recording can be used to evaluate thermal diffusivity of metals by means of Radial Heat Flow Method.

The Axial Heat Flow Method can be used to evaluate thermal diffusivity and thermal conductivity of several kinds of gypsum slurry. It may be expected that this method can be useful in investigations of thermophysical parameters of other similar thermoinsulating materials. To verify this supposition numerous and comprehensive experiments are required.

This work was sponsored by the Polish Ministry of Science and School of Academic Rank under Research Project CPBP 01.06.

REFERENCES

1. W.J. Parker, R.J. Jenkins, C.P. Butler, G.L. Abbott, Flash Method of Determining Thermal Diffusivity, Heat Capacity and Thermal Conductivity, J. Appl. Phys. 32:1679 (1961).
2. A.B. Donaldson, R.E. Taylor, Thermal Diffusivity Measurement by a Radial Heat Flow Method, J. Appl. Phys. 46:4584 (1975).
3. P. Cielo, In Situ Laser Thermal Analysis of Bulk Materials, J. Thermal Anal 30:33 (1985).

4. Y. Tada, M. Harada, M. Tanigaki, W. Eguchi, Laser Flash Method for Measuring Thermal Conductivity of Liquids — Application to Low Thermal Conductivity Liquids, Rev. Sci. Instrum. 49:1305 (1978).

5. R.E. Taylor, L.R. Holland, R.K. Crouch, Thermal Diffusivity Measurements on Some Molten Semiconductors, High Temp.-High Press., 17:47 (1985).

6. M.S. Elchinger, C. Martin, and P. Fauchais, Measurement of the diffusivity of ceramic materials either sintered or sprayed by means of a flash method with a positive or negative pulse, Rev. Int. Htes Temp. et Refract, 16:317 (1979).

7. A.M. Luc, D.L. Balageas, Comportment thermique des composites a renforcement oriente soumis a des flux impulsionnels, High Temp. — High Press. 16:209 (1984).

8. A. Whittaker, R. Taylor, H. Tawil, Thermal Diffusivity of some Fine-Wave Carbon/ /Carbon-Fibre Composites, High Temp. — high Press. 17:225 (1985).

9. E.P. Roth, M.F. Smith, Determination of Thermal Conductivity from Specific Heat and Thermal Diffusivity Measurements of Plasma Sprayed Cermets, Int. J. Thermophys. 7:455 (1986),

10. K. Rozniakowski, T.W. Wojtatowicz, Thermal Diffusity of Building Materials Measured by the Laser Flash Method, J. Mat. Sci Lett. 5:996 (1986).

11. H.L. Lee, D P.H. Hasselman, Comparison of Data for Thermal Diffusivity Obtained by the Laser-Flash Method Using Thermocouple and Photodetector, J. Am. Ceram. Soc. 68:C12 (1985).

12. K. Rozniakowski, A. Drobnik, A. Lipinski, On the Application of the Liquid Crystals to the Visualisation of Temperature Fields, Optica Applicata, 16:253 (1986).

13. R. Singh, N.S. Saxena, D.R. Chaudhary, Simultaneous measurement of thermal conductivity and thermal diffusivity of some building materials using the transient hot strip method, J. Phys. D: Appl. Phys. 18:1 (1985).

14. H.S. Carslaw and J.C. Jaeger, "Conduction of Heat in Solids", Oxford University Press, New York (1959).

15. A.B. Donaldson, Radial conduction effects in the pulse method of measuring thermal diffusivity, J. Appl. Phys, 43:4226 (1972).

16. R.D. Cowan, Pulse Method of Measuring Thermal Diffusivity at High Temperatures J. Appl. Phys. 34:926 (1963).

17. D.A. Watt, Theory of thermal diffusivity by pulse technique, Brit. J. Appl Phys. 17:231 (1966).

18. R.C. Heckman, Finite pulse-time and heat loss effects in pulse thermal diffusivity measurements J. Appl. Phys. 44:1455 (1973).

19. T. Sakai and N. Hirosaki, Properties of hot-pressed barium-doped SiC, J. Mat. Sci. Lett, 5:43 (1986).

20. T. Mitsuhashi, H. Tanaka and Y. Fujiki, Thermal Properties of Sintered Potassium Hexatitanate, Yogyo-Kyokai Shi. 90:676 (1982).

21. A.G. Guy, "Introduction to Materials Science", McGraw-Hill Book. Co., New York (1979).

22. F.I. Chu, R.E. Taylor, A.B. Donaldson, Thermal Diffusivity Measurements at High Temperatures by the Radial Flash Method, J. Appl. Phys. 51:336 (1980).

23. H. Chelminska et al,. "Charakterystyki stali" seria A, tom I, Wydawnictwo Slask", Katowice (1975).

24. K. Rozniakowski, T.W. Wojtatowicz, On the Possibility of the Application of a Laser Flash Method to Evaluate the Influence of the Gypsum Structure on the Thermal Diffusivity, Mat. Sci. Eng. 96:321 (1987).

25. I.K. Kikoin, "Tablicui Fizicheskih Velichin", Atomizdat, Moscov (1976).

26. G.A. Aksielrud, M.A. Altszuler, Ruch masy w ciałach porowatych', Wydawnictwa Naukowo-Techniczne, Warszawa (1987).

A COMPARATIVE HEAT FLOW METER

MEASUREMENT IN STEADY AND VARIABLE

HEAT TRANSFER CONDITIONS

Miroslav Cvetković

Institut for Testing
Materials SR Serbie
Beograd, Yugoslavia

ABSTRACT

Brief explanations are given of the principle of measurements of heat flow with a new method of measurements i.e. with compensated heat flow sensors.

The usual way of measurements of heat flow without compensation, affects the measured sample like an additional thermal insulation. If the sensor is placed only on one part of the sample, the thermal resistance of the sensor produces a non-linear heat flow passing the sample and the heat flow meter. This is opposite to the initial assumption of the method of auxiliary wall i.e. heat flow meter.

The compensation is made with a separate surface heater placed on the meter's surface, which eliminates the influence of the sensor's thermal resistance on the measured in the sample gradient.

The paper gives the results of comparative measurements of heat flow sensors with and without compensation in variable and steady-state heat transfer conditions.

The paper shows that this new kind of measurement with compensation gives very good results in variable heat transfer conditions.

INTRODUCTION

The measurement of heat flow is usually conducted with the aid of an auxiliary wall[1] or heat flow meters[2].

By putting the sample meter on the sample that is being measured, thermal flux is disturbed in two ways. First, the thermal resistance of the heat flow meter affects the sample as additional thermal insulation. Second, assuming that the heat flow meter is placed only on one part of the sample surface, its thermal resistance produces a nonlinear heat flow passing through the sample and the heat flow meter. This phenomenon produces a systematic error in thermal flux measurements, which is known as thermal shunting.

To reduce these errors some refinements of heat flow meters are made (concerning their thickness, sensitivity or use of a guard ring).

To eliminate these errors, a new kind of thermal flux measurement has been developed that uses a heat flow meter that compensates for the disturbance of temperature field through the sample. The compensation is made by a separate surface heater placed on the meter's surface.

PRINCIPLE OF MEASUREMENTS WITH A COMPENSATED HEAT FLOW METER

Fig.1 gives the cross-section of an isotropic and homogeneous plate or sample and the temperature distribution throughout this plate in steady-state heat transfer conditions. The letters mark the temperature on different points throughout this plate.

A new distribution of temperature throughout this sample, a new additional layer (heat flow meter) is placed, when on the warmer side of the whole surface of the sample is shown in Fig.2. The dotted line gives the former temperature distribution given in Fig.1. The temperature t_1 and t_2 of the air are the same as in Fig.1. In plane 2-2 the temperature decreases from θ_2 on θ'_2. The heat flow meter affects the plate like an additional source of cold.

The heat flow meter usually has a smaller surface area than the surface of the measured sample. This case is presented in Fig.3. The average temperature under the heat flow meter is marked as θ_s and under the center of the heat flow meter as θ_m[3,4,5] The line of temperature distribution is curved, that means, that supposed linear heat transfer through the sample and heat flow meter is not valid.

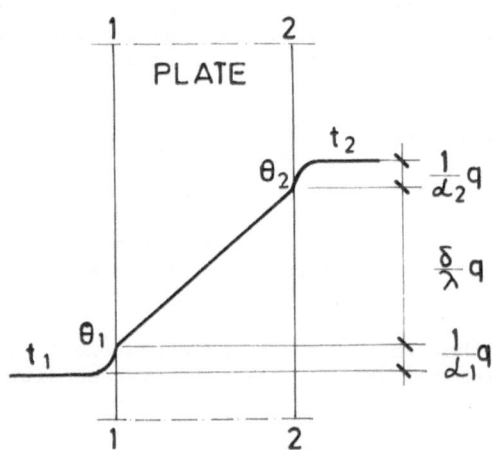

Fig.1. Schematic presentation of temperature distribution.
        Steady state

To compensate for this disturbance caused by the heat flow meter, it is necessary to increase the temperatures of samples to their initial values θ_1 and θ_2, which can be attained by placing an additional source of heat in plane 3-3. This compensating heater, its view and its influence are given in the Fig.4a and Fig.4b.

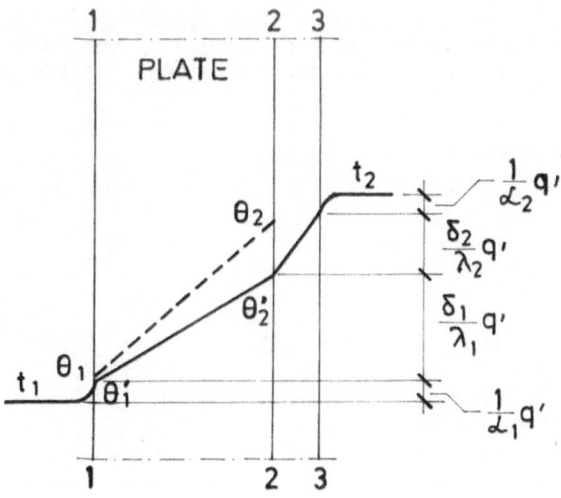

Fig.2. Schematic presentation of temperature distribution
through the plate and heat flow meter. Steady state.

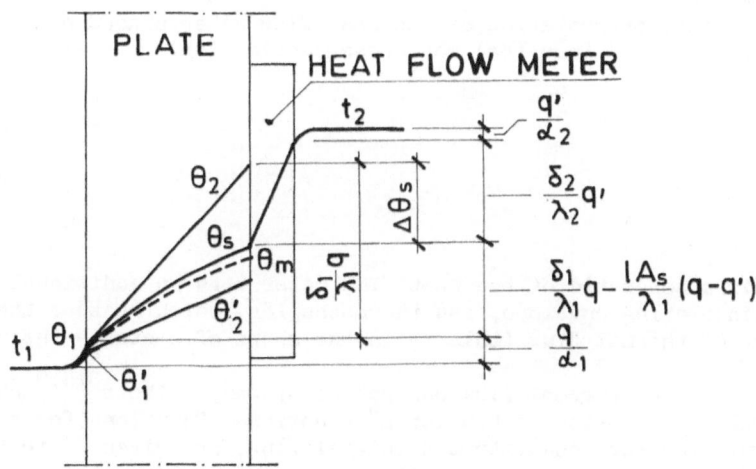

Fig.3. Schematic presentation of temperature distribution.
Steady state.

The control of this heat source has to be adjusted so that the same thermal flux passes through the sample as without the heat flow meter, i.e., through the sample away from the surface covered with heat flow meter (when the sample is homogeneous). Greater or less heating of the heater causes a radial flow of heat through the sample. This radial conduction of heat is the greatest on the surface of the sample covered by the heat flow meter and appears first, so it can be used for the control of the compensation heater*.

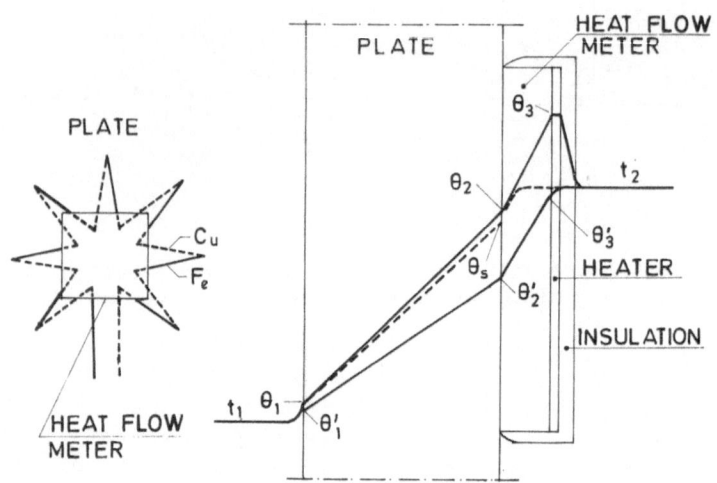

Fig. 4a   Fig. 4b
Schematic presentation of the heat flow meter placed on the sample. A face (4a) and cross-section (4b)

Fig.4 is a schematic of the heat flow meter with an additional source of heat, its insulating envelope, and thermocouples, which measure the radial component of thermal flux (this is in the shape of a star - thermopile).

Radial flow of thermal flux was solved by many authors[3,4,5] on the basis of Maxwell's[5] solution of Fourier's[6] equations. Equations for correction of thermal flux for non-disturbed initial flux, are given in the literature[7,8]. Comparison of the results for measured and corrected values are given in the literature[8] as well.

---

* If the heat flow meter is put on the colder side of the sample, similar relationships are obtained. The heat flow meter acts as a source of heat and for its compensation the "cooling foil" is needed.

MEASUREMENT

To establish the possibility of in-situ heat flow measurement with thermal compensation through the walls of a building, the measurements have been made on the samples in steady and variable states of heat conduction. The same samples and heat flow meters are used for comparative measurements with and without thermal compensation. The same samples are measured with a standard guarded hot plate apparatus.

The measurements with heat flow meters have been made in a square box chamber. Its thermal insulation are panels of expanded polystyrene. A heating aluminium plate is placed under the chamber's cover, and an identical cooling plate is placed on the floor. Thermostats are connected with both plates. A soft rubber plate is placed over the lower cooling plate.

The sample is placed over the plate of soft rubber. The heatflow meter can be lowered of lifted from the sample by a handle, which is manipulated outside the chamber.

The dimensions of the heat flow meter (Schmidt's belt) used in the chamber are 512 x 58 x 6 (mm). The heater for thermal compensation has the same surface as the meter. The meter and heater are put in an insulated little box for measurements with compensation.

The state of heat transfer is made variable by changing the temperature of the water in the lower cooling plate. The following wall samples are used for the measurements:

1. concrete of normal granulation:
   density 2100 kg/m³;
   dimension 750 x 750 x 50 (mm)

2. a wall of hollow brick (36% cavity) with a total of 1 cm of plaster on both sides
   density 1300 kg/m³;
   dimension 750 x 750 x 75 (mm)

3. expanded polystyrene;
   density 15,3 kg/m³;
   dimension 750 x 750 x 20 (mm);

The output signal $K\Delta\theta_{so}$ of differential temperature thermopile is measured in steady state on the surface of the sample, before positioning the heatflow meter. After positioning the heatflow meter the output signal becomes $K\Delta\theta_s$. Then the output signal is adjusted on the value $K\Delta\theta_{so}$ by the compensation heater. In this way the radial component of the thermal flux (caused by surplus of shortage of thermal flux through the heat flow meter, in relation to the largeness of the flux before putting the heat flow meter) in annulled.

The measurements of the thermal flux, the temperature and heating control are made with a data acquisition system. Each of the values is measured or changed every 5 minutes. The measurements with compensation lasted usually 1 to 2 hours with such slow control.

The measurements of heat flow meter with no compensation are made by placing the same heat flow meter on the same position on the sample. The measurements are made in the interval of some hours, until the values become established (this usually happens after more then 24 hours after positioning of the heat flow meter).

It is much more difficult to obtain the initial output signal $K\Delta\theta_{so}$ of differential temperature thermopile during the variable state of heat transfer passing the sample than during its steady state.

An integration of temperature $K\Delta\theta_s$ during at least one full cycle of oscillation of temperature of the sample is made in order to obtain the output signal $K\Delta\theta_{so}$.

EXPERIMENTAL RESULTS

Results of measurements of the heat flow meter with and withount thermal compensation are given in Tab. 1 for steady and variable states of heat transfer. The measurements given in Tab.1 compare the results obtained with heat flow meters with and without compensation in steady and variable state of heat transfer through the corresponding samples and the results obtained with guarded hot plate apparatus (these results serves as an etalon).

Fig.5 to 10 give the diagram of variation of measured values in variable state of heat conduction with and without compensation, as a function of time (start of measurements at time t=0), where:

θ_1, θ_2 – sample's surface temperatures (bottom and upper)
$K\Delta\theta_s$ – value proportional to the difference of sample's surface temperature (due the radial component of heat flux)
$\emptyset$ – heat flux
td – temperature of water in cooling plate

The results of measurements with no compensation in steady and variable state of heat transfer are given in table 1, where:

R – thermal resistance of the sample
$K\Delta\theta_{so}$ – initial value proportional to the difference of the sample's surface temperature (before the heat flow meter is put on the sample)
$K\Delta\theta_s$ – average value proportional to the difference of the sample's surface temperature (during the measurements).

As can be seen in Tab.1, lower thermal resistances are obtained in the measurements with compensation than without compensation. These differences can be considerable. Smaller relative differences are obtained with the same way of measurement (with or without compensation) between thermal resistances in steady and variable states of heat conductivity for the same samples. These differences are practically within the limits of error of measurement with heat flow meters.

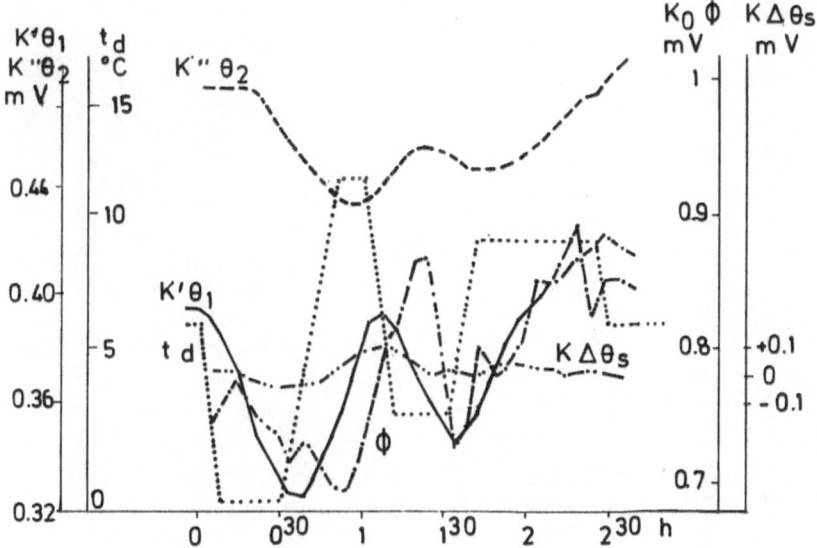

Fig.5. Variation of measured signals as a function of time at variable state for heat flow measurements on a sample of concrete with compensation

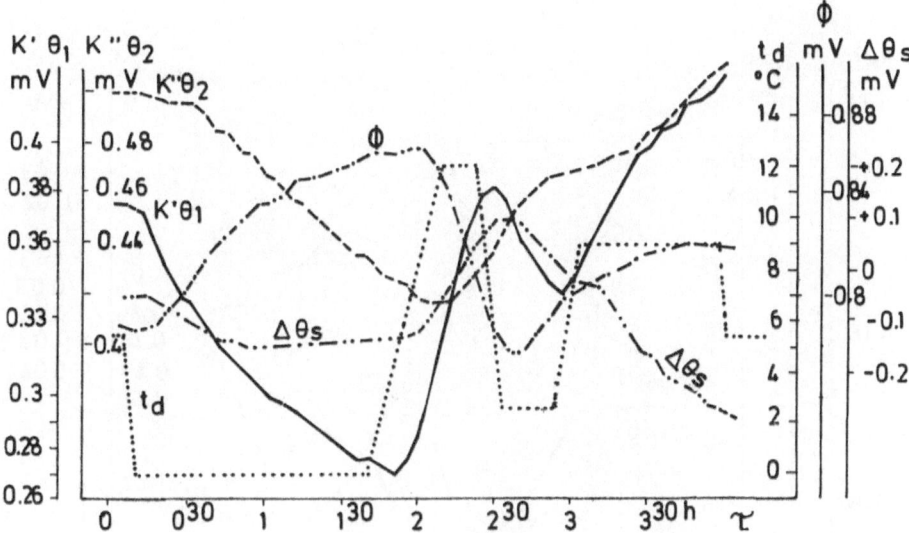

Fig.6. Variation of measured signals as a function of time at variable state for heat flow measurements on a sample of concrete without compensation.

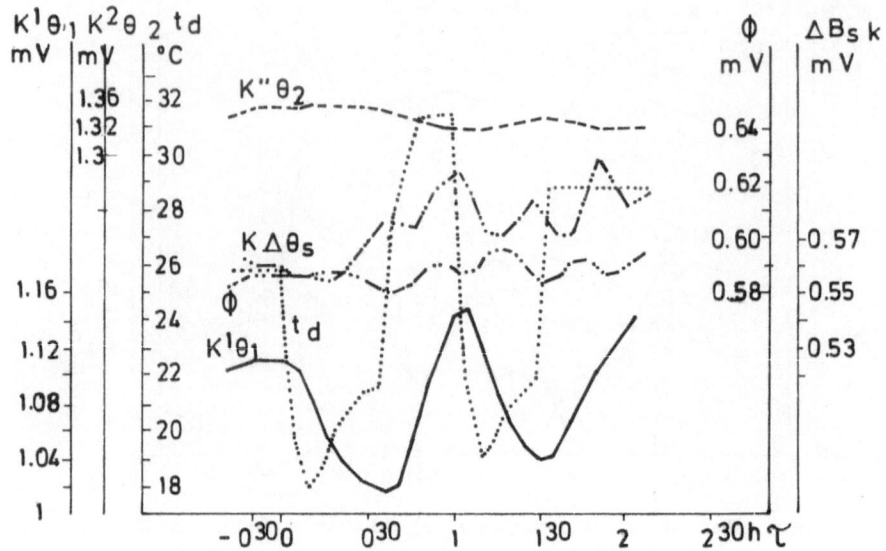

Fig.7. Variation of measured signals as a function of time
at variable state for heat flow measurements on wall
of hollow bricks with compensation

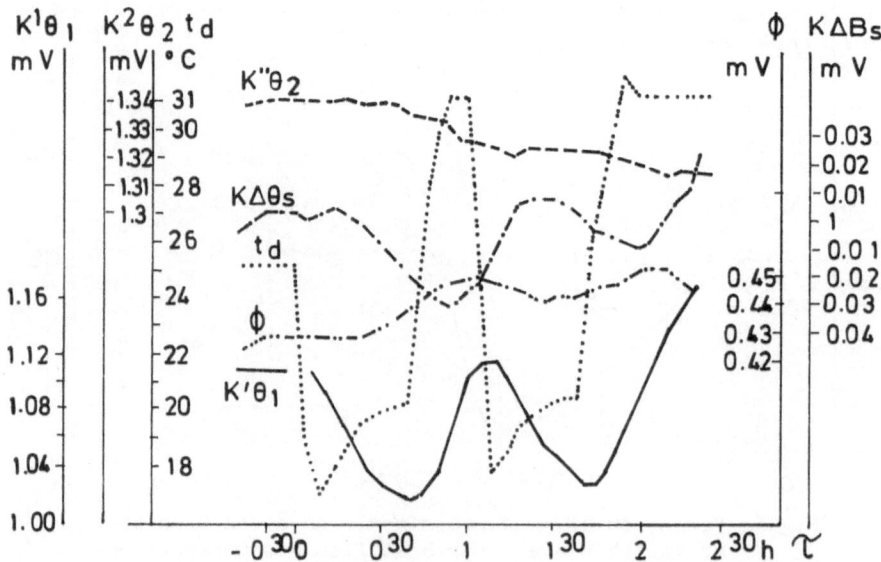

Fig.8. Variation of measured signals as a function of time
at variable state for heat flow measurements on wall
of hollow bricks without compensation

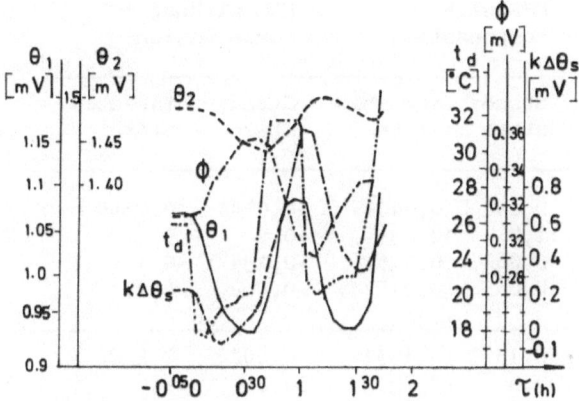

Fig. 9. Variation of measured signals as a function of time at variable state for heat flow measurements on sample of expanded polystyrene with compensation.

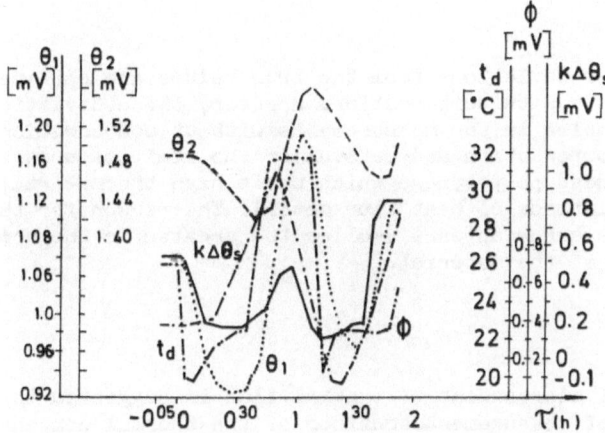

Fig. 10. Variation of measured signals as a function of time at variable state for heat flow measurements on sample of expanded polystyrene without compensation.

Table 1. The heat flow meter measurements with and without compensation under steady and variable state for heat flow measurements and steady state for guarded hot plate apparatus.

Sample	Quantity[*]	Measured value				
		HMF with compensation		HFM without compensation		Guarded hot plate
		Steady state	Variable state	Steady state	Variable state	Steady state
Concrete	$R(m^2 K/W)$	0,042	0,0407	0,0541	0,0544	0,0434
	$\theta(^oC)$	11,8	10,23	10,9	9,8	28,0
	$K\Delta\theta_{so}$ (mV)	0,064	0,016	−0,134	−0,134	−
	$K\Delta\theta_{s}$ (mV)	0,065	0,015	−0,062	−0,071	−
Hollow brick (wall)	$R(m^2 K/W)$	0,1543	0,1531	0,2022	0,2226	0,1531
	$\theta(^oC)$	30,3	29,7	30,3	29,6	32,8
	$K\Delta\theta_{so}$ (mV)	0,236	0,555	−0,070	−0,070	−
	$K\Delta\theta_{s}$ (mV)	0,236	0,556	−0,006	−0,009	−
Expanded polysty-rene	$R(m^2 K/W)$	0,4866	0,4651	0,5155	0,5168	0,5025
	$\theta(^oC)$	31,5	30,9	31,5	30,9	30,2
	$K\Delta\theta_{so}$ (mV)	0,415	0,511	0,065	0,065	−
	$K\Delta\theta_{s}$ (mV)	0,415	0,335	0,208	0,230	−

Relatively few deviations from the true values are obtained in all cases in measurements with compensation. However, these deviations are higher in two of three samples in the measurements without compensation. The results obtained without compensation are outside of the limit of error (except in the sample of expanded polystyrene which has a high thermal resistance relativ to the thermal resistance of heat flow meter). The reason for this is a systematic error. This error appears smaller the greater is the homogeneity and thermal resistance of the material.

CONCLUSION

A new kind of measurement of thermal flux is suggested in earlier[8,9] papers. This kind of measurement consists of the thermal compensation of thermal flux alterations caused by the thermal resistance of the heat flow meter itself.

It is established[8] that the thermal resistance of the heat flow meter affects the sample as a corresponding source of cooling, when put on the warmer side of the partition. An auxiliary source of heat can compensate for the influence of the source of cooling.

---

[*] θ mean sample temperature; see text for other quantities.

The auxiliary source of heat increases the flux through the sample back to what its original value would have been if the heat flow meter were not put on the sample. The method of measurement, in principle, becomes a null method.

The experiments prove that the measurements with compensated heat flow meter in variable state of heat transfer can measure at least as well as without compensation.

The results of measurements with compensation are practically independent of thermal characteristics of the samples and of the mode of heat transfer (zero method). They are inside the limits of error of the equipment with the results obtained in guarded hot-plate apparatus. The results obtained without compensation are (except for expanded polystyrene) outside of these limits. That means that a systematic error exists in the measurements without compensation.

It is not necessary to have a guard-ring arround the measuring surface of the heat flow meter[8]. This surface can be very small and can measure thermal bridges.

It is easier and faster to make a new temperature distribution through the small and thin heat flow meter, than through the much heavier and larger sample. Because of that the measurements by heat flow meter with compensation are faster than without it.

That means that we can measure with compensation practically the instantaneous values of heat flux, and with appropriate mathematical methods find the true values of the in-situ thermal resistance of the samples.

REFERENCES

1. K. Hencky, Ein einfaches praktisches Verfahren zur Bestimmung des Wärmeschutzes verschiedener Bauweisen, Gesundheits Ing.Bd 42, 437(1919).

2. E. Schmidt, Ein neuer Wärmeflussmesser und seine praktische Bedeutung in der Wärmeschutztechnik, Mitteilungen aus dem Forschungsheim für Warmeschutz, e.V. München, Heft 3, 19 (1923)

3. E. G. Loewen and M. C. Shaw, On the Analysis of Cutting-Tool Temperatures, Transaction of ASME, Cambridge, Mass., Vo. 76, 217-231 (1954).

4. Oosterkamp, Philips Res. Rep. 3, 49, (1948).

5. H. S. Carslaw and J. C. Jaeger, Conduction of Heat in Solids, in Oxford at the Clarendon Press (1959).

6. J. B. J. Fourier, Theorie analytique de la chaleur, Paris (1822).

7. E. Raisch und K. Schropp, Die Thermo-elektrische Temperatur-und Wärmeflussmessung, Mitteilungen des Forschungsheimes für Wärmeschutz, e.V.München, Heft 8 (1930)

8. M. Cvetković, Thermal Flux Measurements with a Compensated Heat-Flow Meter, ASHRAE/DOE Conference Proceedings, 749-766 (1982).

9. M. Cvetković, Thermal Flux Measurements with a Compensated Heat Flow Meter at Elevated Radiation Influence, High Temperature-High Pressures, Vol. 18 (1986).

TRANSIENT MEASUREMENTS OF INSULATION MATERIALS

S. C. Bae*

Visiting Professor
Thermophysical Properties Research Laboratory
School of Mechanical Engineering
Purdue University
West Lafayette, IN

ABSTRACT

Several transient techniques were used for measuring the thermal diffusivities of fibrous insulating materials. The transient techniques used in the present study included the constant temperature plate method and the undetermined heat flux method.

The mathematical model of the former is based on a semi-infinite slab with finite thickness, with the front face of the slab modeled at constant temperature and the rear face as being maintained at the initial temperature. At time zero, the sample was placed for several minutes on a hot plate maintained at constant temperature. The temperatures of both sample faces were measured as a function of time to confirm the boundary conditions. Temperatures of an interior position of the sample were measured and were used to calculate the thermal diffusivity. The boundary conditions were satisfied during the time from zero to the time just before the rear face temperature started to rise.

The undetermined heat flux method can be applied to the problem of heat conduction using any kind of heat flux. In this method the transient temperatures inside the materials are measured rather than the heat flux. These actual temperature measurements are utilized as the boundary and the initial conditions in the calculational procedures. A simple projection lamp was used for the input heat flux.

The thermal diffusivities of several insulators were obtained through 300°C. The results obtained by the various techniques were intercompared.

INTRODUCTION

Although the flash method [1] is a well established technique for measuring the thermal diffusivity of many materials, this method is not suitable for the measurement of some materials with poor thermal diffusivity such as fibrous insulating materials. The temperature of the

---

*Dept. of Mechanical Engineering, Dankook University, Seoul, Korea.

surface which received the radiant heat pulse may increase substantially due to the low diffusivity and short pulse time duration, and these materials are often translucent permitting volume energy absorption.

Several methods were suggested in the literature in order to overcome such problems [2-5]. A step-wise heating method for measuring the diffusivity of insulating materials usually requires longer measuring times and larger sample sizes than those for the flash technique. Therefore, the heat losses of sample faces should be considered significant. The surface of any heated material that is not located in a vacuum loses heat to the surroundings by the simultaneous action of convection and radiation mechanisms. When the average film temperature of the surface of a sample and its surrounding temperature are maintained at about 100°C, both free convection and radiation heat loss have the same order of magnitude. But the heat loss by radiation increases very rapidly with increasing average film temperature whereas the heat loss by convection remains almost the same. Under the same conditions, for example, if the average film temperature reaches the neighborhood of 500°C, the heat loss by radiation increases to about three times the convection heat loss. Thus, when the diffusivity measurement is conducted at a high temperature the radiation heat loss should be considered. Hence, the radiation heat loss should be decreased by decreasing the measurement times, sample sizes and the emissivity of surface.

The mathematical analyses of the heat conduction problem are highly dependent upon the heating method. The method allowing a sample to absorb heat through its surface such as the flash method includes a heat source term in the heat conduction equation. In the case of step-wise heating method, the heat flux at the front face is assumed to be constant. However, the emission from the front face may increase as the face temperature is increased during the measurement while the radiation heat flux remains constant. This leads to an obvious error in the assumption. Whatever the heat flux is, the face temperature variations can be measured experimentally and expressed as a function of time. That is, if the temperature boundary conditions of the sample are expressed as a function of time, not only the measurement of the heat flux can be avoided but also there exists closed solutions corresponding to the temperature boundary conditions. The simplest surface condition is the case of constant surface temperature. If the front face of a sample is maintained at a higher constant temperature and the rear surface is maintained at a lower constant temperature, the solution exists in a simple closed form. This is the principle of constant temperature plate method.

The approach to heat conduction problems can be broadly divided into two different procedures. One is termed a direct problem. In this approach, the temperature distribution inside the sample is found after the heat flux and/or temperature histories at the face of sample are determined as functions of time. The other is termed an indirect method [6]. In this method, the temperature distribution inside the body is measured at first, and then the heat flux and/or the temperature as a function of time. The measured surface temperature as a function of time usually becomes nonlinear. So, it is very difficult to find the exact solution of the equation with this kind of boundary condition. One method of finding the solution of such a problem is the parameter estimation method, which is the principle of the undetermined heat flux method. In this method, the transient properties can be found without changing the computer program regardless of the heat flux to the front face.

The present work uses both the constant temperature method and the undetermined heat flux method to measure the thermal diffusivities of several insulating materials. The results are compared with each other to confirm the accuracy of each apparatus. Also a constant heat flux method (step-heat technique) using an infrared lamp is used to compare the accuracy of each method.

MEASUREMENT THEORY

Constant Temperature Plate Method

As shown in Fig. 1 an infinite slab is considered as the mathematical model. Both surface temperatures are maintained constant and the front face temperature is higher than the rear face, i.e., $T_0 > T_\ell$. We assume that at times less than zero, the front and rear surfaces of the specimen are in thermal equilibrium at the constant initial temperature T_i, and there are no heat losses. Under these conditions, we can find the exact solution of the heat conduction equation.

For the one dimensional heat conduction equation, boundary and initial conditions are expressed as follows:

$$\frac{\partial^2 T}{\partial X^2} = \frac{1}{\alpha}\frac{\partial T}{\partial t} \qquad (0 < X < \ell, \ t > 0)$$

$$T = T_0 \qquad (X = 0, \ t > 0)$$

$$T = T_\ell \qquad (X = \ell, \ t > 0) \qquad\qquad (1)$$

$$T = T_i \qquad (0 \le X \le \ell, \ t \le 0)$$

where α, t and ℓ are thermal diffusivity, time and thickness, respectively. The solution of Eq. (1) is obtained as follows [7]:

$$T(X,t) = T_0 + (T_\ell - T_0)\frac{X}{\ell} + \sum_{n=1}^{\infty} A_n \sin\frac{n\pi X}{\ell} \exp(-n^2\pi^2 t/\ell^2) \qquad (2)$$

$$A_n = \frac{2}{\ell}\int_0^\ell T_i \sin\frac{n\pi X}{\ell}\,dX + \frac{2}{n\pi}(T_\ell \cos n\pi - T_0) \qquad (3)$$

where A_n is the Fourier Coefficient.

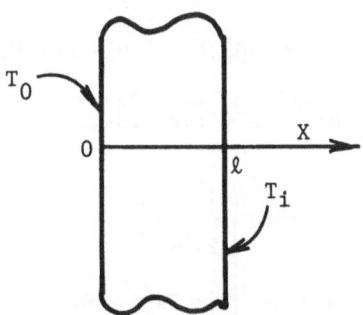

Fig. 1. Mathematical Model of the Constant Temperature Plate Method.

We assume that the rear surface temperature, T_ℓ, is maintained at the initial temperature, and we choose nondimensional quantities as follows:

$$V_T = \frac{T - T_i}{T_0 - T_i}, \qquad X = \frac{x}{\ell}, \qquad F_0 = \frac{\alpha t}{\ell^2} \tag{4}$$

The solution can be expressed in a simpler form.

$$V_T = 1 - X - \frac{2}{\pi} \sum_{n=1}^{\infty} \frac{1}{n} \sin n\pi X \exp (-n^2 \pi^2 F_0) \tag{5}$$

Once the experimental temperature-time response is obtained, the only unknown left in Eq. (5) is the thermal diffusivity, α.

Constant Heat Flux Method (Step-Heating Technique)

A semi-infinite solid which has a constant rate of heat flow, H_0, into the surface at X=0 is considered as the mathematical model in the development of this method. The one dimensional heat conduction equation, boundary and initial conditions are expressed as follows:

$$\frac{\partial^2 H}{\partial x^2} = \frac{1}{\alpha} \frac{\partial H}{\partial t} \qquad\qquad (X > 0, \ t > 0)$$

$$H = 0 \qquad\qquad (X \geq 0, \ t = 0)$$

$$H = H_0 \qquad\qquad (X = 0, \ t > 0) \tag{6}$$

$$T = T_i \qquad\qquad (X \geq 0, \ t = 0)$$

where H is the heat flux.

The temperature distribution in the body can be obtained as follows [8]:

$$T(X,t) - T_i = \frac{2H_0 \sqrt{\alpha t}}{K} \ \mathrm{ierfc} \left(\frac{X}{2\sqrt{\alpha t}} \right) \tag{7}$$

where K is a thermal conductivity and ierfc(z) is the integral complementary error function.

The temperatures at X = 0 and X = ℓ are given by

$$T(0,t) - T_i = \frac{2H_0}{K} \left(\frac{\alpha t}{\pi} \right)^{1/2} = 1.1284 \ \frac{H_0 \sqrt{\alpha t}}{K} \tag{8}$$

$$T(\ell,t) - T_i = \frac{2H_0 \sqrt{\alpha t}}{K} \ \mathrm{ierfc} \left(\frac{\ell}{2\sqrt{\alpha t}} \right) \tag{9}$$

To eliminate the heat flux term from Eqs. (8) and (9), we take a ratio of the two temperatures:

$$V_H = \frac{T(0,t) - T_i}{T(\ell,t) - T_i} = \frac{0.5642}{\text{ierfc}\left[\dfrac{\ell}{2\sqrt{\alpha t}}\right]} \qquad (10)$$

Since the temperature history can be measured, the only unknown left in Eq. (10) is the thermal diffusivity, α.

Undetermined Heat Flux Method

The model used in the development of this method is a semi-infinite solid as shown in Fig. 2, but the input heat flux is a function of time, $H=f(t)$. The boundary conditions are the temperatures at $X=0$ and an arbitrary rear position $X=\ell$. The initial condition is allowed to be $T(X,0) = f_i(X)$. Thus, the problem is modeled as follows:

$$\frac{\partial^2 T}{\partial X^2} = \frac{1}{\alpha}\frac{\partial T}{\partial t} \qquad\qquad (X > 0,\ t > 0)$$

$$T = f_o(t) \qquad\qquad (X = 0,\ t > 0)$$

$$\qquad\qquad\qquad\qquad\qquad\qquad\qquad\qquad (11)$$

$$T = f_\ell(t) \qquad\qquad (X = \ell,\ t > 0)$$

$$T = f_i(X) \qquad\qquad (X \geq 0,\ t = 0)$$

The thermal diffusivity α in Eq. (11) is to be determined. The method of calculating parameters appearing in a differential equation is called the "parameter estimation" method [9]. In general, the parameter estimation is applied to a nonlinear problem. So, the parameter estimation is often called "nonlinear estimation". One way to determine the linearity of an estimation problem is to inspect the sensitivity coefficients (defined as the first derivative of the dependent variable with respect to the unknown parameter). If the sensitivity coefficients are functions of the parameters, then the estimation problem is nonlinear [9].

The least square procedure for estimating α with temperature measurements minimizes

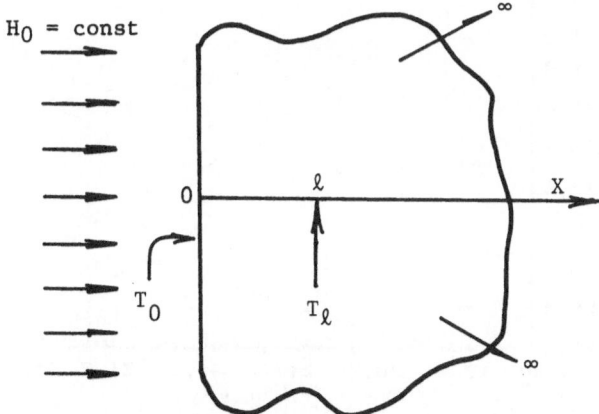

Fig. 2. Mathematical Model of the Constant Heat Flux Method.

$$S = \sum_{i=1}^{n} \sum_{j=1}^{m} [T_{ij}^*(\alpha,X,t) - T_{ij}(\alpha,X,t)]^2 \qquad (12)$$

with respect to α. The first subscript refers to space (thermocouple number) and the second to time. $T_{ij}^*(\alpha,X,t)$ are the discrete experimental temperature values, and $T_{ij}(\alpha,X,t)$ are the discrete calculated temperatures found here by using a finite difference approximation. To evaluate the thermal diffusivity, α, the parameter estimation program PROPPC.FTN of TPRL is used. In the program, a Crank-Nicolson finite element scheme is used to predict the interior temperature history of the sample [9]. A first guess alpha is entered, and then the temperature is found and compared with the measured interior temperature distribution by the minimization of the sum of square function S. The parameter is adjusted and iterated until a best fit is attained. The process is sequential so that the iteration is carried out for each time step.

MEASUREMENT APPARATUS AND PROCEDURES

The constant temperature plate method involves suddenly placing one surface of sample in contact with an infinite heat source at a constant higher temperature, and simultaneously recording the temperature history at three locations, i.e., the front, interior, and rear face. The temperatures of both sample faces are measured to confirm the boundary conditions, and the interior temperatures are used to calculate the thermal diffusivity. The temperature of the hot plate was controlled by a variac and the infinite heat source was achieved by using a relatively massive aluminum block on the plate.

Starting at time zero, the sample was placed for 50-100 seconds on the hot plate maintained at a constant temperature. The temperatures at three locations were measured as a function of time. The measured temperatures and time are used to draw the graphs and calculate the thermal diffusivity. A typical graph is shown in Fig. 3 for the Manville material. As shown in the graph, the hot plate temperature was maintained almost constant during the experiment. The temperature of the rear surface

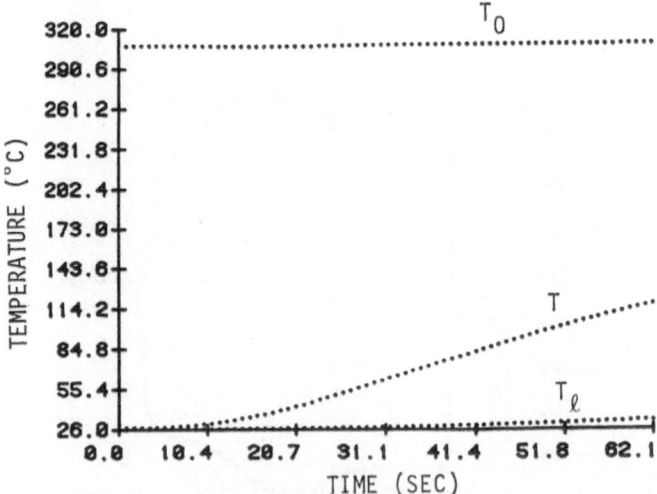

Fig. 3. Temperature History of the Manville Material Measured by the Constant Temperature Plate Method.

started to rise above the initial temperature at 30 seconds. Since the constant temperature plate method assumes that the temperature of rear face is maintained at the constant initial temperature, the data measured after 30 seconds violated the assumption and were not used. The thermal diffusivity values were calculated by using the data for the time region where the interior temperatures of sample increased linearly. As shown in Fig. 3, this region of the time interval was from about 15 second to 30 seconds in this case. The final diffusivity value was an average of all values calculated at each time step in this region. The effective temperature was taken as the arithmetic mean temperature between the hot plate and the initial temperature.

The apparatus for the constant heat flux method coincides with that for the undetermined heat flux method. During the data collection after the shutter is removed, the constant heat flux method involves constantly maintaining the intensity of lamp, while the undetermined flux method need not consider the heating method. As shown in Fig. 4, the apparatus is divided into three parts, i.e., heat source, furnace, and sample holder. A 600 W quartz line projection lamp was selected for the heat source. The lamp intensity was controlled by a variac. The cylindrical furnace was electrically heated with a nichrome wire embedded in a ceramic shell. The window at one end of the furnace is supported by a shutter mount which supports the shutter plate. The ceramic shutter plate is located between the furnace environment and the viewing window. The furnace is rated at 10 amps, and reaches a maximum temperature of about 1200°C. The sample holder consisted of two end rings aligned on threaded rods. The sample is mounted on the sides of the rings and is held by nuts made of Macor machinable ceramic. Both faces of the sample were placed against a thin stainless steel plate of thickness 0.025 cm. The plate on the front face was painted with a black lacquer in order to increase absorption. Three layers of the sample material were stacked together, and a stack of layers was assumed as one continuous medium. K-type thermocouples of 0.01 cm diameter were placed between the layers, and were assumed to cause negligible interference with the heat flow. The thermocouple hot junctions were welded to thin stainless steel foils of 0.005 cm thickness and 0.2 cm x 0.2 cm area. An ice bath was used for the cold junction. The thermocouples used inside the furnace were insulated with tubes made of ceramic textiles.

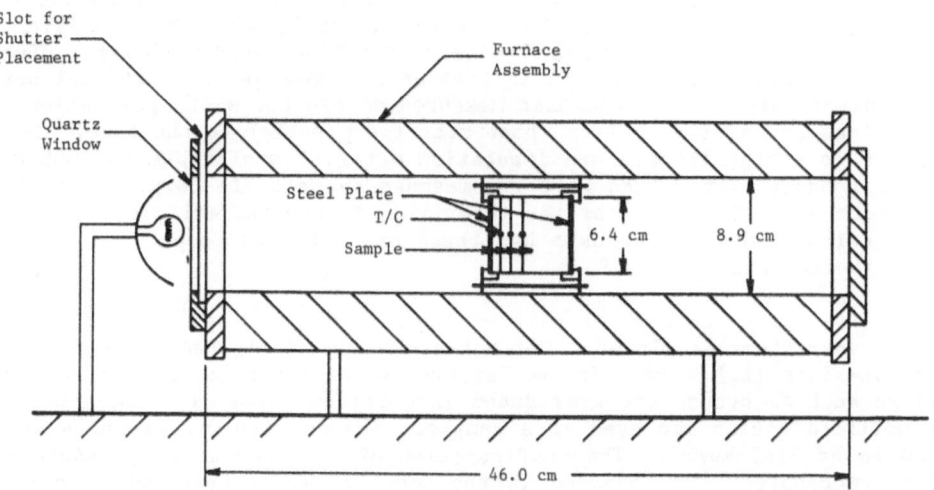

Fig. 4. Cross-Sectional View of Apparatus and Sample.

Once the temperature within the sample was established in a steady state condition, the data collection was started. Data acquisition continued normally for 70-120 seconds, and the time step was 1 second. Data acquisition and analysis were performed using a DEC minicomputer, model PDP 11/34A.

EXPERIMENTAL RESULTS AND DISCUSSION

Experimental results are subjected to many sources of error. Taylor [10] divided broadly the sources of error for the flash technique into two parts, namely, measurement errors and non-measurement errors and examined these. Measurement errors include those associated with determining the effective thickness of the sample, the position of thermocouple, the sensitivity of thermocouple, two-dimensional heat flow, and so forth. Non-measurement errors depend upon the effects of heat losses and heating conditions (the duration, shape, and spatial non-uniformity of heating). Determinations of sample thickness of some insulation materials are often difficult due to the softness and the fragility of materials such as Cerwool, Foam, and Silicon rubber materials. The thermal diffusivity is directly affected by the measurement errors of sample thickness since it involves thickness squared. For example, if the measurement error in the thickness of sample is 2%, the thermal diffusivity value is affected by about 4%. So, it is important to assure accurate measurement of sample thickness. When the sample is inserted into the measuring apparatus, it should be maintained at the same thickness that was measured. In order to do so, the sample was mounted on the sample holder while applying the same weight as was applied when the thickness was determined.

Rooke [5] studied the effects of thermocouple locations. His results showed that the radial effects of thermocouple installation position increased with the radial distance from the center line and decreased as the axial distance from the front face. So, in present work the thermocouples were placed on the center of the sample.

Comparisons of the measured thermal diffusivity values using the three methods, i.e., Constant Temperature Plate (CTP), Constant Heat Flux (CHF), and Undetermined Heat Flux (UHF) method are presented in Figs. 5 and 6. Figures 5 and 6 present the data for Manville and Teflon materials, respectively. The Manville material is Ceraboard 141, manufactured by Manville Corporation. The total sample thickness was 1.347 cm (ℓ_1 = 0.567 cm, ℓ_2 = 0.780 cm) and the density is 235 kg/m^3. The Manville material is a dense clay bonded alumina silica fiberboard with fiber diameters on the order of 3 μm. It is treated as a homogeneous material but it is anisotropic. The properties measured across the width perpendicular to the face are different from those measured parallel to the face. This material is a high temperature insulation material applicable to 1000°C. Thermal conductivity values were recommended by Manville Corp. and Rooke measured the specific heat of the material. So the thermal diffusivities can be calculated from the data and these are referred to as "the recommended thermal diffusivity."

Teflon is the commercial name of the material and its chemical name is "polytetrafluoroethylene". Teflons are being widely used as antifriction materials [11]. In order to improve the antifriction qualities, the fillers such as bronze are introduced into teflon. But in present work, the unfilled teflon was used as a sample. The density was measured and found to be 2145 kg/m^3. The configuration of the sample is cylindrical and thermocouples were embedded in the small diameter holes which were bored from the wall to the center.

As shown in Fig. 5, the experimental thermal diffusivity results for Manville material are compared with values obtained using different methods. Although the experimental results were below the recommended values over the range from room temperature through 300°C, the values obtained by three techniques showed fair agreement in this range. The results obtained by the CTP method had the lowest values, and the results by the UHF method were the nearest to the recommended values. The values

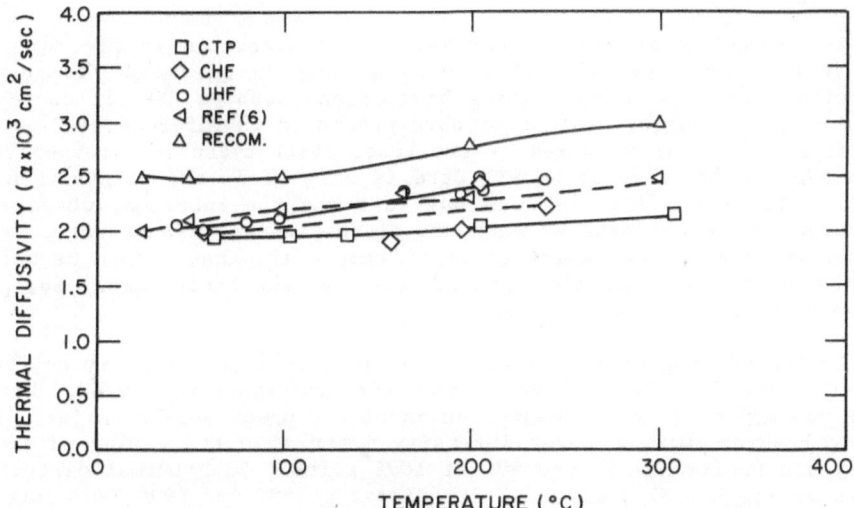

Fig. 5. Comparison of the Thermal Diffusivities Measured by Various Methods for Manville Material.

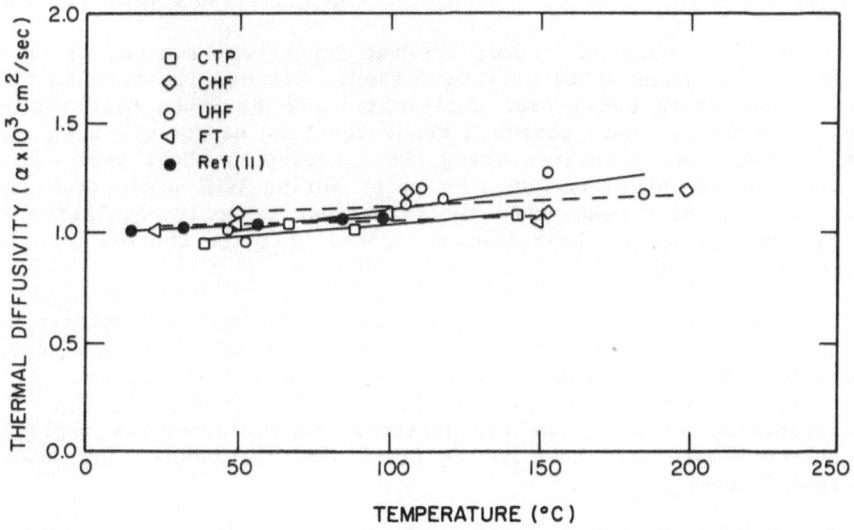

Fig. 6. Comparison of the Thermal Diffusivities Measured by Various Methods for Teflon Material.

obtained by the UHF method and reference [5] agreed well with each other over the range. At 100°C the experimental values were between 1.97×10^{-3} and 2.12×10^{-3} cm^2/s, a seven percent spread. For the Manville material, the results obtained by CTP method were the lowest values. The range of differences between the CTP values and the recommended values was approximately 22-28% and for the UHF values and the recommended values, it was within approximately 18% in this range. If the recommended values are taken for the baseline, the UHF method can be regarded as the best technique. The data measured by the CHF method were somewhat scattered. One of the reasons for this is the response characteristic of the integral complementary error function when the convergent gradient is steep in the small argument region.

The experimental results for Teflon are presented in Fig. 6. All data in the figure agreed well with each other in the range from a room temperature through 200°C. The abbreviations such as REF[11] and FT on Fig. 6 mean the average values of data presented in reference [12] and the experimental results measured by the laser flash technique, respectively. The maximum difference among all data is only 6% in the neighborhood of 150°C. The Teflon material was not soft and the locations where the thermocouples were installed were not discontinuous. Thus, not only the influences of the measurement error of sample thickness could be reduced but also there were not the interferences at the interfaces. So, good agreement data could be obtained.

During testing of the effect of projection lamp intensity on the UHF method, several kinds of fibrous insulating materials were used. The lamp power was adjusted by the manipulation of the power supply variac. Fifty or one hundred percent power intensity means that the variac of power supply was positioned at the 50% or 100% point. The thermal diffusivity values of the Manville material irradiated at 50% and 100% lamp intensity were 2.346×10^{-3} and 2.370×10^{-3} cm^2/sec at about 160°C, and 2.496×10^{-3} and 2.344×10^{-3} cm^2/sec at about 200°C, respectively. For 50% and 70% lamp intensity, the values of foam material were 4.20×10^{-3} and 4.202×10^{-3} cm^2/sec at about 80°C. At about 70°C, the same kind of experiment was conducted for the blanket material. The values irradiated at 50% and 100% projection obtained were 2.561×10^{-3} and 2.554×10^{-3} cm^2/sec, respectively. According to these results, the effect of lamp intensity was small. The error range of the experiments was less than 6%.

In order to find the effect for the input heat sources of the UHF method, several experiments were conducted. The results for the blanket material irradiated using four different kinds of input heat flux are listed in Table 1. Heat source A means the heat source was used continuously at 50% lamp intensity during the experiment. Heat source B means that the shutter was closed at some point during 100% projection experiment. Heat source C means that the variac was randomly oscillated during the data collection and heat source D means that the shutter was opened and closed repeatedly.

For all heat sources, the differences of thermal diffusivity values were within four percent. So, the effect of variable heat source can be neglected. The experimental results for blanket material and silicon rubber material are plotted in Fig. 7. As shown in Fig. 7, the differences generated by using the two different heat sources are negligible. Heat source E means the heat source irradiated the sample continuously at 100% lamp intensity.

The thermal diffusivity values for several fibrous materials were measured by the UHF method and are plotted in Fig. 8.

CONCLUSIONS

The major objective of this study was to compare the accuracy of the apparatuses for the measurement of thermal diffusivity of fibrous insulation materials, and then to recommend the best method. All the methods considered here used apparatuses which can be manufactured cheaply and simply, have fair accuracy and can measure easily the thermal diffusivity.

A weak point of the CTP method is that the original sample temperature should be perhaps 100°C less than the hot plate temperature and it is difficult to achieve this and handle the samples at higher temperatures. At each temperature, the stacked sample should be put on the hot plate and subsequently removed. However, apparatus is the simplest and the cheapest.

Table 1. Thermal Diffusivity of Blanket Material for Various Input Heat Sources at About 70°C

Kind of Heat Source	Thermal Diffusivity (cm^2/sec)
A	2.561 x 10^{-3}
B	2.554 x 10^{-3}
C	2.627 x 10^{-3}
D	2.668 x 10^{-3}

A: 50% continuous projection heat source
B: 100% discontinuous projection heat source
C: Random oscillating heat source
D: Intermittent heat source

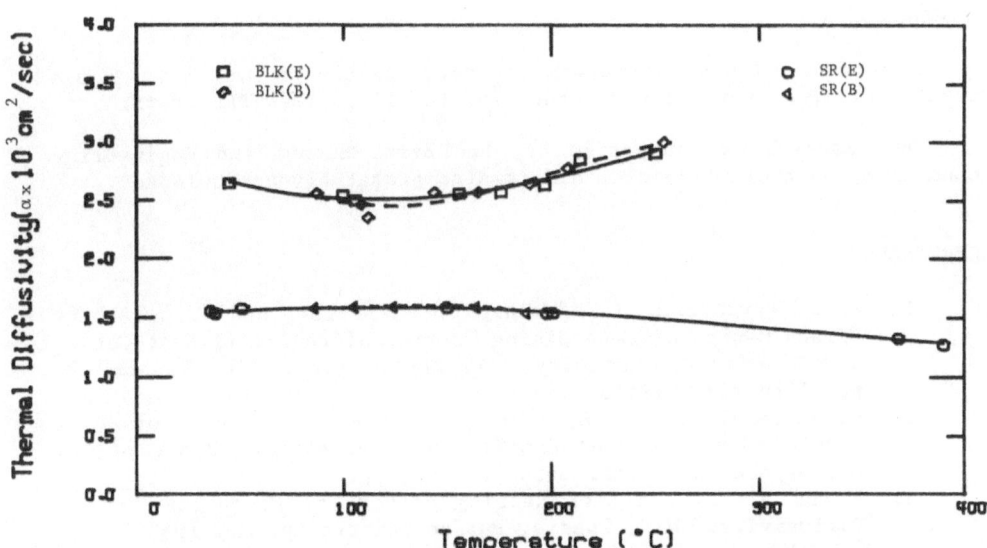

Fig. 7. Comparison of the Thermal Diffusivities Obtained by Using Two Different Heat Sources for Blanket Material and Silicon Rubber Material.

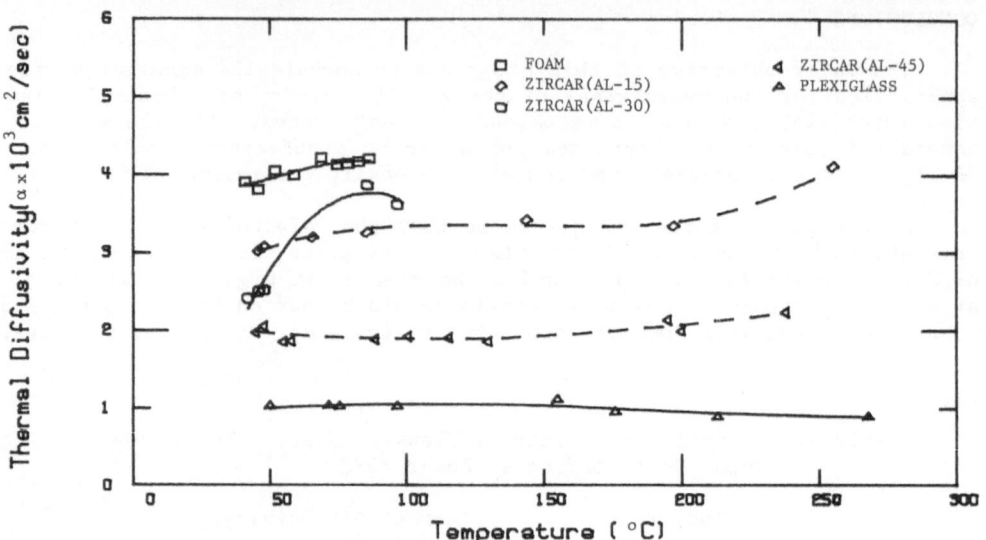

Fig. 8. Thermal Diffusivities for Various Materials.

The data obtained by the CHF method are widely scattered due to the characteristics of the integral complimentary error function. But the computer program for analysis of the thermal diffusivity is much simpler.

Although the computer program for the UHF method is quite big and complicated, the method can obtain the most accurate results, can neglect the effects of heat sources, that is, the intensity, the discontinuity, the duration and the random oscillation of heat source, and easily raise the initial temperature to high temperatures (1200°C).

ACKNOWLEDGMENTS

The author wishes to express his appreciation to Dr. R. E. Taylor, Director of TPRL, for his assistance and to all of the TPRL staff.

The research was supported by the Korea Science and Engineering foundation and their financial aid is also gratefully acknowledged.

REFERENCES

1. W. J. Parker, R. J. Jenkins, C. P. Butler, and G. L. Abbott, "Flash Method of Determining Thermal Diffusivity, Heat Capacity, and Thermal Conductivity," J. Appl. Phys., Vol. 32, No. 9, pp. 1679-1684, 1961.
2. R. R. Bittle and R. E. Taylor, "Thermal Diffusivity of Heterogeneous Materials and Non-Fibrous Insulators," Proc. of 18th Conference on Thermal Conductivity, pp. 379-390, 1985.
3. R. R. Bittle, "A Step Heating Method for Measuring Thermal Diffusivity," M.S. Thesis, Purdue University, May 1983.
4. K. Kobayasi, "Simultaneous Measurement of Thermal Diffusivity and Specific Heat at High Temperatures by a Single Rectangular Pulse Heating Method," Int. J. Thermophysics, Vol. 7, No. 1, pp. 181-195, 1986.

5. S. P. Rooke, "Transient Technique for the Determination of the Thermal Diffusivity of Fibrous Insulators," M.S. Thesis, Purdue University, Dec. 1986.

6. J. V. Beck, B. Blackwell, and C. R. St. Clair, Jr., Inverse Heat Conduction, Wiley-Interscience Publication, New York, NY, 1985.

7. H. S. Carslaw and J. C. Jaeger, Conduction of Heat in Solids, 2nd Ed., Oxford Press, 1959.

8. E. R. G. Eckert and R. M. Drake, Jr., Analysis of Heat and Mass Transfer, McGraw-Hill Book Co., 1972.

9. J. V. Beck and K. J. Arnold, Parameter Estimation in Engineering and Science, John Wiley & Sons, 1977.

10. R. E. Taylor, "Critical Evaluation of Flash Method for Measuring Thermal Diffusivity," Rev. Int. Hautes Temp. Refract., pp. 141-145, 1975.

11. V. M. Baranovskii, V. P. Dushchenko, E. M. Natanson, V. V. Levandovskii, V. M. Chegoryan, and B. P. Dem'yanyuk, "Temperature Dependence of the Thermophysical Properties of a Teflon-Bronze Antifriction Material," Polym. Mech., USSR, Vol. 7, No. 6, pp. 936-939, 1971.

12. Y. S. Touloukian, R. W. Powell, C. Y. Ho, and M. C. Nicolaou, Thermophysical Properties of Matter -- The TPRC Data Series, Vol. 10, IFI/Plenum, New York, NY, pp. 607-608, 1973.

SESSION 7

COMPOSITES

THERMAL DIFFUSIVITY AND CONDUCTIVITY OF

COMPOSITES WITH INTERFACIAL THERMAL

CONTACT RESISTANCE

D.P.H. Hasselman
Department of Materials Engineering
Virginia Polytechnic Institute and State University
Blacksburg, Virginia 24061 USA

ABSTRACT

Experimental data for three composite material systems are presented which indicate that the existence of an interfacial thermal barrier resistance can have a significant effect on the effective thermal conductivity of composites.

By modifications of the original theories of Rayleigh and Maxwell, expressions for the effective thermal conductivity of composites consisting of a continuous matrix phase with dilute concentrations of dispersions with spherical, cylindrical and flat plate geometry with a thermal barrier resistance at the interface between the components were derived.

It was found that for a given composite system and dispersed phase geometry, due to the existence of an interfacial thermal barrier resistance, the effective thermal conductivity depends on the volume fraction of the dispersed phase as well as the dispersion size. This latter result is in contrast with literature predictions that the effective thermal conductivity of composites without an interfacial thermal barrier resistance is independent of the dimensions of the microstructure.

INTRODUCTION

Many engineering materials for structural or other purposes consist of two or more components. The properties of such composites can be measured or predicted from the corresponding properties of the components, their volume fraction and phase distribution. For this purpose, much earlier effort has been devoted to predict theoretically the effective thermal conductivity of composites [1-13]. Generally, in these analyses it was implicitly assumed that at the interface the temperatures within each component were identical, i.e., that the heat flow between the individual components was not affected by the existence of a thermal contact resistance. Experimental results obtained by the author and his co-workers for the thermal diffusivity of three different composite systems indicated that the existence of a thermal contact resistance at the interface between the individual components can have a significant effect on the effective thermal conductivity [14, 15, 16]. The findings provided an impetus for the derivation of analytical expressions for the effective thermal conductivity of composites, which incorporated the effect of an interfacial

contact resistance. It is the purpose of this paper to present a brief
review of these experimental observations and analytical results.

EXPERIMENTAL OBSERVATIONS

1.   Nickel-sodaborosilicate glass composites

The composite samples consisted of a continuous phase of a
sodaborosilicate glass with a dispersed phase of∿ 50μ m diameter spherical
inclusions of nickel, and made by vacuum hot-pressing mixtures of the
appropriate powders at ∿ 725°C. Because of the non-wetting
characteristics, no chemical or mechanical adhesion occurred between the
nickel and glass matrix. The coefficient of thermal expansion of
the nickel greatly exceeded the corresponding value for the
sodaborosilicate glass. As a result, on cooling the composite from the
hot-pressing temperature, the nickel dispersions shrank away from the glass
matrix, leaving a gap at the interface. Figure 1 shows the experimental
data for the thermal diffusivity as a function of temperature for volume
fractions of nickel of 0, 15, 30 and 45 vol. % nickel, as measured by
Powell et al [14] by means of the laser-flash diffusivity method. Included
in fig. 1 is the calculated behavior based on Bruggeman's variable
dispersion (BVD) theory, which assumes perfect thermal contact at the
interface. Two effects are to be noted. Firstly, at the lower temperatures
the experimental data fell below the predicted behavior by about a factor
of two. This is expected as the width of the gap and its effectiveness as
a thermal barrier is expected to increase with increasing degree of
cooling. Secondly, at the higher temperatures, the thermal diffusivity
shows a strong positive temperature dependence. This latter effect is also
expected as with increasing temperature the nickel spheres regain their
original size, so that improved thermal contact between the glass and
nickel is re-established.

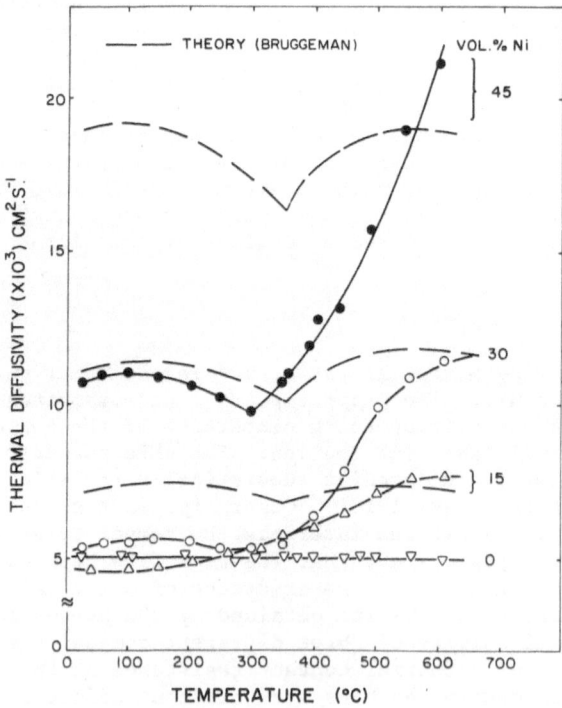

Fig.1.   Thermal diffusivity of soda-borosilicate
         glass matrix with spherical nickel
         inclusions (ref.14).

2.   Uniaxial carbon fiber-reinforced lithia-alumina-silicate glass-ceramic composite

The thermophysical characteristics of this composite material were investigated by Hasselman, et al [15]. Unusual behavior was noted in the temperature dependence of the thermal diffusivity perpendicular to the fiber direction. The experimental results for two successive cycles to ∿ 1000°C are shown in fig. 2. These data indicate that following cooling from the upper temperature, the thermal diffusivity exhibits a permanent decrease, which is absent in the second cycle. This permanent decrease was found to be associated with crack formation in the matrix and interfacial separation between the fibers and matrix, shown in fig. 3. The thermal diffusivity on heating and cooling also appears to exhibit irreversible behavior, especially on heating to 1000°C for which the data on cooling exceed those on heating. It is thought that this latter effect is related to the existence of an irreversible adhesive bond and associated thermal contact resistance at the carbon-matrix interface.

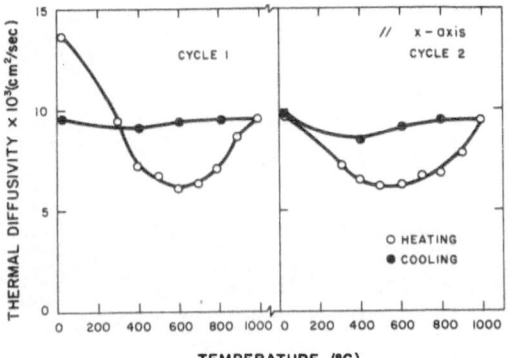

Fig.2.   Thermal diffusivity of uniaxial carbon fiber reinforced lithia-alumina sili-cate glass-ceramic over two successive heating cycle to ∿1000°C (ref.15).

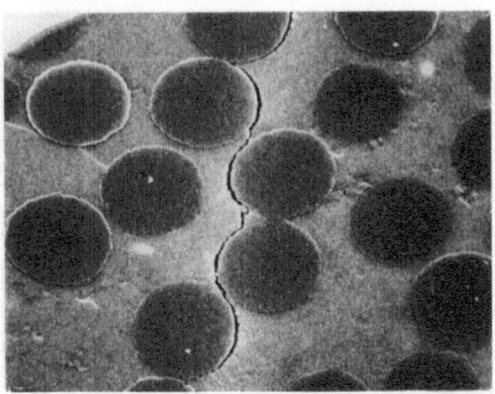

Fig.3.   Evidence for interfacial cracking in carbon fiber-reinforced lithia-alumina silicate glass-ceramic heated to 1000°C. Fiber diameter ∿ 7.3 μm (ref.15).

3. Carbon fiber-reinforced borosilicate glass

These composite samples were made by hot-pressing mixtures of the glass in powder form and the carbon fibers chopped into lengths of a few millimeters. During hot-pressing, the carbon fibers take on a preferred orientation in a plane perpendicular to the hot-pressing direction. Within this plane the fiber orientation is random. The thermal properties of the composite were reported in a detailed study by Johnson et al [16], to which the reader is referred for further detail. For purposes of the present paper the thermal diffusivity perpendicular to the fiber plane above $600^\circ C$ is of primary interest. As shown in fig. 4, during an isothermal anneal at various temperatures > $600^\circ C$, the thermal diffusivity shows a time-dependent decrease, as shown in fig. 4. This decrease, which occurred more rapidly with increasing temperature, appeared to occur in an exponential manner to a constant value. Electron micrography of annealed polished surfaces, as shown in fig. 5, revealed the formation of a gap between the fibers and the matrix. In fact, some of the fibers appeared to stick up above the surface. It is thought that the gap formation resulted from the viscous flow of the glass under the influence of the stresses in

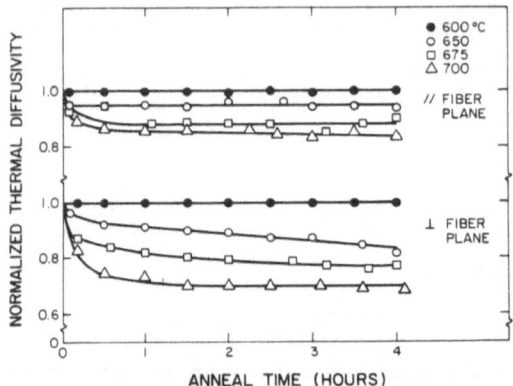

Fig.4. Time dependence of the relative thermal diffusivity of a carbon fiber-reinforced borosilicate glass (ref.16).

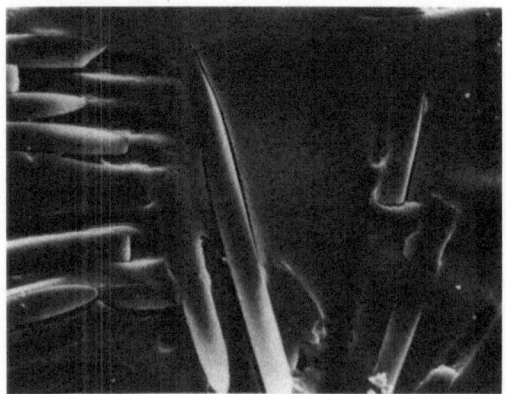

Fig.5. Matrix-fiber separation in carbon fiber-reinforced borosilicate glass annealed at $650^\circ C$ for 4 hr (ref.16).

the fibers which were deformed elastically during hot-pressing and constrained by the rigid matrix at temperatures below 600°C. Above 600°C the viscous state of the glass matrix permitted the relaxation of the stresses in the carbon fibers. For this reason, the specific time dependence of the thermal diffusivity as shown in fig. 4, reflected the thermally activated nature of the viscous relaxation of the glass matrix. The resulting time-dependent gap between the fibers and matrix was responsible for the time-dependent decrease in the thermal diffusivity.

The above observations indicate the significant role interfaces can play in governing the effective thermal diffusivity and conductivity of composites. A review of the literature, however, revealed that this effect appeared not to have been included in the analysis of the effective thermal conductivity of composites. However, a survey of the various theoretical approaches also revealed that for composites consisting of continuous matrices with dilute concentrations of spherical inclusions or circular fibers oriented perpendicular to the heat flow, expressions for the effect of an interfacial thermal barrier resistance on the effective thermal conductivity are easily obtained by quite simple modifications of the original theories of Rayleigh [1] and Maxwell [2]. The general approach and results of this analysis [18] will be presented briefly below.

ANALYSIS

The derivation for the effective thermal conductivity of a continuous matrix with spherical dispersions will be presented in some detail. For a matrix with parallel circular cylindrical inclusions oriented perpendicularly to the heat flow, only the final expressions will be presented.

1.   Matrix with spherical inclusions

The dispersions with thermal conductivity K_d are imbedded in a matrix with thermal conductivity, K_m. The volume fraction of the dispersions is assumed sufficiently dilute that interactions between the local temperature fields of neighboring dispersions are absent. In keeping with usual practice in the theory of heat transfer, the interfacial thermal barrier resistance is expressed in terms of a boundary conductance, h_c. The derivation for the effective thermal conductivity (K_{eff}) first requires obtaining solutions for the temperature distribution in and around a single dispersion, assumed to be of the general form:

$$T_d = rA\cos\Theta \tag{1a}$$

$$T_m = (\nabla T)r \cos\Theta + (B/r^2) \cos\Theta \tag{1b}$$

where (∇T) is the temperature gradient at large distances away from the dispersion, A and B are constants to be determined and r and Θ are spherical coordinates with Θ being the angle between the radius vector r and the temperature gradient.

Eqs. 1a and 1b are subject to the boundary conditions at r=a:

$$K_d (\partial T_d/\partial r) = K_m (\partial T_m/\partial r) \tag{2a}$$

$$T_d - T_m = -(K_d/h_c)(\partial T_d/\partial r) \tag{2b}$$

In the absence of an interfacial thermal barrier resistance (i.e.,

$h_c = \infty$) eq. 2b becomes:

$$T_d = T_m \qquad (r=a) \qquad (2c)$$

which is the boundary condition for the original derivation by Maxwell.

Solving for A and B and substituting into eq. 1, yields:

$$T_p = (\nabla T) r\cos\Theta \; \frac{\left\{1 + \left[\dfrac{K_m}{ah_c} + \dfrac{K_m}{K_d} - 1\right]\left[1 + \dfrac{2K_m}{K_d}\left(\dfrac{K_d}{ah_c} + 1\right)\right]^{-1}\right\}}{\left[\dfrac{K_d}{ah_c} + 1\right]} \qquad (3a)$$

$$T_m + (\nabla T) r\cos\Theta \; + \frac{(\nabla T) a^3 \cos}{r^2} \; \frac{\left[\dfrac{K_m}{ah_c} + \dfrac{K_m}{K_d} - 1\right]}{\left[1 + \dfrac{2K_m(K_d/ah_c + 1)}{K_d}\right]} \qquad (3b)$$

For $h_c = \infty$, eqs. 3a and 3b become:

$$T_m = (\nabla T) r\cos\Theta \; \frac{3K_m}{K_d + 2K_m} \qquad (4a)$$

$$T_m = (\nabla T) r\cos\Theta \; + \frac{(\nabla T) a^3}{r^2} \; \frac{(K_m - K_d)}{(2K_m + K_d)} \qquad (4b)$$

in agreement with the original solutions of Maxwell [2].

The derivation of the effective thermal conductivity of the composite relies on assessing the cumulative effect on T_m of n spheres of radius a within a large sphere of radius b which is considered to exhibit the effective thermal conductivity K_{eff}.

In terms of n spheres of radius a, eq. 3b becomes:

$$T_m = (\nabla T) r\cos\Theta \; + \frac{na^3 (\nabla T) \cos\Theta}{r^2} \; \frac{\left[\dfrac{K_m}{ah_c} + \dfrac{K_m}{K_d} - 1\right]}{\left[1 + \dfrac{2K_m(K_d/ah_c + 1)}{K_d}\right]} \qquad (5a)$$

410

which in terms of the sphere of radius, b and thermal conductivity, K_{eff} becomes:

$$T_m = (\nabla T) r \cos\Theta + \frac{(\nabla T) r \cos\Theta}{r^2} \left[\frac{\frac{K_m}{bh_c} + \frac{K_m}{K_{eff}} - 1}{1 + \frac{2K_m(K_{eff}/bh_c + 1)}{K_{eff}}} \right] \tag{5b}$$

The last terms in eqs. 5a and 5b must be equal. With the volume fraction of spheres, $V_d = na^3/b^3$ and for b very large such that $K_m/bh_c = K_{eff}/bh_c$ o, equating these last terms, yields:

$$K_{eff} = K_m \frac{\left[2\left(\frac{K_d}{K_m} - \frac{K_d}{ah_c} - \right) 1 \; V_d + \frac{K_d}{K_m} + \frac{2K_d}{ah_c} + 2 \right]}{\left[\left(1 - \frac{K_d}{K_m} + \frac{K_d}{ah_c} \right) V_d + \frac{K_d}{K_m} + \frac{2K_d}{ah_c} + 2 \right]} \tag{6}$$

For $h_c = \infty$, eq. 6 agrees with the expression of Maxwell [2] for K_{eff} in the absence of an interfacial thermal barrier resistance.

2. Circular cylinders oriented perpendicularly to heat flow

The temperature, T_d, within the circular cylindrical dispersion with radius a and the temperature T_m in the surrounding matrix were assumed to be generally described by:

$$T_d = rA\cos\Theta \tag{7a}$$

$$T_m = (\nabla T) r\cos\Theta + (B/r) \cos\Theta \tag{7b}$$

where r and Θ now represent a two-dimensional polar coordinate system. Solving for A and B and following the identical approach taken for the spherical phase geometry yields for the effective thermal conductivity:

$$K_{eff} = K_m \frac{\left[\left(\frac{K_d}{K_m} - 1 - \frac{K_d}{ah_c} \right) V_d + \left(1 + \frac{K_d}{ah_c} + \frac{K_d}{K_m} \right) \right]}{\left[\left(1 + \frac{K_d}{ah_c} - \frac{K_d}{K_m} \right) V_d + \left(1 + \frac{K_d}{ah_c} + \frac{K_d}{K_m} \right) \right]} \tag{8}$$

For $h_c \to \infty$, eq. 11 agrees with the solution of Rayleigh [1].

3. Flat plate dispersions oriented perpendicular to heat flow

The composite for this case is considered to consist of a matrix with parallel flat plate dispersions with half-thickness, a/2 oriented perpendicular to the direction of heat flow, with a concentration n dispersions per unit volume. The volume fraction of dispersions, V_d is related to the dispersion size and concentration, n by:

$$V_d = na \tag{9}$$

By considering the composite equivalent to a series-circuit the effective thermal resistivity (K^{-1}_{eff}) is obtained by the addition of the thermal resistivity of each component plus the sum of all the interfacial thermal resistances. The number of such interface equals 2 for each dispersion or a total of 2n per unit-volume. The thermal resistivity of the composite in this manner becomes:

$$1/K_{eff} = V_d/K_d + (1 - V_d)/K_m + 2n/h_c \tag{10}$$

which with the aid of eq. 12 and rearrangement yields :

$$K_{eff} = \frac{K_d}{1 - \dfrac{K_d}{K_m} + \dfrac{2K_d}{ah_c} V_d + \dfrac{K_d}{K_m}} \tag{11}$$

which for $h_c = \infty$ is in agreement with the lower bound thermal conductivity of a two-component composite with perfect thermal contact between the components.

DISCUSSION

The above analytical results indicate that the effect of the interfacial thermal barrier resistance on the effective thermal conductivity is controlled by the non-dimensional parameter, K_d/ah_c and its value relative to the ratio K_d/K_m. For $K_d/ah_c = \infty$ (i.e., $h_c = 0$) the effective thermal conductivity corresponds to the value for a matrix with a dispersed pore phase, irrespective of the absolute value of K_d. This effect arises because for $h_c = 0$, no heat can flow across the boundary into the dispersed phase, which then cannot contribute to the transfer of heat.

Of interest to note also is that because the value of K_d/ah_c depends on the dimensions of the dispersions, the effective thermal conductivity for any value of h_c not equal to infinity, will depend on the size of the dispersed phase particles. Such a size effect is not predicted by the theory for the thermal conductivity of composites without thermal contact resistance.

An interesting condition is encountered for the flat-plate geometry for which $K_{eff} \to 0$ as $h_c \to 0$, regardless of the values of K_d and K_m. This is not the case for the spherical or cylindrical phase geometry, which suggests that a flat-plate geometry for heat flow perpendicular to the plane of the plate will be most effective in lowering the thermal conductivity for $h_c \to 0$.

In general, interfacial thermal barriers are expected to be effective only if they are non-parallel to the direction of heat flow. This suggests that for non-spherical dispersions with preferred orientation, the existence of a thermal barrier resistance can serve as an additional mechanism for introducing anisotropy in thermal conductivity even when $K_d = K_m$.

In summary, experimental data were presented supported by a theoretical analysis, which show that an interfacial thermal contact resistance between components can have a significant effect on the effective thermal conductivity of a composite.

REFERENCES

1. Lord Rayleigh, "On the Influence of Obstacles Arranged in Rectangular Order Upon the Properties of a Medium," Phil. Mag. 34 (1892) p. 481.
2. J.C. Maxwell, A. Treatise on Electricity and Magnetism, 1 3rd Ed., Oxford University Press (1904).
3. H. Fricke, "The Electrical Conductivity of a Suspension of Homogeneous Spheroids," Phys. Rev., 24 (1924) p. 575.
4. D.A.G. Bruggeman, "Dielectric Constant and Conductivity of Mixtures of Isotropic Materials," Annalen Physik 24 (1935) p.636.
5. E.H. Kerner, "The Electrical Conductivity of Composite Media," Proc. Phys. Soc. (London) B69 (1956) p. 802.
6. R.E. De La Rue and C.W. Tobias, "On the Conductivity of Dispersions," J. Electrochem. Soc., 106 (1959) p.827.
7. Z. Hashin, "Assessment of the Self Consistent Scheme Approximation: Conductivity of Particulate Composites," J. Comp. Mat. 2 (1968) p. 284.
8. S.C. Cheng and R.I. Vachon, "The Prediction of the Thermal Conductivity of Two and Three Phase Solid Heterogeneous Mixtures," Int. J. Heat Mass Transfer, 12 (1969) p. 249.
9. B. Budiansky, "Thermal and Thermoelastic Properties of Isotropic Composites," J. Comp. Mat., 4 (1970) p. 286.
10. A. Nir and A. Acrivos, "The Effective Thermal Conductivity of Sheared Specimens," J. Fluid Mech., 78 (1976) p. 33.
11. R.A. Crane and R.I. Vachon, "Prediction of the Bounds on the Effective Thermal Conductivity of Granular Materials," Int. J. Heat and Mass Transfer, 20 (1977) p. 711.
12. S. Nomura and T.W. Chou, "Bounds of Effective Thermal Conductivity of Short-Fiber Composites," J. Comp. Mat 14 (1980) p.120.
13. Hiroshi Hatta and Minoru Taya, "Thermal Conductivity of Coated Filler-Composites," J. Appl. Phys., 59 (1986) p.1851.
14. Bob R. Powell, Jr., G.E. Youngblood, D.P.H. Hasselman, and Larry D. Bentsen, "Effect of Thermal Expansion Mismatch on the Thermal Diffusivity of Glass-Ni Composites," J. Am. Ceram. Soc. 63 (1980) p.581.
15. D.P.H. Hasselman, L.F. Johnson, R. Syed, Mark P. Taylor, and K. Chyung, "Heat Conduction Characteristics of a Carbon Fiber-Reinforced Lithium-Alumino-Silicate Glass-Ceramic," J. Mat. Sc., 22 (1987) p.701.
16. L.F. Johnson, D.P.H. Hasselman, E. Minford, "Thermal Diffusivity and Conductivity of a Carbon Fiber-Reinforced Borosilicate Glass," J. Mat. Soc. (in press).
17. A.E. Powers, Conductivity in Aggregates, Knolls Atomic Power Laboratory Report KAPL-2145, General Electric Company, Schenectady, N.Y. (1961).
18. D.P.H. Hasselman and L.F. Johnson, "Effective Thermal Conductivity of Composites with Interfacial Thermal Barrier Resistance," J. Composite Materials 21 (1987) p.508.

HEAT TRANSFER THROUGH MOIST POROUS MEDIA

Anna M. Schneider*, H.J. Goldsmid**, and B.N. Hoschke*

* CSIRO Division of Textile Physics,
  338 Blaxland Road, Ryde, NSW, 2112, Australia

** School of Physics, University of New South Wales
   P.O. Box 1, Kensington. NSW, 2033, Australia

ABSTRACT

A transient technique has been used to measure the effective thermal conductivity of polyurethane foam containing various amounts of water. In this three-phase porous material, it was observed that the conductivity increases with water content, but not linearly. A model of heat and mass transfer based on (1) conduction, (2) infra-red radiation, and (3) the process of evaporation of water, diffusion of water vapour through the structure and condensation both in the structure and externally has been developed to predict the effective thermal conductivity in this material.

INTRODUCTION

A transient method of measuring the effective thermal conductivity of textile fabrics containing water has been reported elsewhere[1]. Fabrics were sandwiched between a heat source of constant temperature and a smaller, passive guarded heat sink at a lower temperature. The thermal conductivity of the fabric can be calculated from the temperature changes of the heat sink during the first 100 seconds of heat flow.

This method has the advantage over steady state methods, such as the guarded hot plate, in that measurements can be completed quickly. In steady state methods specimens are subjected to a temperature gradient until equilibrium heat flow is reached, and thermal properties are determined from measurements of appropriate parameters. It can take an hour or more to establish equilibrium, and during this period liquid water in the specimen will redistribute itself, possibly even evaporating completely.

The apparatus has been used to study the effective thermal conductivity of textile materials containing water. This paper deals with conductivity of another insulating material, polyurethane foam at a range of moisture levels, from completely dry to containing a mass of water of 1500% of the dry mass of the foam.

PROPERTIES OF POLYURETHANE FOAM

The properties of polyurethane are as follows[2]:

    specific gravity: 1.31
    thermal conductivity: 146.0 mW/(m.K)
    water content at 65%
        relative humidity: 1.2%
    heat of wetting: negligible
    specific heat: 1.89 kJ/(kg.K)

The specimen had a thickness (at a pressure of 10 Pa) of 3.30 mm and mass per unit area of 49.0 g/m^2.

These five properties are very similar to those of non-absorbent textile fibres[2], but the geometrical structure of the material is very different from that of textile fabrics. It does not consist of fibres but has the form of a sponge. The distribution of the solid fraction is random and its mass per unit area is very low. A graph of the effective thermal conductivity of this material plotted against water content (expressed as a percentage of dry weight of the sample) is presented in Figure 1.

Full details of the measurement techniques and means of achieving the appropriate water levels are reported in detail elsewhere[3].

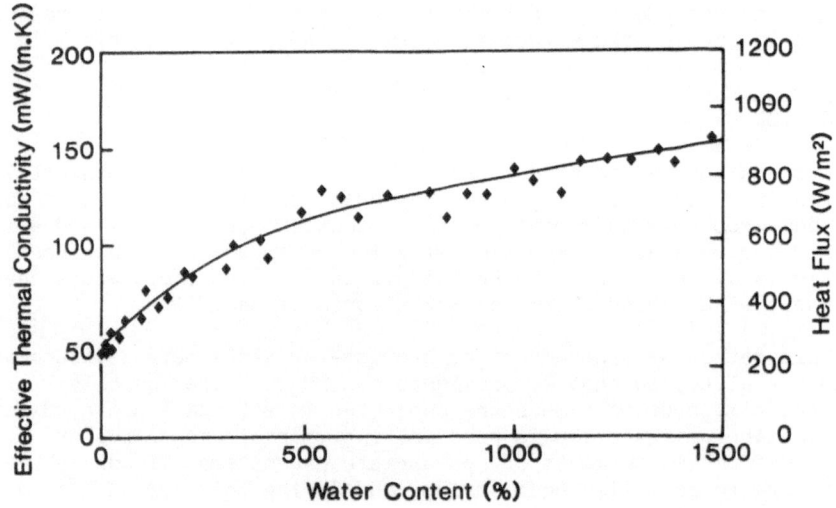

Fig. 1   Effective thermal conductivity of polyurethane foam and total heat flux to the heat sink.

ANALYSIS OF THE HEAT AND MASS TRANSFER THROUGH MOIST FOAM

During a test with the transient conductivity apparatus, water evaporates from within the sample, diffuses through the pores of the material, and condenses on the heat sink, raising its temperature. The other modes of heat transfer are conduction through the mixture of solid

of the radiation heat flux diminishes with water content. The water in the sample was assumed to be opaque, as far as infra-red radiation was concerned, and was assumed to form uniform layers on the surface of the solid. Hence the proportion of the solid fraction increases with water content, and this reduces the radiant heat transfer[4].

Water migration

When the radiant heat flux is subtracted from the total flux at the heat sink (see Figure 1), the remainder is due to the processes of evaporation of water, diffusion of water vapour and condensation, and conduction through air, foam and water.

The total amount of water condensed on the heat sink at each regain tested was determined by absorbing the water into paper tissues at the conclusion of each test and weighing. From this and the latent heat of condensation of water, the total heat absorbed by the heat sink due to condensation during the test can be calculated. To determine the heat flux due to condensation at the heat sink over the course of the test, the ratio of the condensation component of heat flux to the total heat flux was assumed constant throughout the test and equal to the ratio of the heat gain of the heat sink due to the condensation and the total heat gain of the heat sink at the end of the test. This was experimentally verified by repeating the tests using different time spans. Figure 2 shows the heat flux due to the evaporation/diffusion/ condensation process plotted against water content.

Comparison of the condensation heat flux curve of Figure 2 with the total heat flux curve of Figure 1 shows that the former is responsible for the shape of the total effective conductivity curve. There is an initial rise in heat flux in the range of 0-600 % water content, and then the value of the heat flux remains relatively constant up to 1500%.

To our knowledge there is no simple theory available to explain the shape of this curve. A qualitative explanation is as follows. Initially, as the water content in the foam rises from the dry state, it will be preferentially attracted to pores within the structure by surface tension. The condensation heat flux curve levels off at around 600% water content, which suggests that from this point onwards, the amount of water present is in excess of that which can be influenced by surface tension.

When the regain reaches 600%, free water is available in the material. From then on, the heat flux increases only slightly. This suggests that once free water is available, regardless of how much there is, the evaporation rate is limited. There is an upper limit of diffusion of water vapour through the material determined by the vapour pressure gradient and the pore structure. Hence the limitation on vapour migration is probably due to the vapour resistance of the material, and the rate that heat can be supplied to the water distributed through the material to bring about evaporation, that is, its thermal resistance. An analagous model has been proposed for textile materials[7].

Conduction

The heat flux due to conduction is the total heat flux less the radiant and condensation components. This is shown Figure 2.

When porous materials are subjected to a temperature gradient, the gradient within the material is not linear[4], and the heat fluxes associated with both radiation and conduction vary with position. The preceding analysis was based on the heat flux incident at the heat sink.

The conductive heat flux at this point may be used to determine an apparent conductivity due to conductive heat transfer, k*, found by dividing the heat flux from conduction by the temperature difference across the sample and multiplying it by the thickness of the sample. A plot of k* against moisture content for the foam is in Figure 3.

It is instructive to compare this with the value calculated from predictive models such as that of Schuhmeister[8]. It is assumed that water is adsorbed on the surface of the foam, and that both components can be subdivided into components oriented either parallel or perpendicular to the direction of the heat flow. It is also assumed that these components are randomly oriented, and that one third of their volume is oriented parallel to the direction of the heat flow and two thirds perpendicular to it.

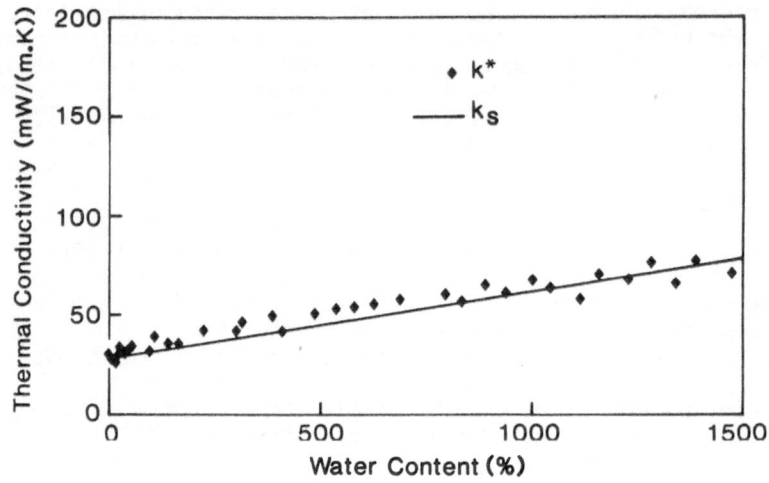

Fig. 3   Comparison of k* and k_s for polyurethane foam.

According to the Schuhmeister model, the expression for the conductivity parallel to the direction of the heat flow is given by:

$$k_{||} = k_f \cdot v_f + k_w \cdot v_w + k_a \cdot v_a$$

and the conductivity perpendicular to the direction of the heat flow is given by:

$$k_{\perp} = \frac{1}{v_f/k_f + v_w/k_w + v_a/k_a}$$

where k_f, k_w and k_a are the conductivities of foam, water and air respectively, and v_f, v_w and v_a are the corresponding volume fractions. The total conductivity of the mixture is expressed as follows:

$$k_s = r \cdot k_{||} + (1 - r) \cdot k_{\perp}$$

419

where r is the proportion by volume of the components parallel to the direction of the heat flow to the total volume, and equal to 1/3.

The conductivity found by Schuhmeister's method k_s is also plotted in Figure 3. There is good agreement between k_s and k*, which suggests that the approach used to derive k* is acceptable.

CONCLUSIONS

A transient technique has been used to measure the effective thermal conductivity of polyurethane foam containing water. It has been established that the effective thermal conductivity of moist foam is substantially higher than that of the dry material; however, the increase in conductivity with water content is not linear as might be expected from a simple proportional model.

The heat and mass transfer in this material has been analysed in terms of its individual components, namely conduction, radiation, and the process of evaporation of water, diffusion of water vapour and condensation. With a dry material, only conduction and radiation are present.

The component of heat flow due to infra-red radiation decreases with increasing moisture content, the conductive component increases in a roughly linear way, whilst the component due to moisture transfer rises until the material reaches a moisture level most probably when surface tension forces no longer limit the evaporation rate and then levels off. The shape of the curve heat flux due to condensation indicates that the evaporation/ diffusion/condensation process is responsible for the characteristic, non-linear shape of the effective conductivity curve.

From the conductive component of the heat flux, an apparent thermal conductivity for the moist foam has been derived. The agreement between theoretically obtained value, using a simple geometrical model, and that of the experimental one, indicates that the above method of analysis of the heat and mass transfer is justified.

Acknowledgements

The Wool Research Trust Fund is thanked for provision of a Junior Research Fellowship to one of us (Anna M. Schneider) to enable this work to be performed. Encouragement in this task, and valuable technical discussions with Dr B.V. Holcombe, are gratefully acknowledged.

LITERATURE CITED

1. A.M. Schneider, B.N. Hoschke, and H.J. Goldsmid, Measurements of thermal conductivity of textile fabrics, Paper presented at The 19th International Thermal Conductivity Conference, Tennessee Technological University, Cookeville, Tennessee 38505, U.S.A., October 20-23, (1985).

2. W.J. Roff, and J.R. Scott, Fibres, films, plastics and rubbers, Butterworth, London, (1971).

3. A.M. Schneider, Heat transfer through moist fabrics, PhD thesis, School of Physics, The University of New South Wales, (1987).

4. I.M. Stuart, and B.V. Holcombe, Heat transfer through fiber beds by radiation with shading and conduction, Text. Res. J. 54:55 (1984).

420

fraction, air and water, and radiation between the heat source, the surfaces of the foam, and the heat sink. All these processes interact, but for the purpose of simplicity they will be treated separately in this analysis.

Radiation

According to Stuart and Holcombe[4] the contribution of infra-red radiation in air/fibre assemblies is a significant component of total heat transfer, even when the temperature in the assembly is low. They developed a mathematical model from which the heat transfer behaviour of fibrous beds may be predicted as a function of conduction and radiation.

If a material is compressed in small steps and the effective thermal conductivity measured at each thickness, this model can be used to determine the 'apparent' fibre diameter of the fibres making up the bed, assuming them to be cylindrical. This method has been described by Holcombe[5] who found that the actual and apparent fibre diameters did not always agree, and the difference he explained in terms of deviation from the assumption in the model of random distribution of the fibres.

Foam has a similar solid/air ratio to that of insulating fibrous materials. On the basis of Holcombe's findings, it is reasonable to assume that, regardless of its actual geometrical structure, the same technique can be used to estimate the apparent fibre diameter of a bed of fibres which behaves in the same way as the foam, in so far as heat transfer by radiation and conduction are concerned. A value of 40 micrometers was found in this case.

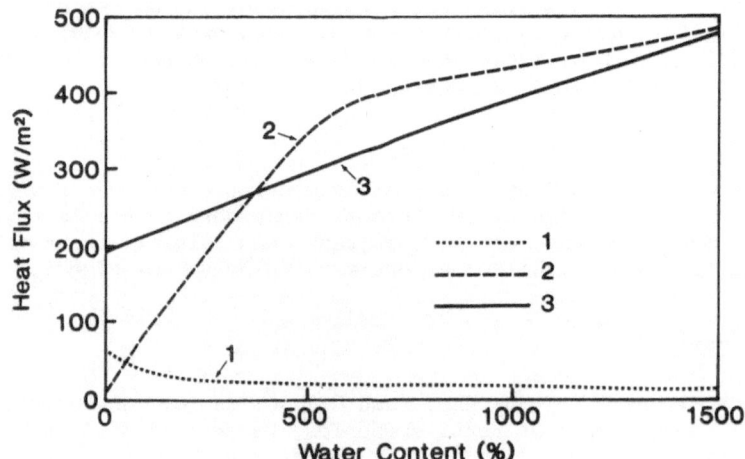

Fig. 2   Heat flux into the heat sink during conductivity tests on polypropylene foam from:
(1)   radiation,
(2)   evaporation/diffusion/condensation process,
(3)   conduction.

By inserting this value of diameter into a suitable computer program[4], the radiation component of the heat flux was calculated for each level of moisture tested, and this is presented in Figure 2. The magnitude

5. B.V. Holcombe, The influence of infra-red radiation on heat transfer through fibrous assemblies, in preparation.

6. B.V. Holcombe, and I.M. Stuart, Method of numerical solution of equations describing heat transfer in fibrous beds., J. Therm. Insul. 7:214 (1984).

7. A.M. Schneider, B.N. Hoschke, and H.J. Goldsmid, Heat transfer through moist fabrics, in preparation.

8. J. Schuhmeister, Ber. K. Akad. Wiss., Wien. Math.- Naturwiss. Klasse, 76:283 (1877).

THE THERMAL CONDUCTIVITY OF BEDS OF SPHERES

D. L. McElroy, F. J. Weaver, M. Shapiro, A. W. Longest,
and D. W. Yarbrough

Metals and Ceramics Division
Oak Ridge National Laboratory
Oak Ridge, Tennessee 37831

ABSTRACT

The thermal conductivities (k) of beds of solid and hollow microspheres were measured using two radial heat flow techniques. One technique provided k-data at 300 K for beds with the void spaces between particles filled with argon, nitrogen, or helium from 5 kPa to 30 MPa. The other technique provided k-data with air at atmospheric pressure from 300 to 1000 K. The 300 K technique was used to study bed systems with high k-values that can be varied by changing the gas type and gas pressure. Such systems can be used to control the operating temperature of an irradiation capsule. The systems studied included beds of 500 µm dia. solid Al_2O_3, the same Al_2O_3 spheres mixed with spheres of silica-alumina or with SiC shards, carbon spheres, and nickel spheres.* Both techniques were used to determine the k-value of beds of hollow spheres with solid shells of Al_2O_3, $Al_2O_3 \cdot 7$ w/o Cr_2O_3, and partially stabilized ZrO_2.** The hollow microspheres had diameters from 2100 to 3500 µm and wall thicknesses from 80 to 160 µm.

INTRODUCTION

Controlled mixtures of two phase systems, such as solids and gases, are the basis of many thermal insulation systems. The performance of an insulation depends on the individual components to heat transfer by convection, conduction by solid and gas, and radiation (1). Several such systems composed of beds of spheres of various materials were investigated in the present study. Poured beds of monosize solid spheres with gas-filled voids typically achieve about 60% of the theoretical density of

* Research sponsored by the Office of Fusion Energy, U.S. Department of Energy, under subcontract DE-AC05-84OR21400 with Martin Marietta Energy Systems, Inc.

**Research sponsored by the Energy Utilization Research, Energy Conversion and Utilization Technologies (ECUT) Program, U.S. Department of Energy, under subcontract DE-AC05-84OR21400 with Martin Marietta Energy Systems, Inc.

the solid. The fraction of theoretical density of the poured bed can be increased to about 85% by filling the voids with smaller spheres (2). As such, these assemblies have the potential to yield thermal properties intermediate to those of the gas and the solid.

A specific use of these intermediate properties is to produce a relatively-high bed thermal conductance that can be varied by changing the gas type and the gas pressure. One primary goal of the present study is to determine if such a variable conductance bed can be applied to the temperature control of irradiation capsules (3). This part of the study was sponsored by the Office of Fusion Energy. The behavior at various exposure temperatures of metals and alloys subjected to neutron irradiation is important to the design of fission and fusion reactors. These irradiation studies have typically been conducted by exposing specimens imbedded in solid cylinders inside capsules that contained a small annular gap for a filling gas. From the estimated heat produced in the solid cylinder due to neutronic heating, and the known thermal conductivity of various gases, the gap could be sized to control the specimen exposure temperature. This can be accomplished through several reactor cycles needed to obtain the total neutron exposure, sometimes exceeding 10^{22} n/cm^2. Such designs have worked well where the gas gap is not too small to obtain accurately.

Table 1 shows the required gap sizes and gap conductances for various specimen irradiation temperatures in a typical irradiation capsule with heat generation rate of 27 kW/m inside the temperature control gap.

Table 1. Control gas gap conductance, C, (k/thickness) required at operating temperature in a typical irradiation capsule with a heat generation rate of 27 kW/m inside the gap.

Specimen Irradiation Temperature (°C)	Control Gas Gap Thickness (mm)	Control Gap Conductance (W/m$^2 \cdot$K)
400	0.117	950
330	0.085	1250
200	0.029	3300

The 200°C design requires a nominal gap of 0.029 mm at a power generation rate of 27 kW/m inside the gap for a reactor coolant near 55°C. One primary problem is machining and assembly of a concentric cylindrical system to maintain the needed gap dimensions and hence the needed gap thermal conductance. Obtaining a bed with higher thermal conductivity offers the possibility of increasing the gap dimension to one which is more easily achievable. The required metallic microsphere bed k for a 0.73 mm gap to match the helium gap conductance of 3300 W/m$^2 \cdot$K is 2.7 W/m$\cdot$K.

A second goal of this study is to model and measure the thermal conductivity of beds of thin-wall hollow spheres produced from ceramic powder slurries using a coaxial nozzle process (4). In this process, slips of ceramics are blown into individual spheres at rates of 50 to 60 per second; these spheres are subsequently dried and fired to provide impervious walls. The modeling effort is to produce working computer programs based on existing, open-literature models for both hollow spheres and solid spheres. The measurement effort provides a base for comparison of thermal properties to commercially available insulation products being used for energy conservation products. This part of the study was sponsored by the ECUT program.

RADIAL HEAT FLOW APPARATUSES

Figure 1 depicts the unguarded radial heat flow apparatus used to measure k of beds of spheres near 300 K with the void spaces filled with argon, nitrogen, or helium at absolute pressures ranging from 5.1 kPa to 30 MPa. This apparatus was designated ORNL-7; its predecessors were described in references 6 and 7. Steady state k determinations are based on the temperature difference across the specimen in the radial direction, ΔT; the electrical power dissipated in the stainless steel core heater, $E \cdot I/L$; and the radial dimensions of the annular space, r_o and r_i. Accordingly, k is calculated from:

$$k = \frac{EI}{2\pi L} \cdot \frac{\ln r_o/r_i}{\Delta T} \tag{1}$$

If significant axial heat flow occurs in the measurment section of ORNL-7, then a correction to k calculated from Eq. (1) is necessary. The length-to-diameter ratio of ORNL-7 is large, 22.5, which minimizes the amount of correction to k. The effect of axial heat flow for an unguarded radial measurement apparatus becomes increasingly important as the specimen k decreases. Two-dimensional thermal modeling with the HEATING5

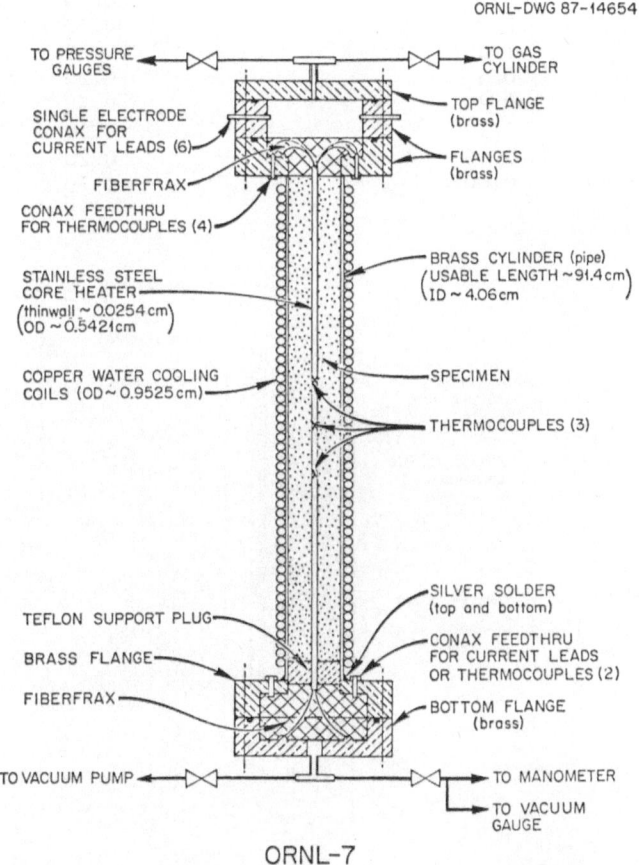

ORNL-DWG 87-14654

ORNL-7

Figure 1.   THE 300 K UNGUARDED RADIAL HEAT FLOW APPARATUS, ORNL-7

computer program (8) was performed to obtain the corrections required for a range of specimen k-values (see Table 2). A determinate error analysis (9) of ORNL-7 indicated a total error in k of ±3%.

Figure 2 depicts the unguarded radial heat flow apparatus (ORNL-8) used to measure k of beds of spheres from 300 to 1000 K in air. ORNL-8 is similar to ORNL-7 except the specimen annulus is formed by an instrumented stainless steel core heater tube and an instrumented outer cylindrical nichrome screen wire heater. Both are heated by dc power supplies. The core heater is instrumented with two Pt vs Pt-10% Rh thermocouples to measure the voltage drop in the central 10 cm of the heater and the surface temperature of the heater. Five Pt vs Pt-10% Rh thermocouples are welded to Pt studs welded near the central plane of the nichrome screen heater. The length-to-diameter ratio of ORNL-8 is about 17, which is sufficient to minimize the corrections to k calculated from Eq. 1. A determinate error analysis (9) of ORNL-7 indicated a total error in k of ±2.4%.

Table 2. Results of Thermal Modeling of ORNL-7 Using the HEATING5 Computer Program

Specimen Thermal Conductivity (W/m-K)	Percent Error in k Due to Axial Heat Flow %
5.0	0.20
2.0	0.20
1.0	0.20
0.5	0.20
0.2	0.22
0.1	0.20
0.05	0.20
0.007	0.70

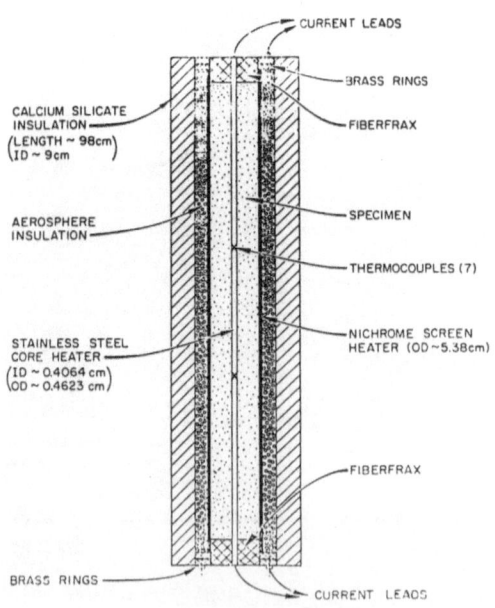

Figure 2. THE HIGH TEMPERATURE UNGUARDED RADIAL HEAT FLOW APPARATUS, ORNL-8

The characteristics of the beds of solid and hollow spheres tested in ORNL-7 and ORNL-8 are given in Table 3.

Table 3A provides the solid sphere materials, nominal diameters, and the density of the beds that were tested using ORNL-7. Table 4 lists the corresponding k-results obtained with N_2 and He gases at absolute pressures of 0.1, 1.0, and 30 MPa at 300 K.

The k-values of each bed increase with gas pressure; the percentage increase is much greater in helium than in nitrogen. The k-values with helium are 2.7 to 5 times larger than those with nitrogen. The k-values of the binary (coarse/fine) beds with Al_2O_3 (500 µm) are greater than the k-values of the beds of Al_2O_3 (500 µm) only, and of glassy carbon (355-425 µm) only. The bed of Al_2O_3 and SiC yielded the highest k-values for the ceramic materials. The binary bed of nickel spheres yielded k-values slightly higher than those of the bed of Al_2O_3 and SiC. Price (11) provides data for SiC that shows neutron irradiation decreases its k-value. Since similar decreases are expected for Al_2O_3 and glassy carbon, but not for metals, future work for application to irradiation capsule temperature control systems will focus on beds of metallic spheres.

Table 3B provides the sphere diameter and wall thickness for the ceramic hollow spheres forming the beds tested in ORNL-7 and ORNL-8. Figure 3 shows that the k-values increase with nitrogen gas pressure and depend on the material forming the hollow sphere and the sphere diameter. The bed k-values decrease when the Al_2O_3 is alloyed with Cr_2O_3 and are lowest for spheres of partially stabilized ZrO_2.

Figure 4 shows that the bed k-values in helium for Al_2O_3 shells increase as bed density increases.

Table 3. Characteristics of Beds of Spheres Tested

A. Solid Sphere Beds

Bed Designation	Material	Diameter (µm)	Bed Density (kg/m³)	Apparatus (ORNL-7 or 8)
A	Al_2O_3	500	2375	7
AZ6	Al_2O_3 plus 600 Zeeospheres(a)	500 3-25	2410	7
AZ8	Al_2O_3 plus 850 Zeeospheres(a)	500 18-107	2440	7
ASC	Al_2O_3 plus 320 grit SiC	500 48	2590	7
C	Glassy Carbon	355-425	848	7
NN	Nickel (b) Nickel	700 200	5260	7

B. Hollow Sphere Beds (9)

Bed Designation	Material	Diameter (μm)	Wall (μm)	Bed Density (kg/m^3)	Apparatus (ORNL-7 or 8)
11	Al_2O_3	3448	78	260	7
12	Al_2O_3	2809	104	480	7, 8
13	Al_2O_3	2229	92	590	7
14	Al_2O_3	2289	132	790	7
15	Al_2O_3	2106	130	900	7
21	$Al_2O_3 \cdot 7$ w/o Cr_2O_3	2852	91	410	7, 8
22	$Al_2O_3 \cdot 7$ w/o Cr_2O_3	3498	70	240	7, 8
31	Partially Stabilized ZrO_2	2250	55	560	7, 8

(a) Zeeospheres is an alloy of $Al_2O_3 \cdot SiO_2$ hollow spheres, Zeelon Industries, Inc., St. Paul, MN

(b) Federal Mogul Metal Powder

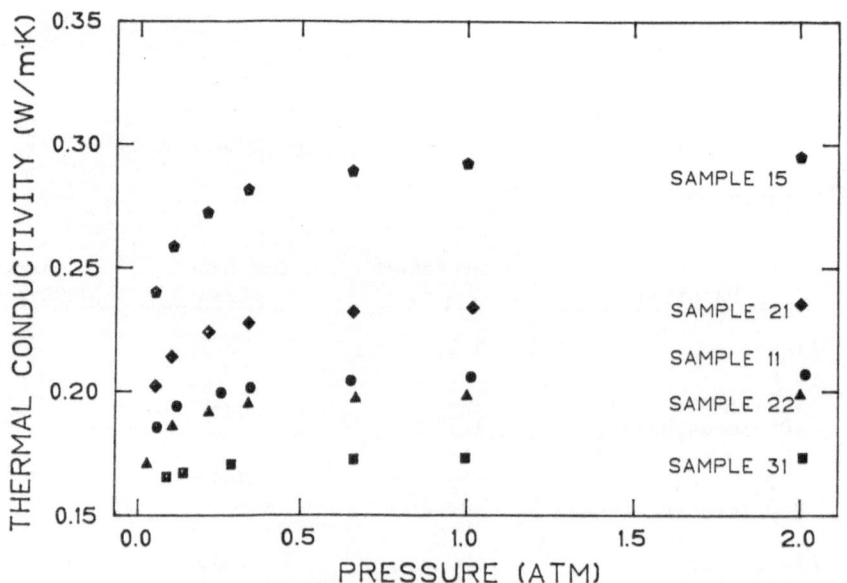

Figure 3. THE THERMAL CONDUCTIVITY AT 300 K OF VARIOUS HOLLOW CERAMIC SPHERE BEDS AS A FUNCTION OF NITROGEN GAS PRESSURE

Table 4. Thermal Conductivity Results for Various Beds of Solid Spheres in Nitrogen and Helium as a Function of Gas Pressure at 300 K.

Bed Designation	Gas	Thermal Conductivity, W/m•K 0.1 MPa	1.0 MPa	30 MPa
A	N_2	0.274	0.293	0.30
	He	1.19	1.40	1.44
AZ6	N_2	0.45	0.47	0.50
	He	1.23	1.74	1.91
AZ8	N_2	0.40	0.45	0.47
	He	1.30	1.61	1.66
ASC	N_2	0.43	0.47	0.57
	He	1.42	2.36	2.46
C	N_2	0.39	0.40	–
	He	1.08	1.34	–
NN	N_2	0.49	0.51	0.53
	He	1.83	2.38	2.52

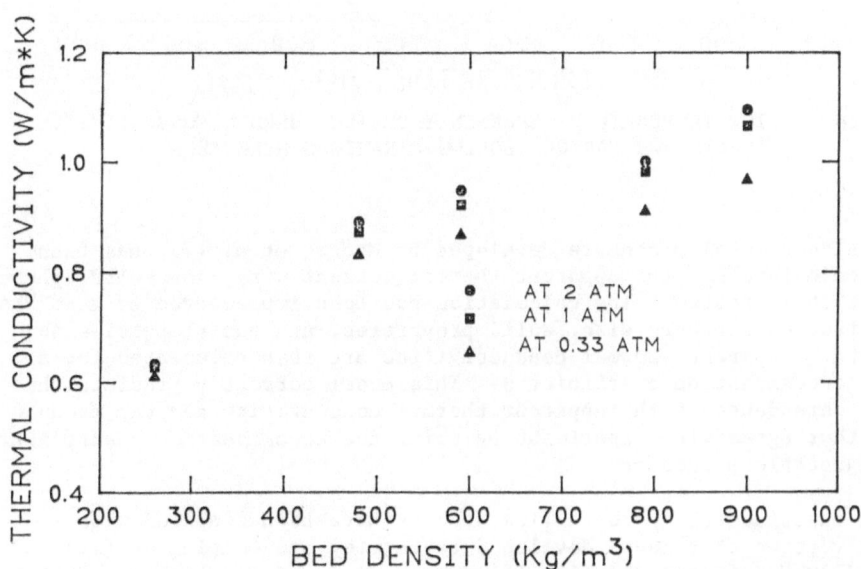

Figure 4. THE THERMAL CONDUCTIVITY AT 300 K OF VARIOUS HOLLOW AL_2O_3 SPHERE BEDS IN HELIUM AS A FUNCTION OF BED DENSITY

Figure 5 shows that the k-values measured for ceramic hollow spheres in air, using ORNL-8, increase with increasing temperature in the range 300 to 1000 K. These results were empirically fitted to equations of the form $a + bT + CT^3$. If the radiation contribution is associated with the cubic term, then extinction coefficients in the range 800 to 1200 m^{-1} are obtained using the Rosseland approximation:

$$k_r \simeq \frac{16}{3} \frac{n^2 \sigma T^3}{E},$$

where n is the index of refraction, σ is the Stefan-Boltzmann constant, E is the extinction coefficient, and T the absolute temperature.

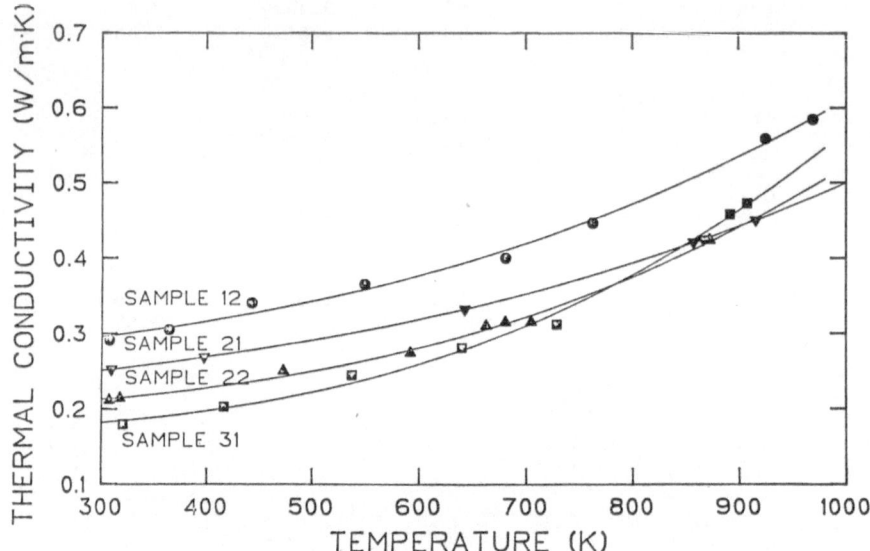

Figure 5. THE TEMPERATURE DEPENDENCY OF THE THERMAL CONDUCTIVITY IN AIR FOR VARIOUS HOLLOW CERAMIC SPHERE BEDS

MODELS

A calculational procedure developed by Moore, et al (2), has been used to calculate k_a, the apparent thermal conductivity, for solid sphere beds like those tested. The calculation has been implemented by a FORTRAN program that uses sphere size, solid properties, and gas properties as input data. Apparent thermal conductivities are then calculated for a range of accommodation coefficients. This model correctly predicts the pressure dependence of the apparent thermal conductivity and can produce results that agree with experiment by using the accommodation coefficient as an adjustable parameter.

The experimental k_a for hollow spheres have been compared with a modified version of a model developed by Parmley and Cunnington (12) and discussed by Cunnington and Tien (5). The model treats k_a as a sum of solid conduction, gas conduction, and radiation.

$$k_a = k_{sc} + k_{gc} + k_r \tag{2}$$

The solid conduction term, k_{sc}, is the product of a contact coefficient, a pressure term, and a bulk-solid thermal conductivity. The absolute pressure of the gas phase has been used in the k_{sc} calculation rather than the external compressive load used by Parmley and Cunnington. Agreement between calculated and measured k_a required k_{sc} proportional to P^n with $n = \frac{1}{2}$ or less.

The gas conduction term, k_{gc}, includes the pressure dependent bulk gas thermal conductivity and the effective thermal conductivity of the hollow sphere. The material properties of the gas and the accommodation coefficient are included in the calculation of k_{gc} (12).

The radiative component, k_r, requires boundary emittances and an extinction coefficient for the sphere bed. An extinction coefficient of 1200 m^{-1} was obtained from the T^3 coefficient obtained for the Al$_2$O$_3$ spheres on the interval 300 to 1000 K.

The model can be viewed as having at least two adjustable parameters; the contact coefficient and the pressure exponent. The sphere bed void fraction was taken to be the ideal value m = 0.4, but subsequent analysis shows that the calculation of k_a is sensitive to this parameter.

Figures 6 and 7 show calculated k_a-values compared with measured k_a-values. Figure 6 is typical of the results obtained for hollow spheres in N$_2$ using $n = \frac{1}{2}$ for the pressure exponent while Figure 7 shows similar results for hollow spheres in He. The agreement between calculated and experimental values can be improved to better than ±10% shown in the figures by adjusting the pressure exponent or the void fraction or both.

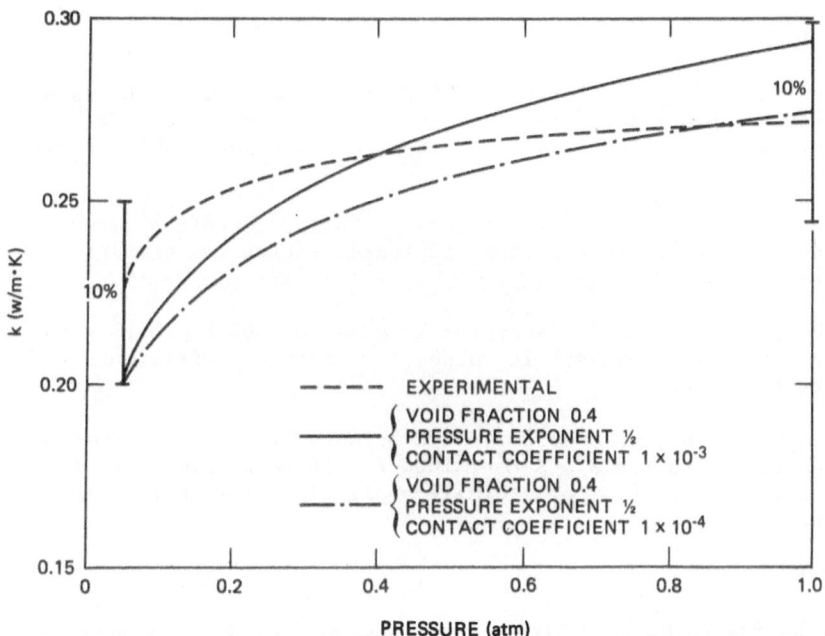

Figure 6. MEASURED k_a FOR 2229 μm-diam. HOLLOW SPHERES WITH 126 μm WALL THICKNESS IN NITROGEN COMPARED WITH CALCULATED VALUES. (a) EXPERIMENTAL DATA (b) CALCULATIONS WITH VOID FRACTION 0.4, PRESSURE EXPONENT 0.5, AND CONTACT COEFFICIENT 0.001 (c) SAME AS (b) EXCEPT CONTACT COEFFICIENT 0.0001.

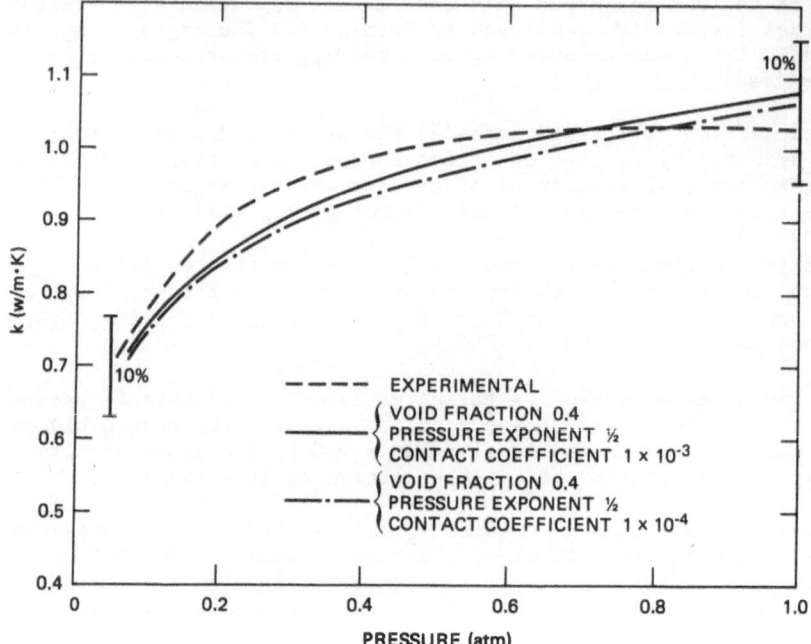

Figure 7. MEASURED k_a FOR 2106 μm-diam. HOLLOW SPHERES WITH 157 μm WALL
THICKNESS IN HELIUM COMPARED WITH CALCULATED VALUES.
(a) EXPERIMENTAL DATA (b) CALCULATIONS WITH VOID FRACTION 0.4,
PRESSURE EXPONENT 0.5, AND CONTACT COEFFICIENT 0.001 (c) SAME
AS (b) EXCEPT CONTACT COEFFICIENT 0.0001.

CONCLUSIONS

The thermal conductivity at 300 K of beds of solid spheres are
dependent on sphere material and increase with gas pressure and bed
density. Beds of metallic spheres are promising gap media for irradiation
capsules.

The thermal conductivity of beds of hollow ceramic spheres are
dependent on the sphere material and increase with temperature, gas
pressure, and bed density.

The pressure dependence of bed k-values at 300 K can be predicted to
±10% by models with adjustable values for contact coefficients and
pressure exponent.

Bed k-values in air at 300 K are between 0.2 and 0.3 W/m·K and
increase to 0.5 to 0.6 W/m·K near 1000 K. These k-values are about twice
those of lower-density, commercially-available, fibrous thermal
insulations.

REFERENCES

1. R. E. Pawel, D. L. McElroy, F. J. Weaver, and R. S. Graves, High
   Temperature Thermal Conductivity of a Fibrous Alumina Ceramic, Paper
   to be published in Thermal Conductivity 19 (October 1985, Tennessee
   Technological University).

2. J. P. Moore, R. J. Dippenaar, R. O. A. Hall, and D. L. McElroy, Thermal Conductivity of Powders with UO_2 or ThO_2 Microspheres in Various Gases fromn 300 to 1300 K, ORNL/TM-8196 (June 1982).

3. A. W. Longest, J. E. Corum, and K. R. Thoms, "Design and Fabrication of HFIR-MFE RB* Spectral Tailoring Irradiation Capsules," Fusion Reactor Materials Semiannual Progress Report for Period Ending March 31, 1987, pp. 8-9, DOE/ER-0313/2 (September 1987).

4. A. T. Chapman, J. K. Cockran, J. M. Britt, and T. J. Hwang, Thin-Walled Hollow Ceramic Spheres from Slurries, Draft Report to ORNL from Georgia Institute of Technology (86X-2204 3C) (January 19, 1987).

5. G. R. Cunnington and C. L. Tien, Heat Transfer in the Presence of a Gas, Thermal Conductivity 15, pp. 325-333 (1981).

6. D. W. Yarbrough, F. J. Weaver, R. S. Graves, and D. L. McElroy, Development of Advanced Thermal Insulation for Applicances Progress Report for the Period July 1984 through June 1985, ORNL/CON-199 (May 1986).

7. G. L. Copeland, D. L. McElroy, R. S. Graves, and F. J. Weaver, Insulations with Low Thermal Conductivity, Thermal Conductivity 18, pp.367-377, ed. by T. Ashworth and D. R. Smith (Plenum, 1985).

8. W. D. Turner, D. L. Elrod, and I. I. Siman-Tov, HEATING5 - An IBM 360 Heat Conduction Program, ORNL/CSD/TM-15 (March 1977).

9. M. J. Shapiro, "An Experimental Investigation of the Thermal Conductivity of Thin-Wall Hollow Ceramic Spheres," M.S. Thesis, Georgia Institute of Technology (March 1988).

10. S. H. Jury, D. L. McElroy, and J. P. Moore, "Pipe Insulation Testers," Thermal Transmission Measurements of Insulation, ASTM STP 660, R. P. Tye, Ed., Americal Soceity for Testing and Materials, pp. 310-326 (1978).

11. R. J. Price, Properties of Silicon Carbide for Nuclear Fuel Particle Coatings, GA-A14061 (January 1977).

12. R. T. Parmley, and G. R. Cunnington, Jr., "Evacuated Load-Bearing High-Performance Insulation Study," Lockheed Missiles and Space Company, Inc., NASA CR-135342 (December 1977).

SUBJECT INDEX